Advances in
Carbohydrate Chemistry and Biochemistry

Volume 65

Advances in Carbohydrate Chemistry and Biochemistry

Editor
DEREK HORTON

Board of Advisors

DAVID C. BAKER
GEERT-JAN BOONS
DAVID R. BUNDLE
STEPHEN HANESSIAN
YURIY A. KNIREL

TODD L. LOWARY
SERGE PÉREZ
PETER H. SEEBERGER
NATHAN SHARON
J.F.G. VLIEGENTHART

Volume 65

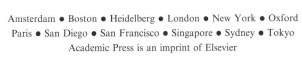

Amsterdam • Boston • Heidelberg • London • New York • Oxford
Paris • San Diego • San Francisco • Singapore • Sydney • Tokyo
Academic Press is an imprint of Elsevier

Academic Press is an imprint of Elsevier
Radarweg 29, PO BOX 211, 1000 AE Amsterdam, The Netherlands
Linacre House, Jordan Hill, Oxford OX2 8DP, UK
32 Jamestown Road, London NW1 7BY, UK
225 Wyman Street, Waltham, MA 02451, USA
525 B Street, Suite 1900, San Diego, CA 92101-4495, USA

First edition 2011

Copyright © 2011 Elsevier Inc. All rights reserved

No part of this publication may be reproduced, stored in a retrieval system or transmitted in any form or by any means electronic, mechanical, photocopying, recording or otherwise without the prior written permission of the publisher

Permissions may be sought directly from Elsevier's Science & Technology Rights Department in Oxford, UK: phone (+44) (0) 1865 843830; fax (+44) (0) 1865 853333; email: permissions @elsevier.com. Alternatively you can submit your request online by visiting the Elsevier web site at http://elsevier.com/locate/permissions, and selecting *Obtaining permission to use Elsevier material*

Notice
No responsibility is assumed by the publisher for any injury and/or damage to persons or property as a matter of products liability, negligence or otherwise, or from any use or operation of any methods, products, instructions or ideas contained in the material herein. Because of rapid advances in the medical sciences, in particular, independent verification of diagnoses and drug dosages should be made

ISBN: 978-0-12-385520-6
ISSN: 0065-2318

British Library Cataloguing in Publication Data
A catalogue record for this book is available from the British Library

Library of Congress Cataloging-in-Publication Data
A catalog record for this book is available from the Library of Congress

For information on all Academic Press publications
visit our website at books.elsevierdirect.com

Printed and bound in USA
11 12 13 14   10 9 8 7 6 5 4 3 2 1

Working together to grow
libraries in developing countries
www.elsevier.com | www.bookaid.org | www.sabre.org

ELSEVIER   BOOK AID International   Sabre Foundation

CONTENTS

CONTRIBUTORS vii

PREFACE ix

Jean Montreuil 1920–2010
JEAN-CLAUDE MICHALSKI

Laurance David Hall 1938–2009
BRUCE COXON

Potential Trehalase Inhibitors: Syntheses of Trehazolin and its Analogues
AHMED EL NEMR AND EL SAYED H. EL ASHRY

I.	Introduction	46
II.	Natural Occurrence and Characterization of Trehazolin	47
III.	Biosynthesis of Trehazolin	48
IV.	Retrosynthetic Analysis of Trehazolin and Its Analogues	49
V.	Synthesis of the Sugar Isothiocyanate Fragment	50
VI.	Synthesis of the Trehazolamine Fragment	53
	1. Synthesis from D-Glucose	54
	2. Synthesis from D-Mannose	60
	3. Synthesis from D-Ribose	63
	4. Synthesis from D-Arabinose	65
	5. Synthesis from D-Mannitol	66
	6. Synthesis from Sugar 1,5-Lactones	66
	7. Synthesis from *myo*-Inositol	68
	8. Synthesis from D-Ribonolactone	73
	9. Synthesis from Cyclopentadiene	77
	10. Synthesis from Acylated Oxazolidinones	81
	11. Synthesis from *cis*-2-Butene-1,4-diol	82
	12. Synthesis from Norbornyl Derivatives	85
	13. Synthesis from 3-Hydroxymethylpyridine	85
VII.	Synthesis of Trehazolins	87
	1. Trehazolins Having Trehazolamines at the Anomeric Center	87
	2. Trehazolines with Trehazolamines on Positions Other than the Anomeric Center	100

3.	Synthesis of 1-Thiatrehazolin.	104
4.	Deoxynojirimycin Analogues.	106
	Acknowledgments.	108
	References.	108

Polysaccharides of the Red Algae
Anatolii I. Usov

I.	Introduction.	116
II.	Storage Carbohydrates: Floridean Starch.	117
III.	Neutral Structural Polysaccharides: Cellulose, Mannans, and Xylans.	119
IV.	Sulfated Galactans.	122
	1. General Information and Nomenclature.	122
	2. Isolation of Galactans.	127
	3. Chemical Methods of Structural Analysis.	129
	4. Physicochemical Methods of Structural Analysis.	145
	5. Enzymatic and Immunological Methods.	155
	6. Polysaccharides of the Agar Group.	162
	7. Polysaccharides of the Carrageenan Group.	165
	8. DL-Hybrids.	168
	9. Biological Activity of Sulfated Galactans.	169
V.	Sulfated Mannans.	171
VI.	Glycuronans.	173
VII.	Heteropolysaccharides of Unicellular and Freshwater Red Algae. Miscellaneous Polysaccharides.	174
VIII.	Algal Systematics and Polysaccharide Composition.	177
	References.	179

Toward Automated Glycan Analysis
Shin-Ichiro Nishimura

I.	Introduction.	220
	1. Background.	220
	2. Proteomics and Glycomics: Glycan Expression is Not Template-Driven.	220
	3. Unmet Needs in Glycobiology and Glycotechnology.	221
II.	Emerging Glycomics Technologies.	223
	1. Progress in Mass Spectrometry.	223
	2. The Rate-Determining Stage in Glycomics and Glycoproteomics.	234
III.	Toward Automated Glycan Analysis.	261
	References.	265

AUTHOR INDEX.	273
SUBJECT INDEX.	297

LIST OF CONTRIBUTORS

Bruce Coxon, The Eunice Kennedy Shriver National Institute of Child Health and Human Development, National Institutes of Health, Bethesda, Maryland 20892, USA

El Sayed H. El Ashry, Chemistry Department, Faculty of Science, Alexandria University, Alexandria, Egyptc, and Fachbereich Chemie, Universität Konstanz, Fach M 725, Konstanz, Germany

Ahmed El Nemr, Environmental Division, National Institute of Oceanography and Fisheries, Kayet Bey, El Anfoushy, Alexandria, Egypt

Jean-Claude Michalski, Unité de Glycobiologie Structurale et Fonctionnelle, Université de Lille 1, 59655 Villeneuve d'Ascq, France

Shin-Ichiro Nishimura, Field of Drug Discovery Research, Faculty of Advanced Life Science, Hokkaido University, N21, W11, Kita-ku, Sapporo, Japan

Anatolii I. Usov, N. D. Zelinsky Institute of Organic Chemistry, Leninskii Prospect 47, Moscow, Russian Federation

PREFACE

This volume pays tribute to the life and work of two notable figures in the carbohydrate scene, each of whom made major contributions in quite different areas. Jean Montreuil, who lived for almost ninety years, was a world pioneer in the field now known as glycobiology, and he was recognized as one of the great leaders in French biochemistry. He was a colorful and influential personality who held a lifetime attachment to the north of France and his native city of Lille, where he created the Lille Glycobiology School. The lively account presented here by Michalski, one of his associates in Lille, illustrates Montreuil's far-reaching scientific contributions, and his work in organizing many national and international conferences on glycoconjugates.

The career of Laurance David (Laurie) Hall took a quite different course; he was a mercurial figure who crisscrossed the Atlantic and played a major role in the development of NMR spectroscopy of sugars following on from the seminal work of Lemieux, and he later became a pioneer in what became known as magnetic resonance imaging (MRI). Setting out originally in synthetic carbohydrate chemistry in Hough's laboratory in Bristol, UK, he soon turned to the new field of NMR and made astonishingly prolific contributions in a career that developed in Vancouver at the University of British Columbia. Bruce Coxon (Bethesda, Maryland), his close friend from their days together in Bristol, and himself an NMR spectroscopist, traces in thorough technical detail the innovative work of Laurie Hall on the NMR of sugars bearing a wide range of different magnetic nuclei, and on to the last phase of Hall's work conducted back in the UK at Cambridge, where again he pioneered the use of magnetic resonance, especially MRI, in a wide range of physiological and technological applications. Indeed, had Laurie Hall not died at a relatively young age, the article here recorded by Coxon might have been a chapter written by Hall himself, as a sequel to his two earlier noteworthy contributions to *Advances*.

In the early days of this *Advances* series, authors were offered the opportunity to record experimental procedures as part of their contributions, but this has been rarely used in recent years. However, the practice is revived here in the article by El Nemr and El Ashry (Alexandria) on the synthesis of trehazolin, a natural nitrogen-containing pseudodisaccharide that is an inhibitor of the enzyme trehalase, together with various analogues. These synthetic targets can be approached in a wide variety of ways, and the authors present for each synthetic sequence sufficient detail to permit an experienced synthetic chemist to repeat the procedure.

The red algae are lower plants, mostly from marine sources, that contain a wide diversity of sulfated polysaccharides in their cell walls and intercellular matrices, and many of these have enjoyed long practical use as a consequence of their ability to form strong gels in aqueous solution. Such examples as agar and carrageenan are well known; their structures were surveyed by Mori in Volume 8 of this series and their physicochemical properties were the subject of a landmark article by Rees in Volume 24, but only with the advent of sophisticated new structural techniques has the vast range of different structures in these algae been recognized. Usov (Moscow) has been a leading contributor to the literature in this field, and his article in this volume provides a comprehensive overview of the known chemical structures, along with interesting new perspectives on the classification of the *Rhodophyta* according to the data of molecular genetics.

The terms glycomics and glycobiology are now firmly established as major themes of carbohydrate science, notably in relation to cell-surface glycoproteins and glycosphingolipids and their relation to human disease. The detailed analysis of glycans in these components offers high potential for medical/pharmaceutical applications. However, as Nishimura (Sapporo, Japan) points out in this volume, there are major problems involved in separating and analyzing the complex mixtures obtained from whole-serum glycoproteins, and this limits the current therapeutic potential. The author details the rapid evolution of automated methodology to overcome these difficulties, and offers promise for major advances in glycan profiling by employing the "glycoblotting" technique in conjunction with mass-spectrometric analyses.

John M. Webber, who died on January 4, 2011, was the author of an article on chitin in Volume 15 and higher-carbon sugars in Volume 17 of this series, and was a member of the editorial team that founded the journal *Carbohydrate Research*.

Nathan Sharon, a leading carbohydrate biochemist and long-time member of the Advisory Board of this series, died June 17, 2011. A detailed account of his life and work will appear in a future volume of *Advances*.

DEREK HORTON

Washington, DC
May, 2011

JEAN MONTREUIL

1920–2010

Jean Montreuil, a "glycomaniac" and pioneer glycobiologist, was an emblematic and charismatic figure of French biochemistry. He passed away on July 16, 2010, just a few weeks short of his 90th birthday, at "Vasile Goldis," Western University of Arad, Romania, where he was organizer of the 16th meeting of French-speaking Summer Schools "Molecular pathologies and pharmacology; Biotechnology." Jean Montreuil was among the early pioneers in the field of carbohydrate chemistry and glycobiology, and one of the fathers of what is now termed "glycomics." He made seminal contributions to the subject and created the "Lille Glycobiology School," where he trained a number of leaders in this field and mentored a new generation of glycobiology and biochemistry researchers. He was one of the prime organizers of the Second International Symposium on Glycoconjugates, held in Lille in 1973; served as president (1977) and member of the International Glycoconjugate Organization (IGO); and was a member of the International Steering Committee for Carbohydrate Symposia (ICS).

Lille and His Early Life

Jean Montreuil was born in Lille, in the north of France, on October 11, 1920, into a family of modest origins, and he lost his mother at a young age. Jean was very proud of his "Flemish" heritage and was extremely attached to the North of France region and to the friendly and "welcoming" character of people from this area. He studied at the "Lycée Wallon" in Valenciennes, in northern France, from where he obtained his bachelor degree in philosophy in 1939.

During the period of political unrest and the beginning of the Second World War, Jean Montreuil elected to join the French army and participate in the "Campagne de France." Throughout his life, Jean carried a lasting impression of this difficult experience, which contributed greatly in molding many features of his character, such as respect for others and a sense of responsibility and discipline.

Lille has a long history in the field of chemical biology, this specialty having been created as early as 1854 by Louis Pasteur, who performed in Lille his work on beer preservation and where he discovered the lactic fermentation. Pasteur was also dean of the Lille Faculty of Sciences from 1854 to 1857. Although Pasteur subsequently returned to Paris, his school of chemical biology remained in Lille. Jean Montreuil was always extremely impressed by the work of those scientists who succeeded to the chair of chemical biology, most significantly Eugene Lambling (1857–1924) who wrote in 1911 one of the first biochemistry essays "Précis de Biochimie" (Masson Eds., Paris), and Michel Polonowski (1889–1954) whose pioneer work on human milk oligosaccharides led to the discovery of the so-called gynolactose and his work on plasma proteins with the discovery of haptoglobin. In his initial research work, Jean Montreuil followed the "research avenue" traced by Michel Polonowski.

Upon his return from military service, Jean Montreuil started pharmaceutical studies at the Medicine and Pharmacy, University of Lille, and obtained his diploma in pharmacy in 1945. This "pharmaceutical" background ensured that Jean always combined his interests for both chemistry and biology, and he was fascinated by the chemistry of life. After obtaining his pharmacy diploma, he first opened a small drugstore in a small village located near Saint Quentin (northeast of Paris), together with his wife Jeannine, who was also a pharmacist. He and Jeannine were to have four children, three daughters, and a son.

From Nucleic Acids to Glycoproteins of Body Fluids

But Jean had already caught the "virus for research," and he spent his time travelling between Saint Quentin and the Cancer Institute in Lille, mainly in the laboratory directed at that time by Professor Paul Boulanger in the Department of Chemical Biology. He then joined the Department of Chemistry of the Lille Cancer Institute, where under the supervision of Professor Boulanger, and like many others at this time, he started working on the structure of nucleic acids, with a special interest in the so-called pentose-containing nucleic acids. He was awarded the D.Sc. degree in 1952. Immediately after his thesis, Jean Montreuil received strong support from his supervisor to move into "less explored" territories concerning the isolation and chemical characterization of carbohydrate molecules from human biological fluids, notably urine and serum.

In 1957, Jean was appointed as assistant professor at the University of Sciences of Lille and became full professor of biological chemistry in 1963. At that time, he decided to move with his group to the brand new University of Sciences and Techniques, located on a new campus located in Villeneuve d'Ascq, on the outskirts of Lille, and created there a new research lab of chemical biology "Laboratoire de Chimie Biologique," where he became head of the Department of Biochemistry. We will remember Jean Montreuil as an outstanding teacher, particularly skilled at giving clear lectures, illustrated with the most topical data taken from his own research experiences.

The very first contributions of Jean Montreuil were in the study of human milk glycoproteins, which led to the discovery in 1959 of lactotransferrin (lactoferin) and milk IgAs in 1960. These glycoproteins, as well as serotransferrin, quickly became his "fetish" proteins, serving as models for the characterization of their glycan moieties. Together with his coworkers Geneviève Spik, Michel Monsigny, Bernard Fournet, and Gérard Strecker, his first achievements were in the application of various methods for the isolation, assays, and structural characterization of carbohydrates by such procedures as hydrazinolysis, acetolysis, methylation analysis, sequencing by exo-glycosidases, and lectin fractionation. During those early days, relatively little was known about the exact structures of N-linked glycans on glycoproteins. In 1969, he was one of the first to elucidate the "N-(β-aspartyl)-N-acetylglucosaminylamine nature of the glycan–protein bond" in human transferrin and elucidated the complete primary structure of its glycan. This was followed by the characterization of the glycan in many other glycoproteins.

Everyone involved remembers the astonishing complexity of Jean's transparencies at different meetings, listing the glycan heterogeneity in diverse glycoproteins at a time when the term "glycomics" had not yet been invented! Sharing the expertise of his collaborators, Jean Montreuil proposed that all N-glycans have the same core, to which are attached various branches that he named "antennae" to suggest their mobility and potential physicochemical and biological roles. Jean has always been a kind of "visionnaire"; he was convinced that the "informative" message of glycans was not only written in terms of sequence but also supported by their conformation. Since no crystallographic or NMR data concerning glycans were then available, only the building of molecular models, he proposed interconvertible conformations for glycans, and he was the first to coin the terms "Y" and "T" conformations, later called "umbrella" or also "broken wing" conformations.

The "Dutch Connection"

Jean was continuously looking for and applying new methodologies for structural characterization .In the early 1970s, he established contact with Hans Vliegenthart's group at the University of Utrecht. They were involved in the study of biological molecules by NMR. This was the beginning of a long adventure the aim of which was to analyze glycans by high-field NMR. This collaboration led to the first full characterization by NMR of the glycosylation pattern of many glycoproteins and oligosaccharides. This effort was further continued in Lille by Gérard Strecker.

Human Milk and Urine Oligosaccharides

In addition to the study of protein glycosylation, Jean's interests also lay in the free oligosaccharides occurring in such biological fluids as human milk in regard with their biological function in nutrition and protection from intestinal bacteria. This work continued the earlier initiative of Richard Kuhn on oligosaccharides in human milk, convinced that they would serve as biomarkers for much pathology. Together with Gérard Strecker, he greatly contributed to the characterization of urinary storage materials in the urine of patients suffering from lysosomal storage disorders, and they defined the scheme for catabolism of glycans within the lysosomes. A visitor to his laboratories during this period could not fail to be struck by the finesse by which these separations were performed. A temperature-controlled room was equipped with many cabinets for preparative paper chromatography on thick paper. Slow elution over days or weeks allowed beautiful separations of the pure components in quantities sufficient for analysis by the newer techniques, especially NMR.

The "Red Blood Cells Aging" Period

At the age of 70, Jean Montreuil retired from his responsibilities as director of the Lille Glycobiology Institute and started a "new life," devoting most of his time to the development of education and research in Romania, which became his "second homeland," and where he organized congresses, meetings, and summer schools for promoting glycobiology and biochemistry among young scientists, continuing until the last day of his life. Together with Professor Daniela Bratosin, from Arad University, he also initiated a new research program on aging of red blood cells, leading to a new provocative concept about "red blood cells apoptosis," with practical applications for preservation of these cells in blood banks. His last paper appeared in the *Cytometry. Part B, Clinical Cytometry* journal in January 2011.

Dedicated Teacher, Editor, and Organizer

As already mentioned, Jean Montreuil was a respected "Big Boss" and, up to his very last moment, was an inveterate teacher, imparting his knowledge and enthusiasm to his colleagues and students and making Lille a nationally and internationally recognized place for training in glycosciences. Physically, Jean was "built like an oak." He had all the attributes of Flemish people, huge and strong, eating a lot and appreciating good meals and wines. He was also extremely seductive, with his clear blue eyes, cultivating a "long-haired appearance." All through his life, he had a great deal of success with the female gender! At the same time, he also exerted this seduction and natural authority on his students and coworkers. As a manager, he was appreciated, but also feared, cultivating the nickname of "Mandarin" by his coworkers.

Jean was encouraged by his biochemist friend Roger Jeanloz to organize the Second International Symposium on Glycoconjugates in Lille in 1973, and he reorganized the IXth edition of this meeting in 1987. One of his main enterprises, together with André Verbert, was certainly the creation and organization of the "International Course in Glycoconjugates," which was held every two years for two weeks in Villeneuve d'Ascq, gathering together internationally recognized leaders and students around a theme of lectures and practical works. This school contributed largely to promote glycobiology in Europe, and many of us remember the friendship and convivial atmosphere and the Saturday evening "Genièvre (jenever) and cheese party."

In his public addresses, he was a most impressive presence on the lecture podium. He would build up his story to a crescendo of hand-waving animation, employing two projectors, two screens, and an impossible number of slides, but would still manage to finish within his allotted time slot and leave the audience gasping at his virtuosity. Most comfortable, naturally, in his native French, he was nevertheless little deterred when making his presentations in English.

Jean Montreuil wrote several influential reviews that synthesize much information on glycan structures, among them the one published in Volume 37 of *Advances in Carbohydrate Chemistry and Biochemistry*, entitled "Primary Structure of Glycoprotein Glycans: Basis for the Molecular Biology of Glycoproteins." He also edited several books, the most notable of which is a three-volume treatise: *Glycoproteins*, *Glycoproteins II*, and *Glycoproteins and Diseases* (1995–1997) in the New Comprehensive Biochemistry series, coedited with Hans Vliegenthart and Harry Schachter. Jean Montreuil also contributed to the peer-review process, serving on the Editorial Board of *Glycoconjugate Journal*, *Biochemical Journal*, and *Carbohydrate Research*. His name is associated with more than 750 scientific papers. The community of French biochemists and glycobiologists is greatly indebted to him for his service as a central figure in the "Société Française de Biochimie et Biologie Moléculaire" (SFBBM), being one of the symbolic figures of the Society, and chief editor of the journal "Regards sur la Biochimie," which served as a link between French biochemists. He was also promoter and president from the "Groupe Français des Glucides," a society that brings together French carbohydrate chemists and glycobiologists.

Jean has also served in many national and international science institutions throughout his career. He was deeply engaged in scientific administrations in France. He was a CNRS delegate for the North of France region (1981–1990), president of the Life Sciences CNRS National Department (1987–1991), council member of French Minister for Higher Education and Research (1986–1989), member of the French Academies of Medicine and Sciences, honorary member of the Romanian Academy, and Doctor Honoris Causa of the Free University of Brussels (ULB), the University Al. I. Cuza of Iasi, and the University Vasile Goldis of Arad. He received numerous awards, among them, the "Foundation Jaffé Award" and the "Charles Leopold Mayer Award" of the French Academy of Sciences, the Silver Medal of the Royal University of Utrecht, and the Gold Medal of the V. Goldis Western University of Arad.

As a director, laboratory chief, and mentor, Jean Montreuil had a sharply critical mind and showed little patience for poor research. He was a man with principles, and a capacity for hard work constituted for him a most important merit. At the same time, he was extremely open and displayed much attention to people and to novelty. He was extremely attentive to young scientists, never hesitating to help them and sharing his long-term experience or influence. At the same time, he was jovial, good-natured, a "bon vivant" who much enjoyed also "partying" and joking. Overall, he loved his "North of France" region, its history, and people and enjoyed speaking in the local dialect.

Jean was a tireless traveler; he had a special attraction for countries of eastern Europe, their history and culture. During his Romanian period, he became fascinated by orthodox religion and the richness of the old monasteries. His last wish was to have both orthodox and religious celebrations for his funeral. This final wish was indeed fulfilled.

Jean gives to us a wonderful example of a rich and extremely well-filled scientific life, following his passion for science and teaching up to the very last moment, and he certainly realized his unconfessed wish: dying, like Molière, on the stage. Jean Montreuil's influence and scientific contributions will continue to have a lasting impact on current and future scientists. He leaves us a very substantial legacy.

JEAN-CLAUDE MICHALSKI

LAURANCE DAVID HALL

1938–2009

Early History and Education, 1938–1963

Laurance Hall (affectionately known as Laurie) was born on March 18, 1938 in Wapping, in the dockland area to the East of the City of London, United Kingdom. Fortunately, he was evacuated to the countryside during the war years because Wapping was soon to be devastated by German bombing, and the docks never recovered as such. He was the eldest of six children raised by Daniel William Hall and Elsie Ivy Hall. Laurie attended Leyton County High School for Boys on the northeast side of London. Like many British grammar schools, the school underwent significant changes over the years, adopting the comprehensive model in 1968, and becoming the coeducational, Leyton Sixth Form College in 1985. A prescient teacher at the school encouraged him to attend the Christmas Lectures at the Royal Institution. Laurie was so inspired by this experience that he resolved to return one day to give his own lectures, an ambition that was fulfilled in 1991 and 1997, in the form of Friday Evening Discourses. Laurie appears to have inherited his very strong work ethic from his paternal grandmother, who determinedly managed the family furniture business until she died at age 92.

In due course, Laurie was accepted to read Honours Chemistry at the University of Bristol, with physics and mathematics as minor subjects. It was customary at the University of Bristol for honours chemistry students to be assigned to a research group during their third (senior) undergraduate year, where they pursued a small research project chosen by the leader of the group. I first met Laurie in this way in 1958, when as an undergraduate, he joined the carbohydrate research group of Dr. (later Professor) Leslie Hough, who occupied joint positions in the Departments of Chemistry and Biochemistry, and who was subsequently to be a coinventor of the widely used noncaloric sweetening agent, Splenda. Les had a variety of sugar research projects in progress, including synthetic monosaccharide and sucrose chemistry, periodate oxidation studies, polysaccharide structure, and the biosynthesis of polysaccharides. The system of undergraduate research at Bristol had the advantage that it assessed the research potential of students before they graduated, it enhanced their academic performance by exposing them at a personal level to more senior research personnel, and it gave them a sense of belonging to a working research organization so that in many cases, they chose to stay on and undertake graduate research in the same group. The selection of, or assignment to, a research group was an important turning point in an undergraduate's career, which tended to influence it indefinitely. Public financial support was available if a high level of academic performance was established during the undergraduate years, which amounted to obtaining a first class, or upper-second class, bachelor's degree. As an undergraduate, Laurie was an extremely bright and industrious student, who had no trouble securing a first-class honours B.Sc. in chemistry, thus guaranteeing that he was accepted for graduate research, supported by a Department of Scientific and Industrial Research Studentship. In those days at Bristol, there were virtually no graduate level academic courses, so that Ph.D. students were free to concentrate on their research full time. An initial research project was always suggested to the beginning research student by Leslie Hough, but if the project did not work out for one reason or another, he believed in giving his students complete freedom to follow their noses into more profitable areas.

Laurie's initial graduate research project was the chemical synthesis of sugars having sulfur in the pyranose ring. It was believed that comparison of their reactions and properties with those of the oxygen analogues would give considerable insight into the behavior of carbohydrates, and that the sulfur analogues might have useful medicinal properties. 5,6-Dideoxy-5,6-epithio-1,2-O-isopropylidene-α-D-glucofuranose was prepared from 5,6-anhydro-1,2-O-isopropylidene-α-L-idofuranose by reaction with thiourea, and similarly, 5,6-dideoxy-5,6-epithio-1,2-O-isopropylidene-α-L-idofuranose, from 5,6-anhydro-1,2-O-isopropylidene-α-D-glucofuranose. I recall that Laurie in those days was somewhat subject to alternating moods of euphoria and mild depression, especially if a chemical reaction went wrong. Unfortunately, as it seemed at first, but, fortunately, as it turned out later, Laurie's synthesis of key intermediates for the preparation of 5-thio sugars was prepublished by Creighton and Owen in 1960. Being "scooped" on his main research project was extremely disappointing to Laurie, and he resolved to find an alternative research area that was "less busy." In the departmental library, he found a book on the then new subject of nuclear magnetic resonance (NMR) spectroscopy and could visualize the enormous potential that the technique had for the analysis of carbohydrates. At this point in time, Ray Lemieux had collaborated with NMR spectroscopists at the National Research Council in Ottawa, and had in 1958 demonstrated experimentally the angular dependence of vicinal proton–proton coupling constants of sugars on proton–proton dihedral angle, and described by the theoretical Karplus relationship, that was to revolutionize their conformational analysis. However, the NMR spectroscopy of sugars was still virtually unexplored. With permission from Les Hough, Laurie contacted Keith McLauchlan, a recent Ph.D. in physical chemistry at the University of Bristol, who was working in the Magnetic Resonance Section headed by David Whiffen in the Basic Physics Division at the National Physical Laboratory (NPL) in Teddington, Middlesex. A collaboration was arranged to work on the conformational analysis of sugars, which turned out to be very productive, enhanced by the fact that the group at Teddington also included a number of other talented NMR people, such as Ray Abraham, Ray Freeman, and K. G. R. Pachler. Eventually, all of the people with whom Laurie associated at the NPL left this government research institute for prestigious academic or research positions, Ray Abraham to the University of Liverpool, Ray Freeman to Varian Associates and both Oxford and Cambridge Universities, Keith McLauchlan to Oxford University, and David Whiffen to a Chair of Chemistry at the University of Newcastle. As I learned during my own visit to NPL in 1965, Dr. Whiffen could be quite a hard taskmaster. A typical task for new employees in his section might be to fix the NMR spectrometer, which certainly taught them how it worked, and also how it could be modified. Thus, Laurie came to abandon his initial synthetic research project for a life in NMR, and derivatives of 5-thio-D-glucopyranose were soon synthesized by Milton Feather and the late Roy Whistler, by acetolysis of the 5,6-epithio sugars that Laurie and L. N. Owen and coworkers had prepared. It was discovered later that 5-thio-D-glucopyranose prevents the uptake of D-glucose by cells and inhibits sperm development in mice.

Laurie obtained his Ph.D. from the University of Bristol in 1962, and on August 1st of that year, married Winifred Margaret Golding, a 1961 graduate of the French Department at the University, whom he had met in the University Ski Club. It was common for freshly minted Ph.D.s in Les Hough's group to apply for a postdoctoral position in North America. A number of these people went to work with Professor J. K. N. Jones at Queen's University in Kingston, Ontario, Canada, and others applied to the more limited number of carbohydrate laboratories in the United States. Instead, Laurie chose to take up a postdoctoral appointment with Professor F. A. L. Anet, an NMR spectroscopist at the University of Ottawa. However, there was still a carbohydrate connection, as Professor Anet is the brother of the late Dr. E. F. L. J. Anet, who was a well-known carbohydrate chemist in Australia. The newly married couple departed Bristol for Ottawa, but on arriving, found that Professor Frank Anet was to be in residence for only 6 weeks, before departing for a sabbatical year at the University of California, Los Angeles (UCLA). Evidently, Professor Anet liked the location of his sabbatical year so much that he soon moved permanently to UCLA and remained there until his retirement.

Professor Anet's absence from Ottawa afforded Laurie the opportunity to achieve independence at a young age, and during that period, he wrote the first review article on the NMR of sugars, which was published in the *Advances in Carbohydrate Chemistry*. Laurie and Margaret visited my wife and me at our apartment in Silver Spring, Maryland at that time, which was an opportunity to renew an old friendship. During our overlapping years in Bristol, I had attended student parties at Laurie's flat, at which there was swing dancing, as the student music of the time was traditional jazz in the New Orleans style. Laurie was fully aware of the value of the dancing (known as jiving) for breaking the ice between the party-goers, and his awareness of jazz music was later to produce a most startling publication in NMR. Notwithstanding Professor Anet's absence from Ottawa, he and Laurie published two papers together on cyclohexane ring inversion and amino-cyclohexanediols. A concurrent collaboration with Professor Hans Baer allowed the assignment of configuration by proton NMR to three aminodeoxy-anhydroheptuloses.

The Vancouver Period, 1963–1984

After 1 year in Ottawa, Laurie accepted an offer of a position as instructor level II in the Department of Chemistry at the University of British Columbia (UBC) in Vancouver, thus launching his independent career at the age of 25. He was promoted to assistant professor in 1964, to associate professor in 1969, and to full professor in 1973. In this location, he benefited from excellent students and sabbatical visitors, attracted by the prestige of the University, its beautiful location on the Pacific coast of Canada, and Laurie's great drive and enthusiasm for science. A large amount of innovative, high-quality research on the NMR and chemistry of carbohydrates was conducted during this period, giving Laurie a fine international reputation in the area.

Laurie's approach to research has sometimes been characterized as "jumping on the bandwagon" or for not exploring certain fields fully enough, but he explained it this way: "I intend to cream off the best results from new fields as I go." By this means, he usually managed to stay far ahead of the competition. My personal impression of him at the time was that he was continually running and jumping toward new projects and scientific challenges. It has been said (by Ray Abraham, if my memory serves me correctly) that one can either read the scientific literature or contribute to it. Amazingly, Laurie managed to do both and always seemed to be aware of the latest developments in his science. He developed a facile, conversational style of writing, which explained his important scientific results without too much jargon, excessive hubris, or flowery justification. Laurie could write a preliminary communication in a short evening, and he did this frequently in the early years.

He often described battles that he had with his Department Chairman, Professor Charles McDowell over the allocation of space and resources, but due to Laurie's drive and determination, his wish usually prevailed, Professor McDowell doubtless realizing that he had a winner on his hands. Laurie had an open, attractive personality, which served him well in his interactions with students and peers.

In the early 1960s, Laurie soon realized the limitations of proton NMR, which were its low sensitivity (compared with what was to come later), and limited spectral dispersion, owing to the small magnetic field strengths (proton resonance at 60 or 100 MHz) that were available from electromagnets or permanent magnets at the time. He therefore resolved to pursue investigations of the heteronuclear NMR of carbohydrates, and other organic, and also inorganic molecules. A convenient starting point was the ^{19}F NMR of fluorinated carbohydrates because the ^{19}F nucleus is 100% abundant, and the technique has almost equivalent sensitivity to that of proton NMR. This field of research was wide open, but the NMR instrumentation of the time was not well suited to heteronuclear NMR, a situation which Laurie remedied by constructing his own instrumental modifications, with the assistance of a talented electronics expert, R. Burton, in the department.

The fluorine NMR project started well when the late Christian Pedersen generously provided a suite of glycosyl fluoride derivatives, in both the pyranosyl and furanosyl series. However, Laurie also initiated a program for the synthesis in-house of deoxyfluoro sugars. Glycosyl fluorides were prepared initially by classical methods, involving either treatment of peracylated sugars with anhydrous hydrogen fluoride or reaction of acylated glycosyl bromides with silver fluoride. The program yielded a wealth of new data for the NMR characterization of fluoro sugars, and these studies had the further advantage that proton NMR could be used for the initial definition of the configuration and conformation of these sugars, as a basis for assigning the stereospecific dependencies of the ^{19}F NMR parameters. Dispersion of the ^{1}H NMR spectra of the glycosyl fluorides was assisted initially by the late Norman Bhacca, who provided 220 MHz ^{1}H NMR spectra from his position at Varian Associates. For D-glucopyranosyl fluoride derivatives, equatorially oriented fluorine nuclei showed a ^{19}F chemical shift of 131.8–137.8 ppm from internal CFCl$_3$ in chloroform, whereas axially disposed ^{19}F displayed 147.8–149.9 ppm. Equatorial fluorine nuclei resonated an average of 14 ppm to low field of their axial counterparts, thus paralleling the behavior of the ^{1}H chemical shifts of anomeric and other ring protons (in the absence of other shielding effects), but with a much larger dependence of the ^{19}F shifts on anomeric configuration. Laurie was quick to point out that, for cyclohexyl fluoride, the axial and equatorial fluorine chemical shifts measured by Bovey *et al.* are 45 times more sensitive to the steric environment than are the shifts of the protons attached to the same carbon atoms, corresponding to the measurement of ^{1}H shifts at a resonance frequency of 4 GHz, compared with ^{19}F shifts at 94 MHz. As we have seen, immense efforts in magnet development and other technical advances in NMR instrumentation over the past 50 years have led to measurement of ^{1}H NMR spectra still only at 1 GHz.

When glycopyranosyl fluorides having an axial O-substituent at C-2 (typically D-*manno* derivatives) were examined, a striking dependence of the ^{19}F shift on the configuration of the adjacent carbon atom was revealed. Axial (α) anomeric fluorine nuclei resonated in the range 137.6–138.8 ppm, which is similar to the range observed for β-D-glucopyranosyl fluorides. A marked temperature dependence of the ^{19}F shift of β-D-glucopyranosyl fluoride tetraacetate was also observed, which over the range of 190–330 K changed by −8 ppm. This was interpreted in terms of conformational mobility, caused by the strong anomeric effect of the fluorine atom, a factor that was found to be conformationally dominant in the pentopyranosyl fluoride series. The vicinal ^{19}F–^{1}H coupling constants of the glycopyranosyl fluorides displayed a marked angular dependence so that $J_{trans} \sim 24$ Hz, whereas $J_{gauche} = 1.0$–1.5 Hz for $J(F_a,H_e)$ and 7.5–12.6 Hz for $J(F_e,H_a)$. Geminal ^{19}F–^{1}H couplings were found to depend on the orientation of the substituent at C-2, being ~ 53.5 Hz when this substituent was equatorial, and ~ 49 Hz when axial.

^{1}H and ^{19}F NMR studies of a series of fully acylated pentopyranosyl fluorides indicated that all of these derivatives adopt primarily the conformation in which the fluorine atom is axial. This conclusion was supported by interpretation of the values of geminal ^{19}F–^{1}H and ^{1}H–^{1}H coupling constants, long-range (^4J) ^{19}F–^{1}H and ^{1}H–^{1}H couplings, and ^{19}F chemical shifts. The pentopyranosyl fluoride derivatives showed the same types of angular dependence of the ^{19}F–^{1}H coupling constants as had the hexopyranosyl analogues, once more paralleling the Karplus dependence of ^{1}H–^{1}H couplings, albeit with totally different magnitudes. The importance of the anomeric effect in determining the conformation or thermodynamic stability of glycosyl halides in general had also been recognized by others.

In an extension of the work, treatment of 2-deoxy-β-D-*arabino*-hexopyranose tetraacetate with anhydrous HF yielded a major proportion of 3,4,6-tri-*O*-acetyl-2-deoxy-α-D-*arabino*-hexopyranosyl fluoride, with some of the β-anomer. However, a similar reaction of 2-deoxy-β-D-*arabino*-hexopyranose tetrabenzoate afforded exclusively 3,4,6-tri-*O*-benzoyl-2-deoxy-α-D-*arabino*-hexopyranosyl fluoride. The α anomeric configuration of the major products followed from the large values (38.0–38.3 Hz) of the F_1,H_{2a} coupling constants, the F_1,H_{2e} couplings being much smaller (5.0–5.3 Hz).

As the project progressed, a major study was made of the electrophilic addition of various fluorinated species to the double bonds of peracylated carbohydrate glycals, particularly addition of the elements of "XF," where X = I, Br, or Cl. Additions of "XOR" to glycals had been reported earlier by Lemieux and Fraser-Reid. Although Laurie and coworkers D. L. Jones and J. F. Manville showed that the addition of XF to cyclohexene proceeds exclusively trans, reactions of pyranose glycals yielded up to four products, from both cis- and trans-addition, but with the trans adducts predominating. For example, the reaction of 3,4,6-tri-O-acetyl-D-glucal with Br_2-AgF yielded three adducts, 3,4,6-tri-O-acetyl-2-bromo-2-deoxy-α-D-mannopyranosyl fluoride (70%), 3,4,6-tri-O-acetyl-2-bromo-2-deoxy-α-D-glucopyranosyl fluoride (9%), and the corresponding $β$ anomer (21%). The same three products were obtained when "BrF" was generated *in situ* from N-bromosuccinimide and HF. However, under these conditions, 3,4,6-tri-O-acetyl-2-deoxy-α-D-*arabino*-hexopyranosyl fluoride was also formed in ~7% yield. The 2-chloro-2-deoxy and 2-deoxy-2-iodo analogues were prepared by using Cl_2-AgF and I_2-AgF in a similar manner, the structures of several of the products being confirmed by independent synthesis. In the case of the Cl_2-AgF reaction, all four configurational possibilities were formed, but with a different product distribution to that from the other halogens, the major outcome being 2-chloro-2-deoxy-$β$-D-glucopyranosyl fluoride triacetate. Also, the reaction of D-glucal triacetate with N-iodosuccinimide and HF afforded the three 2-deoxy-2-iodo-hexopyranosyl fluorides of corresponding structure, the literature assignment of the $β$-D-*manno* configuration to the major product being corrected to α-D-*manno*, in the process. ^{1}H NMR studies confirmed the configurations and $^4C_1$ conformations of the 2-deoxy-2-halo derivatives, and the vicinal $^{19}F-^1H$ coupling constants showed the same type of angular dependence as that of $^3J_{H,H}$ so that $F_{1a},H_{2a} = 24.0$–27.8 Hz, $F_{1a},H_{2e} = 1.9$–4.3 Hz, and $F_{1e},H_{2a} = 9.3$–10.6 Hz. These coupling constants displayed an approximately linear relationship with the Huggins electronegativity of the substituents at C-1 and C-2.

During his fluoro sugar period, Laurie was invited by Professor Stephen Angyal to spend a mini-sabbatical at the University of New South Wales, financed by a Royal Society and Nuffield Foundation Commonwealth Bursary. During his stay, Laurie industriously wrote up a large quantity of the fluoro sugar work. Along the way, Laurie and his students demonstrated that since reactions of the 2-deoxy-2-halo sugars at C-1 yield exclusively trans products, such reactions must involve participation of the halogen atoms via 1,2-halonium ions, even for the less polarizable chlorine atom.

The ^{19}F NMR studies of Laurie and J. F. Manville attracted the attention of others who were engaged in the synthesis of fluoro sugars, including P. W. Kent, N. F. Taylor, Christian Pedersen, and Allan Foster. A productive collaboration ensued with Allan Foster's group at the Chester Beatty Cancer Research Institute, with some participation from John Brimacombe. Together, they pursued investigations of sugars substituted with fluorine at all of the remaining positions on the sugar chain, including

2-deoxy-2-fluoro-D-glucose and 2-deoxy-2-fluoro-D-mannose, 2-deoxy-2-fluoro- and 3-deoxy-3-fluoro-D-hexopyranosyl fluorides, 4-deoxy-4-fluoro- and 6-deoxy-6-fluoro-D-glucose and their anomeric glycosyl fluoride derivatives, 4-deoxy-4-fluoro-α- and β-D-galactopyranosyl fluoride triacetates, and 4-deoxy-4-fluoro-D-sorbose and -tagatose, 5-deoxy-5-fluoro-L-sorbose, and derivatives of 3,6-anhydro-5-deoxy-5-fluoro-1.2-O-isopropylidene-α-L-idofuranose. Other types of electrophilic addition reactions of sugar glycals were also investigated. Reaction of fluoroxytrifluoromethane with D-glucal triacetate afforded not only two trifluoromethyl glycoside triacetates, in the 2-deoxy-2-fluoro-α-D-*gluco* and 2-deoxy-2-fluoro-β-D-*manno* series, but also, somewhat uncharacteristically, 2-deoxy-2-fluoro-α-D-glucopyranosyl fluoride triacetate and 2-deoxy-2-fluoro-β-D-mannopyranosyl fluoride triacetate, involving addition of two fluorine atoms to the double bond and the generation of carbonyl fluoride. For such electrophilic addition reactions to activated double bonds, cis-addition is very strongly favored, leading to products having only the α-D-*gluco* and β-D-*manno* configurations. Acid hydrolysis of the trifluoromethyl glycosides or the glycosyl fluorides yielded the parent 2-deoxy-2-fluoro-D-glucose and 2-deoxy-2-fluoro-D-mannose. These methods provided a synthesis of 2-deoxy-2-fluoro-D-glucose from D-glucose, in only four steps, which was expected to be useful for the preparation of decagram quantities of this key fluoro sugar for antitumor testing. For all of these fluoro sugars, the strong spin-coupling properties of the ^{19}F nucleus provided a treasure trove of coupling constants over one to five chemical bonds.

With Foster's group, it was demonstrated that the angular dependence of long-range $^4J_{F,H}$ coupling constants is similar to that of $^4J_{H,H}$ couplings so that the stereochemistry that generates the largest coupling is when the coupled nuclei and their intervening bonds are in a planar "M" ("W") arrangement, which is common for 1,3-diequatorial orientations. For example, β-D-allopyranosyl fluoride tetraacetate showed a coupling of 3.6 Hz between F-1e and H-3e, whereas β-D-glucopyranosyl fluoride tetraacetate displayed no resolvable coupling between F-1e and H-3a. As for $^1H-^1H$ couplings, $^4J_{e,a}$ can sometimes be resolved for fluorine and a coupled proton (typically 0.7–1.1 Hz) but is generally smaller than $^4J_{e,e}$ (3.6–4.2 Hz).

Seven pentofuranosyl fluorides were studied in collaboration with Chris Pedersen and graduate student P. R. Steiner. Unexpectedly, large ^4J coupling constants of 5.5–7.9 Hz were measured between F-1 and H-4 when these atoms were trans, and smaller values, 1.0–1.8 Hz, when they were cis. However, the ^4J couplings between F-1 and H-3 were smaller (<0.7 Hz) when they were trans than when they were cis (1.7–2.4 Hz). In those days, the possibility of multiple pathways for spin coupling was considered but not seriously explored, as was also the question of through-space F,H or F,F coupling. Nevertheless, these authors started the difficult task of defining the multiple conformational possibilities for the furanosyl ring, which were found to be different for the anomeric tri-*O*-benzoyl-D-ribofuranosyl fluorides.

Concurrently with his studies of fluoro sugars, Laurie published a series of papers on the application of nuclear magnetic double-resonance methods to sugars and steroids. One of the key applications was determination of the signs of coupling constants, using traditional selective spin decoupling or spin tickling of parts of a spin–spin multiplet, or a new method involving irradiation off-resonance from the ^{19}F multiplet, for example, while observing the ^{1}H transitions. The latter method was convenient in that it required less precise setting of the ^{1}H or ^{19}F frequencies. Each method involved an initial determination of the *relative* signs of coupling constants, but these were put on an *absolute* basis by reference to a coupling constant of known absolute sign; for example, the geminal coupling $^2J_{1,F}$ was known to be absolutely positive.

Naturally, sign determinations were applied to many of the ^{1}H–^{1}H and ^{19}F–^{1}H coupling constants that Laurie *et al.* measured for the fluoro sugars, and the availability of newly synthesized dideoxydifluoro sugars allowed the opportunity for extension of the studies to ^{19}F–^{19}F couplings. The work provided some of the first indications that ^{19}F–^{19}F couplings are anomalous in several different respects. First, the vicinal ^{19}F–^{19}F coupling constants of 2-deoxy-2-fluorohexopyranosyl fluorides were shown to be absolutely negative, in contrast to vicinal ^{1}H–^{1}H and ^{19}F–^{1}H couplings, which are all positive. Second, in marked contrast to the usual situation for ^{1}H–^{1}H and ^{19}F–^{1}H couplings, where $^3J_{trans} > {}^3J_{gauche}$, the trans coupling of ^{19}F–^{19}F was -20.0 Hz, and was about the same as the gauche couplings, -18.8, -15.8, and -13.5 Hz. Thus, there is no simple, Karplus-type $\cos^2\varphi$ dependence of vicinal ^{19}F–^{19}F couplings on dihedral angle φ. Additionally, the magnitudes and signs of $^4J_{F,F}$ couplings of 3-deoxy-3-fluoro-D-pyranosyl fluorides were found to be unusual, showing *ax,ax* $+10.4$, *ax,eq* $+1.0$, and *eq,eq* -3.0 Hz, where the large *ax,ax* value probably reflects the operation of a through-space, spin-coupling mechanism. However, the results for the anomeric 2,3,6-tri-*O*-acetyl-4-deoxy-4-fluoro-D-glucopyranosyl fluorides indicated a partial stereospecificity for both $^5J_{F,H}$ and $^5J_{F,F}$ couplings, since $^5J_{F-4e,H-1e}$ $+3.0$ to $+3.3$ Hz were found to be larger than $^5J_{F-1e,H-4a}$ 0.8 Hz and $^5J_{F-4e,H-1a}$ $+0.5$ Hz, and $^5J_{F-1e,F-4e}$ $+3.1$ Hz was larger than $^5J_{F-1a,F-4e}$ -0.6 Hz. Moreover, in the analogous anomeric 2,3,6-tri-*O*-acetyl-4-deoxy-4-fluoro-D-galactopyranosyl fluorides, $^5J_{F-1a,F-4a}$ -3.7 Hz and $^5J_{F-1e,F-4a}$ -1.3 Hz were observed, thus indicating that similar magnitudes of couplings may be measured, but with different signs. These studies emphasized the importance of sign determinations (as well as the magnitudes) for interpreting the stereochemical dependence of the couplings. With Diane Miller, Laurie investigated the synthesis and reactions of carbohydrate triflates. For example, the reaction of 1,2:3,4-di-*O*-isopropylidene-α-D-galactopyranose with triflic anhydride in pyridine yielded the 6-C-(pyridinium triflate) salt directly, and the corresponding pyridinium derivatives substituted on secondary carbon atoms could also be prepared, with or without isolation of the intermediate triflate ester. Fluorinated carbohydrates were also synthesized stereoselectively by addition of acetyl hypofluorite to vinyl ethers of sugars. In summary, Laurie and coworkers made a monumental contribution to the chemistry and NMR spectroscopy of fluoro sugars.

Following the success of his fluoro sugar work, Laurie extended his studies of NMR coupling constants to additional heteronuclei. His general procedure was to synthesize compounds that were either C-substituted with the heteroatom or contained this atom in the ring. The dihedral angles subtended by such atoms with nearby protons or other magnetic nuclei were of prime interest for studies of angular dependence of the couplings, and in pyranose, furanose, or other heterocyclic derivatives of known conformation, were established by the ring geometry. ^{31}P NMR studies were initiated by synthesis of C-phosphonates. Reaction of 1,2:5,6-di-O-isopropylidene-α-D-*ribo*-hexofuran-3-ulose with dimethyl phosphite and catalytic sodium methoxide afforded a major yield of 3-C-(dimethoxy)-phosphinyl-1,2:5,6-di-O-isopropylidene-α-D-allofuranose, together with a minor proportion of its D-*gluco* epimer, the assignment of configurations being supported by analysis of $^3J_{HCCH}$ and $^3J_{PCCH}$ coupling constants. Similarly, reaction of methyl 4,6-O-benzylidene-2-O-tosyl-α-D-*ribo*-hexopyran-3-uloside with dimethyl phosphite gave methyl 4,6-O-benzylidene-3-C-(dimethoxy)-phosphinyl-2-O-tosyl-α-D-allopyranoside, and 1,2-O-isopropylidene-5-O-tosyl-α-D-*erythro*-pentofuran-3-ulose yielded 3-C-(dimethoxy)-phosphinyl-1,2-O-isopropylidene-5-O-tosyl-α-D-ribofuranose, which showed cis and trans couplings $J_{P,2}$ 8.1 Hz and $J_{P,4}$ 26.8 Hz, respectively. Studies of cyclic phosphates such as 2-oxo-1,3,2-dioxaphosphorinanes having phosphorus in the ring reported $^3J_{POCH}$ coupling constants in the ranges 21.0–22.8 Hz (J_{trans}) and 1.2–3.3 Hz (J_{gauche}), whereas the related phosphonate derivatives having one less phosphorus bound oxygen atom showed similar J_{trans} and J_{gauche} couplings of 18.2–20.2 and 3.2–4.7 Hz, respectively, thus confirming the utility of these stereospecific dependencies. Vicinal ^{31}P–O–C–^{1}H couplings were also measured for a series of acyclic organophosphate esters and were rationalized in terms of varying populations of rotamers. Data from variable-temperature experiments were interpreted in relation to enthalpy and entropy differences between trans and gauche rotamers. A further series of organic phosphonate and hydroxy-phosphonate derivatives was synthesized that contained examples of rigid cyclic systems and more flexible acyclic molecules. The values of $^3J_{PCCH}$ couplings measured for these systems corresponded to nominal PCCH dihedral angles of 0°, 60°, 120°, and 180° and allowed the graphical plotting of an approximate, Karplus-type dependence. In related work, a reaction of phenyl phosphorodichloridate with pentane-2,4-diol yielded a mixture of three isomeric cyclic phosphates of the 2-oxo-2-phenoxy-1,3,2-dioxaphosphorinane type. Analysis of the ^{1}H and ^{31}P NMR spectra of these derivatives confirmed the stereospecificity of several sets of ^{31}P–^{1}H coupling constants, including a number of long-range $^4J_{POCCH}$ values that were larger when the coupled proton was equatorial.

In studies directed toward measurement of tin coupling constants, C-stannylated carbohydrates were prepared by reaction of epoxide or O-tosyl derivatives of sugars with triphenyltin lithium. For example, the reaction of 1,2:3,5-di-O-methylene-6-O-tosyl-α-D-glucofuranose afforded (1,2:3,5-di-O-methylene-α-D-glucofuranos-6-yl)-triphenylstannane, and methyl 2,3-anhydro-4,6-O-benzylidene-α-D-allopyranoside and its D-*manno* isomer yielded (methyl 4,6-O-benzylidene-α-D-altropyranoside-2-yl)triphenylstannane and (methyl 4,6-O-benzylidene-α-D-altropyranoside-3-yl)-triphenylstannane, respectively, both in good yield. The compounds were studied by ^{1}H and ^{13}C NMR, but attempts to measure tin–proton couplings ran into problems at ^{1}H resonance frequencies of 100 and 270 MHz. It was expected that the couplings would appear in the ^{117}Sn and ^{119}Sn satellites in the ^{1}H spectra, but these transitions turned out to be mostly obscured by the stronger signals of protons attached to nonmagnetic tin isotopes. The only tin couplings that could be clearly resolved were $^3J_{Sn-2,1}$ 25.8 Hz and $^2J_{Sn,2}$ 73.4 Hz for the altropyranoside-2-yl stannane, and $^2J_{Sn,3}$ 56.2 Hz for the corresponding 3-yl stannane, and the particular tin isotopes (^{117}Sn or ^{119}Sn) causing these splittings were not identified. Detection of the tin–proton couplings by direct measurement of either ^{117}Sn or ^{119}Sn NMR spectra should have revealed the individual isotopic couplings more easily (since ^{117}Sn and ^{119}Sn are both spin 1/2 nuclei and have different Larmor frequencies) but does not appear to have been investigated, possibly because of instrumental limitations. The ^{13}C NMR spectra of the stannylated sugar derivatives were sufficiently well dispersed (even at 20 MHz) for a number of tin–carbon coupling constants to be measured. However, only the one-bond Sn–C couplings could be resolved for the separate tin isotopes: $^1J(^{119}Sn-^{13}C)$ 340–392 Hz, and $^1J(^{117}Sn-^{13}C)$ 323–376 Hz, for the three sugar stannanes studied. For the two- and three-bond Sn–C couplings, only averaged values for the two Sn isotopes could be measured: $^2J(Sn-^{13}C)$ −9.6, −10.1, −14.8, −24, and −32 Hz (signs assumed) and $^3J(Sn-^{13}C)$ 0, 15.7, 32.8, and 60 Hz.

Several groups (Inglis and Schwarz; Takiura and Honda; Horton *et al.*) had prepared mercurated sugar derivatives, principally by the methoxymercuration of acetylated glycals, especially those of D-glucal, D-galactal, D-xylal, and D-arabinal. The reaction procedure involved treatment of the peracetylated glycal with mercuric acetate, followed by sodium chloride, yielding products such as methyl

3,4,6-tri-O-acetyl-2-chloromercuri-2-deoxy-β-D-glucopyranoside. Laurie initiated a study of the ^{1}H and ^{13}C NMR spectra of a series of such products and measured the couplings of these nuclei to the only important magnetic isotope of mercury, namely, ^{199}Hg, which has a spin of 1/2 and, therefore, introduces only doublet multiplicity into mercury-coupled spectra. He and Victoria Gibb found that both $^3J_{HgCCH}$ and $^3J_{HgCCC}$ coupling constants showed Karplus-type dependences on dihedral angle, electronegativity, and on the orientation of adjacent substituents. Thus, trans-antiparallel atom pairs ($\varphi_{HgCCH} \sim 180°$) showed $^3J_{HgCCH}$ 410–460 Hz, whereas gauche atoms ($\varphi_{HgCCH} \sim 60°$) displayed $^3J_{HgCCH}$ 110–131 Hz. Similarly, $^3J_{HgCCC}$ couplings were observed to be 144–225 Hz for a nominal 180° dihedral angle, and 36–53 Hz for nominally 60° angles, but these couplings were thought to be very sensitive to the orientations of nearby substituents. One-bond $^1J(^{199}Hg-^{13}C)$ couplings were found to be very large (1825.2–1879.7 Hz) but appeared to be relatively insensitive to their surroundings. A number of long-range ^4J and ^5J couplings were also observed, for both ^{199}Hg–^{1}H and ^{199}Hg–^{13}C. Once again, the stereochemical dependences paralleled those of ^{1}H–^{1}H couplings so that a planar W-arrangement of the coupling nuclei and intervening atoms and bonds favored the observation of larger ^4J values, and a large value, $^5J_{Hg-2,C-6}$ 11.2 Hz was detected only when Hg-2 and C-6 were both equatorial.

Impressed by the facile reactions of glycal triacetates under methoxymercuration conditions, Gibb and Hall extended this type of reaction to 2-C-diacetoxythalliation of D-galactal triacetate, which in anhydrous methanol with thallium(III) triacetate yielded a single product that was shown to be methyl 3,4,6-tri-O-acetyl-2-deoxy-2-(diacetoxythallio)-α-D-talopyranoside. $trans$-Diaxial addition to the double bond of the glycal occurred, as it had for the corresponding methoxymercuration reaction. In most cases, separate resolution of the ^{1}H and ^{13}C couplings to the two isotopes of thallium (^{203}Tl, 29.5% and ^{205}Tl, 70.5%, both spin 1/2) was impossible at 100/25 MHz, and only averaged values could be measured. An exceptionally pronounced Karplus dependence was observed for vicinal Tl–^{1}H couplings, since the values for $trans$-diaxial Tl-2 and H-3 [$^3J(^{205}Tl-2,H-3)$ 2646 Hz and $^3J(^{203}Tl-2,H-3)$ 2622 Hz] were almost seven times that for $gauche$ Tl-2 and H-1 [$^3J(Tl-2,H-1)$ 386 Hz, average]. Also noteworthy, were the one-bond Tl–^{13}C couplings, $^1J(^{205}Tl-2,C-2)$ 6466 Hz and $^1J(^{203}Tl-2,C-2)$ 6404 Hz, and the thallium nuclei were also coupled to C-1, C-3, C-4, C-5, C-6, and the methoxyl carbon.

Laurie's studies of the heteronuclear NMR parameters of sugars and other molecules did much to extend the experimental generality of the Karplus dependence of vicinal coupling constants on the dihedral angles and other properties of the coupled nuclei.

In the early days of NMR, it was realized that the full exploitation of ^{13}C NMR for the structural analysis of organic molecules would require spin decoupling of all of the protons in the molecules from their ^{13}C neighbors. This was so because the proton coupled, ^{13}C NMR spectra of most organic molecules, including carbohydrates, are extremely complex, due to the presence of ^{13}C–^{1}H coupling constants over one, two, and three bonds. Although such spectra can be analyzed, the presence of the couplings increases the amount of spectral analysis and coupling constant assignment work by an order of magnitude. Also, it was anticipated that the collapse of ^{13}C–^{1}H coupled multiplets to single resonances by wide-band proton spin decoupling would increase the sensitivity of ^{13}C NMR, which was already much less than that of proton NMR, because of the lower natural abundance and inherent sensitivity of the ^{13}C nucleus. A further enhancement of ^{13}C sensitivity during ^{1}H decoupling was also expected from the operation of the nuclear Overhauser effect. In a landmark paper, Nobel Laureate-to-be Richard Ernst demonstrated that modulation of the decoupling carrier frequency with random noise produces a broadband, heteronuclear spin decoupling effect. A practical implementation of this "noise decoupling" technique was to use a pseudo-random sequence generator to phase modulate the radiofrequency (RF) carrier with either 0° or 180° phase shifts. However, the generator equipment was relatively expensive, and in the 1960s, commercial NMR probe technology was somewhat primitive in that the probes did not include multinuclear observation and were not designed for heteronuclear decoupling of a range of nuclei. Laurie's NMR studies of ^{19}F and ^{31}P containing compounds and other heteronuclei required heteronuclear spin decoupling for the assignment of chemical shifts and the identity of heteronuclear coupling constants, using either selective or broadband irradiation. The contributions of his group to this area were first to modify commercial NMR probes to support a range of heteronuclear decoupling frequencies, such as ^{2}H, ^{19}F, or ^{31}P, while observing ^{1}H, and second, to irradiate ^{1}H, ^{2}H, or ^{31}P, during the detection of ^{19}F NMR signals. The work showed considerable sophistication in electronics, as it involved the construction of double-tuned RF circuits for a number of pairs of nuclei, technology that was also being used by other research groups. An inexpensive solution to the problem of performing broadband decoupling was devised by Laurie that involved the use of commonly available, high-fidelity audio equipment. Audio noise signals of varying bandwidths, for example, 50–500, 50–1000, and 50–2000 Hz were stored on continuous tape loops on a high-fidelity tape recorder, they were played back through a high-fidelity audio amplifier and were used to modulate the output of an RF synthesizer, via an RF mixer, thus allowing heteronuclear decoupling over a defined bandwidth.

In a surprising corollary to this work, it was demonstrated that jazz music could be used as the noise source for spin decoupling, specifically by using the worldwide smash hit "Midnight in Moscow," played by Kenny Ball and his Jazzmen, and taken from the loudspeaker output of a portable record player. The player was borrowed from one of Laurie's graduate students, which could only have served to increase the corporate spirit in his research group! As one might expect, the editor of the journal had some difficulty in accepting the manuscript for publication, but in the end, Laurie's persuasive abilities prevailed. At the time, I was immodest enough to think that the project might have been a joke directed at me, as Laurie knew of my strong interest in jazz. However, it definitely illustrates his impish sense of humor. In this work, Laurie and Burton demonstrated that for many heteronuclear decoupling experiments, the precise source of the audiofrequency noise is almost irrelevant.

Later, Freeman and coworkers investigated broadband decoupling systematically and discovered supercycles of phase-shifted, composite pulses that decoupled a wide bandwidth at low RF power. The latter properties became extremely important as the increasing magnetic field strengths in NMR spectrometers required the decoupling of an ever larger frequency range, and as the Larmor frequency moved closer to the microwave range in which RF sample heating effect became more significant than at lower frequencies. The acronymic sequences WALTZ-16 and GARP are now legendary in NMR history and are very widely used for broadband ^{1}H and ^{13}C/^{15}N decoupling, respectively. For various reasons, the basic sequences were modified to WALTZ-17, WALTZ-64, and WALTZ-65, GARP-1 and GARP-2, etc. Since then, the sample heating problem has been alleviated further by the development of efficient, adiabatic decoupling methods, in which the irradiating frequency is swept repetitively over the chemical shift range to be decoupled. Line narrowing in the NMR of solids had been investigated earlier by Waugh, Huber, and Haberlen (1967) leading to the WAHUHA multiple pulse sequence, and its derivatives.

In addition to determining the relative and absolute signs of a host of coupling constants, Laurie applied magnetic double-resonance methods in many other ways, including identification of hidden signals in the complex envelopes of the ^{1}H NMR spectra of sugars by internuclear double resonance (INDOR), the measurement by INDOR of ^{2}H chemical shifts of deuterated solvents and carbohydrates, and the ^{13}C chemical shifts of methyl group-containing substituents on sugars, including suitable instrumental modifications for the implementation of these techniques. A pulse Fourier transform (FT) equivalent of the ^{1}H–^{1}H INDOR experiment was also developed, and a combination of nuclear Overhauser enhancement (NOE) difference and decoupling-difference spectroscopy was applied to resolve and assign every ^{1}H resonance of 1-dehydrotestosterone and 11β-hydroxyprogesterone.

As Laurie's work on fluoro sugars came to an end, he moved into new areas. With Ian Armitage, A. G. Marshall, and others, he systematically studied the application of lanthanide NMR shift reagents, which had been discovered by Conrad Hinckley in 1969.

Initially, the ^{1}H NMR spectra of 1,2:5,6-di-O-isopropylidene-α-D-allofuranose and 1,2:5,6-di-O-isopropylidene-α-D-glucofuranose and their 3-O-acetyl derivatives were recorded in the presence of various concentrations of the tris(dipivaloylmethanato) (DPM)$_3$ complexes of europium, thulium, and praseodymium. The europium and thulium complexes induced shifts to lower field, whereas the praseodymium analog produced shifts to higher field. However, the use of the lanthanide (DPM)$_3$ complexes as shift reagents was soon replaced by experiments with tris(2,2-dimethyl-6,6,7,7,8,8,8-heptafluoro-3,5-octanedionato)lanthanide(III) chelates [Ln(FOD)$_3$] which had the advantages of higher solubility, and stronger complexing ability, due to the electron-withdrawing fluorine atoms. Detailed investigations of the equilibria between lanthanide shift reagents and a variety of substrates were performed, including simple aliphatic compounds, rigid bicyclic structures, and a number of sugar derivatives, such as the aforementioned di-O-isopropylidene-hexofuranoses, 1,2:3,5-di-O-methylene-α-D-glucofuranose, and methyl 4,6-O-benzylidene-2-deoxy-α-D-*arabino*- and α-D-*ribo*-hexopyranoses. The studies involved the measurement and interpretation of binding constants, bound chemical shifts, lanthanide-substrate stoichiometry, including its variation with solvent, and an examination of various theoretical models. Using ^{1}H–{^{13}C} INDOR, the lanthanide shift reagents were applied to measurement and identification of the ^{1}H and ^{13}C chemical shifts of 2,2-dimethyl-1-propanol. The lack of shifts when gadolinium (III) (DPM)$_3$ was used supported the view that the pseudo-contact mechanism predominates for lanthanide shift reagents.

Nevertheless, the widespread interest in shift reagents was to be short-lived, as the increasing availability and use of higher field spectrometers made the dispersion of NMR spectra by chemical means unnecessary. Using shift reagents undoubtedly had some disadvantages too, including a requirement for anhydrous conditions, increased chemical manipulation, and they tended to reduce the relaxation times of the nuclei in the substrate, because of the diminished tumbling rate of its complex, thus leading to spectral broadening that obscured useful coupling information. By 1981, Laurie *et al.* had observed that Eu(DPM)$_3$-induced line broadening in the ^{1}H NMR spectra of cholesterol and *n*-alkanols was much more severe at 270 MHz than at 100 MHz, due to the effects of chemical exchange being field dependent. Nevertheless, Bulsing, Sanders, and Hall provided a method for removing broadened peaks from the spectra, selectively and controllably, without distorting the remaining signals. This was based on the Carr–Purcell–Meiboom–Gill spin-lock sequence:

$$90°_x - (-t - 180°_y)_n - t - \text{acquire}$$

where t is a delay of ~ 1 ms, n is a variable number, and where x to y represents an RF phase change of 90°. A final nail was driven into the coffin of shift reagents when the development of multidimensional NMR methods provided a much more sophisticated means for increasing the dispersion of the spectra, by separating homo- and heteronuclear frequencies, coupling constants, and subspectra, etc., into different dimensions.

Together with Christopher Grant and Caroline Preston, Laurie pioneered the measurement of ^{1}H spin-lattice relaxation times (T_1) of sugars in carefully deoxygenated solutions. Initially, an audiofrequency pulse method was used with a home-built spectrometer based on a design by Freeman and Wittekoek (1971). However, measurements of complex spectra were tedious by this method, and of limited selectivity. Fortunately, the widespread availability and adoption of the pulse-FT technique was just around the corner, and when the FT method became available to Laurie, he made full use of it to speed up and simplify the measurement of nuclear relaxation times. ^{1}H T_1 values of the anomeric protons of a series of simple pyranoid sugars and derivatives were studied first, using the three-pulse sequence: $180° - t - 90° - T - 90°$ of Freeman and Hill. Later, the more standard two-pulse, inversion recovery sequence of Robert Vold et al. was used, $180° - t - 90°$, depending somewhat on the manufacturer of the spectrometer.

An immediate and important discovery was made; axial protons have shorter T_1 values than equatorial protons. Laurie immediately suspected that this difference was due to dominance of the intramolecular, ^{1}H–^{1}H dipole–dipole relaxation mechanism, which depends on the inverse, sixth power of the internuclear distance. This proved to be the case, based on a theoretical analysis and measurement of T_1s and NOEs in collaboration with NMR experts Freeman, Hill, and Tomlinson at Varian Associates, and further work in Laurie's group. Nonselective, inversion experiments were used throughout the work, in which a short, strong, initial 180° pulse inverted the magnetization of all of the ^{1}H nuclear spins in the sample. However, a class of selective, inversion recovery experiments was developed, using a weak, long 180° pulse to invert the spins for only one chemically shifted nucleus. Double-selective and triple-selective experiments were also explored. Although the results of relaxation measurements were expressed initially as relaxation times in seconds, it became clear that relaxation rates ($R_1 = 1/T_1$ s^{-1}) were more convenient because different contributions to relaxation rates are numerically additive. Laurie showed that the assumption of a 100% dipole–dipole relaxation mechanism could be tested (a method of quality control) by taking the ratio of nonselective to selective relaxation rates, which according to theory was 1.5. This value was indeed observed for a large variety of simple sugar derivatives, thus confirming the dominance of the intramolecular, dipole–dipole relaxation mechanism.

Two important relaxation mechanisms were identified for pyranose carbohydrates: (a) the syn-diaxial effect, in which axial H-1, H-3, and H-5 mutually relax each other; (b) a vicinal gauche effect, for example, in which equatorial H-1 is relaxed by a gauche proton at C-2. The importance of the latter mechanism was demonstrated by the observation that the H-1 T_1 values of both anomers of 2-deoxy-D-*arabino*-hexose are relatively short. These phenomena reflect the close physical proximity of the pairs of protons involved, and the operation of the through-space, dipole–dipole relaxation mechanism. Complete sets of T_1 values could not be obtained for all of the protons in the simple sugars at the time, because of the limited dispersion of their ^{1}H NMR spectra at 100 MHz. However, the spectra of certain acylated derivatives were well dispersed at this frequency. For example, T_1 values could be measured for all of the protons in the anomeric 3,4,6-tri-*O*-acetyl-1-benzoyl-2-chloro-2-deoxy-D-glucopyranoses. The data indicated that the relaxation of H-5 is less influenced by interaction with H-1 and H-3 because its dominant relaxation pathway is via the H-6 nuclei.

The same stereospecific effects were observed for anomeric protons at both the reducing and nonreducing ends of oligosaccharides, namely the α:β T_1 ratio was invariably $\sim$2:1, thus offering a new method for identification of anomers. For monosaccharides and disaccharides, the addition of small proportions of gadolinium nitrate was found to have a leveling effect on the ^{1}H T_1 values, as stereospecific relaxation was replaced by the nonstereospecific effect of the paramagnetic metal ions. In polysaccharides, the stereospecific effects of intramolecular, dipolar relaxation tend to be replaced by the effect of increased correlation time, due to slower molecular tumbling. Comparison of T_1 ratios was thought to be a more reliable method than comparing actual T_1 values because it was demonstrated that the latter numbers depend on concentration by as much as $\sim$35%, and that a temperature change of $\sim$30° can produce a greater than twofold change in the T_1 value. Nevertheless, Laurie was sufficiently thrilled by the results, that in his Tate and Lyle Lecture delivered in Birmingham, United Kingdom in April 1974, he extolled spin-lattice relaxation as a fourth dimension in proton NMR spectroscopy (in addition to the use of chemical shifts, coupling constants, and integrated areas). In one sense, this presaged the development of multidimensional NMR techniques a few years later, but no experiments having spin-lattice relaxation times on the axes have been invented.

Although Allerhand's group had started reporting the ^{13}C T_1 values of sugars in 1971, Laurie extended such studies to the methyl α- and β-D-glucopyranosides and their tetraacetates, in collaboration with sabbatical visitor Klaus Bock, who also expanded the ^{1}H relaxation investigations to the nucleosides uridine and its $2',3'$-O-isopropylidene derivative, to furanose derivatives, and a set of eight isomeric 2,3,4-tri-O-acetyl-1,6-anhydro-β-hexopyranoses (both of the latter with Christian Pedersen). For the methyl glucoside derivatives, the ^{13}C T_1 values of C-1–C-5 were closely similar, indicating a similarity in correlation times. The ^{13}C T_1 values for C-6 were about half those of the ring carbons, reflecting the fact that C-6 is relaxed by interaction with two attached protons, and suggesting that its correlation time is not significantly affected by rotation about the C-5–C-6 bond. For the furanose derivatives studied, it was demonstrated that epimeric pairs may be distinguished by comparison of their nonselective ^{1}H spin-lattice relaxation rates, the important effect being the substantially increased rate caused by the proximity of cis, vicinal protons. The well-dispersed ^{1}H NMR spectra of the 1,6-anhydro-β-hexopyranose triacetates allowed a complete set of nonselective R_1 values to be measured for the ring protons of all eight isomers. These values were analyzed by regressional methods to generate specific, interproton relaxation rate contributions (ρ values): $\rho_{e,e}$ 61, $\rho_{e,a}$ 76, $\rho_{a,a}$ 23, $\rho_{3,6}$ 54, $\rho_{4,6}$ 37, $\rho_{5,6}$ 71, $\rho_{2,4}$ 18 (1,6-anhydrotalose derivative), and $\rho_{2,4}$ 5 ms^{-1} (1,6-anhydroidose derivative), thereby giving a deeper understanding of relaxation mechanisms in carbohydrates. In a multifrequency study, the ratios of R_1 values measured at 90, 270, and 300 MHz to those determined at 100 MHz were found to be constant, thus suggesting the reliability of comparisons of data from different spectrometers. Relaxation contributions from intermolecular, solvent–solute interactions were shown to be small. Hall and Yalpani also used measurements of the ^{13}C T_1 values of pyridine to demonstrate its association in solution with saccharidic substances such as sucrose and alginate, values of the rotational diffusion constants of pyridine being derived by application of the tumbling, rigid ellipsoid model of Woessner.

Collaborations with visiting Professor Laurie Colebrook (and others) helped expand Laurie Hall's horizons into non-carbohydrate areas. Together, they worked on measuring the ^{1}H relaxation rates of a number of heterocyclic systems and natural products, and demonstrated that for many purposes, especially qualitative ones, the quick and simple T_1-null method is almost as accurate as inversion recovery employing a large set of delay times and fitting of the intensity data by nonlinear regression. In the T_1-null method, the time t taken for the magnetization to reach the zero-crossing point (after the initial 180° pulse) is determined by interpolation between straddling data sets. The spin-lattice relaxation rate R_1 is then calculated from the simple expression: $R_1 = 0.69/t$. A highly informative plot was published for the relative relaxation efficiency between two protons versus the inverse sixth power of their internuclear distance. Laurie also made full use of the nulled magnetization phenomenon for the selective suppression of solvent resonances before pulsed field gradient, water-suppressing methods were invented, and for discriminating against sugar proton signals of differing spin-lattice relaxation times.

In a collaboration with student Liane Evelyn and sabbatical visitor John Stevens, the ^{1}H R_1 measurements were extended to the ring protons of all eight D-pentopyranose tetraacetates and methyl α-D-xylopyranoside triacetate, the ^{1}H NMR spectra of which were reasonably well dispersed at 100 MHz. A detailed analysis of the ^{1}H R_1 values of the peracetyl derivatives in terms of proton steric relationships, preponderant chair conformations, and inter-hydrogen distances estimated for a tetrahydropyran model, demonstrated the diagnostic utility of these values for configurational assignments. Hall, Wong, and Hill showed how interproton distances in the model compound 1,2,3,4,7,7-hexachloro-6-*exo*-benzoyloxy-bicyclo-[2.2.1]hept-2-ene could be determined by a combination of nonselective, single-selective, double-selective, and triple-selective relaxation rate experiments, using the inverse sixth-root dependence and a geminal proton distance $r_{2,2}$ of 1.80 Å. For example, the various experiments gave $r_{1,2}$ in the range of 2.27–2.29 Å, as compared with the values 2.28 and 2.27 Å estimated from Dreiding molecular models and computer simulation, respectively. These distance determinations were found to be remarkably precise for several different reasons, not the least of which is the fact that the inverse sixth-root calculation means that a 10% measurement error in the specific relaxation contribution $\rho_{i,j}$ for protons i and j results in only a ~1.7% uncertainty in the distance $r_{i,j}$. In a study of 1,2,3,4-tetra-*O*-trideuterioacetyl-β-D-arabinopyranose, its 5,5-dideuterio derivative, and two isomeric 5-deuterio derivatives at 400 MHz, Hall, Wong, Hull, and Stevens demonstrated how measurements of nonselective ^{1}H relaxation rates could be applied to the determination of the geometry of diamagnetic carbohydrates. Specific, interproton relaxation contributions $\rho_{R,J-j}$ were calculated from rate differences between the protiated and deuterated compounds by application of the equation:

$$\rho_{R,J-j} = 0.6959 \left[R_1^R(\text{ns, H-j}) - R_1^R(\text{ns, D-j}) \right]$$

where R refers to a relaxation receptor proton, J-j is a donor proton, R_1^R(ns, H-j) is the nonselective relaxation rate for the 5-protio species, and R_1^R(ns, D-j)] is that for the 5-deuterio compound. Calculation of interproton distances from the pairwise relaxation contributions, using the inverse sixth power relationship gave results that were in good agreement with those determined by neutron diffraction on the nondeuterated β-D-arabinopyranose tetraacetate, suggesting that it has the same $^1C_4$ conformation in C_6D_6 solution and the solid state. Again, the calculation was referenced to a geminal $r_{5,5'}$ distance of 1.80 Å (from neutron diffraction). These methods were summarized in an ACS monograph with Berry, Welder, and Wong in 1979.

During the 1970s, Laurie was fascinated by Paul Lauterbur's new zeugmatography technique, which together with the work of others, became the basis for the revolutionary magnetic resonance imaging (MRI) method, the word "nuclear" being dropped in clinical radiology, to avoid alarming the public. Laurie built his own superconducting NMR spectrometer based on an Oxford 270 MHz magnet, a Bruker WP-60 console, and other commercial components. He then converted this instrument to perform imaging of phantoms, and a combined NMR tomographic-spectroscopic experiment that mapped out the distribution of chemical substances in an object. The latter technique was referred to as "chemical microscopy." These experiments gave Laurie considerable insight into the MRI technique. This instrument was frequently upgraded by the substitution of new consoles and other components.

Laurie was one of the first to apply two-dimensional (2D) NMR to carbohydrates. With students Subramaniam Sukumar and Michael Bernstein, and visiting scientist Gareth Morris, he demonstrated and extended the application of many of the directly observed 2D NMR techniques of the time. These included the homo- and heteronuclear 2D J-resolved techniques, delayed proton J-resolved NMR that allowed broad resonances to be suppressed, for example, those of dextran in the presence of methyl β-xylopyranoside, proton–proton chemical shift correlation spectroscopy (COSY), nuclear Overhauser enhancement spectroscopy (NOESY), proton–carbon chemical shift correlation (known later as HETCOR), and spin-echo correlated spectroscopy (SECSY). Trideuteriomethyl 2,3,4,6-tetrakis-O-trideuterioacetyl-α-D-glucopyranoside served as a commonly used model compound for these studies.

In the 1970s, we were all impressed by the ability of the proton 2D J-resolved technique to provide (after tilting of the twice-Fourier-transformed data matrix) a proton-decoupled proton spectrum comprised of a singlet at each of the ^{1}H chemical shifts, and resolved J-multiplets by taking slices parallel to the F_1 axis. These J-multiplets displayed all of the coupling constants for a resonance (no chemical shift) and were remarkable in that they were sharper than the spin multiplets in 1D ^{1}H NMR spectra, due to the refocusing effect of the 180° pulse in the J-resolved sequence.

The proton-decoupled proton spectra allowed a distinction to be made between homo- and heteronuclear spin couplings, and Laurie and coworkers also demonstrated nulling of residual solvent resonances during the 2D J-resolved NMR of uridine in aqueous solution, wrote software for 45° tilting of the 2D spectra, and developed experimental protocols for multiple data-acquisition and processing, and a method for acquisition of the 2D J-resolved spectra in phase-sensitive mode. Lately, the 2D J-resolved technique has been less used, as it yields little evidence for spectral assignments.

During his graduate work with Ray Freeman, Gareth Morris was a codeveloper of several of the other 2D NMR techniques and helped Laurie considerably with the proton and carbon chemical shift assignments for a number of monosaccharides and oligosaccharides, including measurement of the one-bond $^{13}C-^1H$ coupling constants of these saccharides by the 2D heteronuclear J-resolved method, a parameter that was already known to be an excellent determinant of anomeric configuration as a result of the work of Perlin and Bock et al. Laurie and Michael Bernstein demonstrated how 2D proton NMR could be used for the de novo sequencing of oligosaccharides, and that a combination of such methods with 2D NOESY and HETCOR allowed complete assignment of the ^1H spectrum of brucine, most of its ^{13}C assignments, and characterization of its ring forms. With sabbatical visitor Jeremy Sanders, Laurie applied the 2D NMR methods and spectral difference techniques to complete assignments for several steroids, particularly 1-dehydrotestosterone and 11β-hydroxyprogesterone, as well as to the conformation of the alkaloid vinblastine (also with Brian Hunter), and to structural proof and conformational deductions for acetaldehyde–enkephalin adducts. At this time, Laurie also commenced a foray into the complex world of zero-quantum NMR (with G. Pouzard and S. Sukumar), which was to be explored in even greater detail during his Cambridge period.

Another interest of Laurie's was to discover additional uses for the abundant, but intractable polymer chitin, and its partially N-deacetylated product, chitosan. Preliminary studies with M. J. Adam demonstrated the strong copper-complexing ability of Schiff's bases constructed from glucosamine derivatives and aromatic aldehydes, for example, methyl 3,4,6-tri-O-acetyl-2-amino-2-deoxy-β-D-glucopyranoside hydrobromide and 3-formyl-2-hydroxybenzoic acid. Then, in an initial demonstration of the reductive amination reaction as applied to chitosan, he and Mansur Yalpani condensed it with salicylaldehyde to give the Schiff base, and reduced the azomethine function with sodium cyanoborohydride, thereby enhancing the metal-chelating properties of chitosan. The copper complex of this product was of interest for demonstrating the utility of ESR spectroscopy for analysis of the metal complexes of glycosaminoglycans. This facile and versatile process was shown to be generally applicable to sugars having a potential aldehyde group, and to keto and lactone sugars. Water-soluble, branched-chain polysaccharides were obtained by reaction of chitosan with D-glucose, D-galactose, N-acetyl-D-glucosamine, cellobiose, lactose, maltose, melibiose, maltotriose, fructose, D-glucoheptonic 1,4-lactone, streptomycin sulfate, C^6-aldehydocyclomaltohexaose, and dextran, in which each (reduced) sugar was substituted on the amino group of the chitosan. Reductive amination of lactose with the small percentage of free amino groups in chitin yielded a water-insoluble, branched-chain derivative. The customary NMR studies were supplemented by an examination of the physical structures of the branched-chain chitosans by scanning electron microscopy (SEM). The products were found to have a wide range of ultrastructures; for example, xerogels derived from 1-deoxymelibiit-1-yl chitosan displayed a highly ordered microfibrillar ultrastructure. However, the SEM results did not indicate any simple relationship between the microarchitecture and the chemical structures of the side chains of the chitosan derivatives. Shear, viscosity, and rheologic properties of solutions, sols, or gels of the chitosan derivatives were also assessed, and the solubility of the compounds in both water and organic solvents could be tailored by the reductive amination of admixtures of sugars and alkyl and aryl aldehydes and ketones. Chitosan derivatives bearing pendant 1-thio-β-D-glucopyranosyl groups that have potential as affinity chromatography materials for β-D-glucosidases were also synthesized and evaluated, as well as the tailored rheology of such derivatives (with K. R. Holme). With Michael Bernstein, Laurie also demonstrated a general synthesis of model glycoproteins, using reductive amination of aldehydes prepared by ozonolysis of alkenyl glycosides. In a collaboration with Jacques Defaye and Andrée Gadelle, reductive amination was applied to 3-oxy-cellulose and 6-O-triphenylmethyl-2-oxy-cellulose. With glucosamine, the 3-oxy-cellulose yielded a highly branched, water-soluble polymer, with 4-amino-2,2,6,6-tetramethylpiperidine-1-oxyl, a spin-labeled cellulose derivative was obtained, and reductive amination with ammonium acetate afforded 3-amino-3-deoxy-cellulose. A similar reaction of the 2-oxy-cellulose derivative with p-toluidine chromium tricarbonyl yielded a novel organometallic cellulose conjugate, and from 1,10-diaza-18-crown-6, an unusual cellulose-crown ether adduct was synthesized.

Although Laurie did not use solid-state NMR to a large extent, he did demonstrate the application of cross-polarization/magic-angle spinning ^{13}C techniques (with M. Yalpani and T. K. Lim) to determination of the primary structures of polysaccharides and their derivatives, inclusion complexes of cyclodextrins, and metal–sugar complexes. For the latter compounds, it was shown that paramagnetic ions such as Cu(II) caused severe broadening of the solid-state ^{13}C NMR spectra, whereas diamagnetic ions like Zn(II) did not.

Together with Michael Adam, John Aplin, Michael Bernstein, Liane Evelyn, Mansur Yalpani, John Waterton, and others, Laurie made an extensive investigation of the application of nitroxide spin labeling to the chemical and physical characterization of polysaccharides, and a host of biologically important materials. The spin-labeled reagents employed were usually 4-amino-2,2,6,6-tetramethylpiperidin-1-oxyl, 3-amino-2,2,5,5-tetramethylpyrrolidin-1-oxyl, or their derivatives. A variety of chemical attachment methods were used, including activation by cyanogen bromide, reductive alkylation and amination, carbodiimide coupling to give amides, reductive amination of galactose residues that had been oxidized to an aldehydo group at C-6 by galactose oxidase, or of bis-aldehydes produced by periodate oxidation of diols, and coupling and substitution using s-triazine derivatives of carbohydrates. A vast array of biomaterials was examined, including agarose, alginate, bovine and human serum albumin, bovine submaxillary mucin, calf thymus DNA, cellulose, chitin, chitosan, fetuin, guar gum, human erythrocyte surface glycoconjugates, locust bean gum, monosaccharide derivatives, sections of human colon, sialic acid residues, starch, sucrose, wood, and xanthan gum.

The nitroxide adducts were studied by ESR spectroscopy, the sensitivity of the technique being such that only a very small proportion of nitroxide was sufficient for good signal:noise in ESR spectroscopy. An important result from these studies was the calculation of motional correlation times τ_c, which permitted assessment of the local dynamic environments of the spin labels in the macromolecular structures. Polysaccharide supports for affinity chromatography were also investigated, particularly the binding selectivity resulting from spacer arms of various lengths. Application of this technology to materials in different physical states was feasible, including solutions, solids, sols, and gels. The relationship of microscopic and macroscopic solution viscosities to τ_c was studied, and the melting and setting of gels. The τ_c values of the spin-labeled macromolecular systems were generally shorter (1–2 ns) than expected, thus suggesting that the nitroxide groups have additional degrees of motional freedom.

During this period, Laurie also investigated the preparation of radiolabeled, iodo and bromo sugars via carbohydrate boranes (with J.-R. Neeser), fluorescent probe–sugar conjugates (with M. Yalpani), sugar ferrocene derivatives (with M. J. Adam), and extended his studies of the conformations of polycyclic sugar derivatives.

With S. Sukumar and/or V. Raganayagam, a long series of papers was published on NMR imaging-related technologies, including chemical microscopy using a high-resolution NMR spectrometer, with evaluation of field gradient magnitudes and B_1 RF field homogeneity, rapid data-acquisition for NMR imaging by the projection-reconstruction method, chemical shift-resolved tomography using frequency-selective excitation and suppression of specific resonances, 3D FT NMR imaging by a high-resolution, chemical shift-resolved planar technique, a new image processing methodology for NMR chemical microscopy, chemical imaging with measurement of proton spin-lattice relaxation rates of model phantoms, evaluation of the magnitudes of magnetic field gradients by line-shape analysis of NMR spectra of simple objects (with S. D. Luck), chemical shift-resolved tomography using 4D FT imaging, construction of a high-resolution NMR probe for imaging with submillimeter spatial resolution, visualization of chromatography columns, mapping of pH and temperature distribution using chemical shift-resolved tomography (with S. L. Talagala), and MRI of wood (with Wendy Stewart and Paul Steiner), particularly evaluation of the distribution of water by 3D proton NMR volume imaging, and the detection of hidden morphology (also with S. Chow).

On a lighter note, one of Laurie's hobbies in Vancouver was wine making. He would order a ton of grapes from California, and using his chemical skills, produce a most presentable wine, which guests were invited to sample at his home on the east side of the park bordering the UBC campus. In this activity, he appears to have been influenced by his fellow chemistry faculty member, the late Guy Dutton (see *Advances*, Volume 56).

List of Professor Hall's Graduate Students at UBC

Ph.D.: Michael Adam, John Aplin, Ian Armitage, Julius Balatoni, Michael Bernstein, Joffrey Berry, Jacques Briand, Christopher Grant, Kevin Holme, T. K. Lim, Jonathan Lee, Stanley Luck, John F. Manville, Caroline Preston, Vasanthan Rajanayagam, Paul Steiner, Wendy Stewart, Subramaniam Sukumar, Lalith Talagala, Kim Wong, Mansur Yalpani.

M.Sc.: Liane Evelyn, Victoria Gibb, Ben Malcolm, Diane Miller, Terry Murname, Joyce Schachter.

Postdoctoral Fellows and Sabbatical Leaves

John Campbell, Klaus Bock, Laurie Colebrook, Gareth Morris, G. Pouzard, Jeremy Sanders, John Stevens, John Walsh, John Waterton.

The Cambridge Period, 1984–2005

In 1982, Laurie received an offer that he could not refuse. His excellent scientific reputation, drive, and enthusiasm had impressed a private benefactor, Dr. Herchel Smith so much, that Laurie was invited to apply for a position in which he would build a new Department of Medicinal Chemistry at Cambridge University in the United Kingdom, and be the first occupant of a new Herchel Smith Chair of Medicinal Chemistry. Dr. Smith had made his fortune from the birth control pill and provided sufficient funding for an entire new department, the creation of which at Cambridge was said to happen only once in 100 years. Laurie was given a green field near the Addenbrooke's Hospital, and with careful planning and attention to furnishings, a new department was built, with facilities personalized for his extensive, future research plans. On moving to Cambridge, Laurie's main focus became MRI. He applied this new technology not only to medical problems in keeping with the mission of his department but also to materials characterization, process control, food analysis, the fundamental technology of MRI, and the elucidation of physical phenomena.

From Cambridge, Laurie published a long series of papers with the late Timothy Norwood on the technology and application of zero-quantum NMR methods. The attraction of this technique was its insensitivity to magnetic field inhomogeneity, which allowed high-resolution NMR spectra having natural line-widths to be detected, as opposed to those broadened by inhomogeneity. They applied this technology to both the traditional 2D NMR methods, such as *J*-resolved, chemical shift correlation, and SECSY, but also to imaging, including the publication of a book chapter in "New Imaging Methods for the Brain." Regrettably, Dr. Norwood died on November 12, 1999 at the age of 37, as a result of injuries suffered in a road accident.

During the middle part of his career in Cambridge, Laurie was ably assisted and supported by Dr. Adrian Carpenter, and together with their many students, postdoctoral fellows, and collaborators, they published a large number of papers on the practical applications of MRI, and its fundamental technology. Laurie Hall also continued his collaboration with Laurie Colebrook during the Cambridge period, working with him on applications of gradient-selected COSY and DQCOSY to brucine, gibberellic acid, and 2-deoxy-D-*arabino*-hexose, estimation of the spin-lattice relaxation times of metabolites in urine, and selection of multiple-quantum spectra of molecules in liquid-crystalline solution using pulsed magnetic field gradients.

The medical applications of MRI included detection of experimental allergic encephalomyelitis in primates, the use of paramagnetic pharmaceuticals for functional studies, mathematical models, instrumentation, and techniques for cerebral blood flow, the delivery and degree of oil/water partition of paramagnetic contrast reagents, spectral editing techniques for ^{31}P NMR of blood, paramagnetic metal complexes of polysaccharides as contrast agents (for enhancement of the relaxation of water), imaging of the brain and brain stem in elderly patients with dizziness, lipid characterization in animal models of atherosclerosis, mechanical output in ^{31}P MRS muscle studies, axillary fibrosis or recurrent tumor in breast cancer studies, localized, *in vivo* spectroscopy of the human kidney, cardiac tagging in the rat using a DANTE pulse sequence, imaging of the head or spine, measurement of regional left and right ventricular function and geometry, relative metabolic efficiency of concentric and eccentric exercise determined by ^{31}P MRS, microscopy and moisturization processes of living human skin, physiological monitoring of the anesthetized rat, and a simple device for respiratory gating of laboratory animals. Other topics studied included schizophrenia, anatomy and induced ischemic damage in lab animals, MRI planes for the 3D characterization of human coronary arteries, visualizing changes in cerebral blood flow produced by pharmaceuticals, animal-brain studies, visualization of the natural state of water in human stratum corneum, magnetic resonance in hyperthermia and oncology, design of biplanar gradient coils for MRI of the human torso and limbs, simultaneous evaluation of the effects of RF hyperthermia on intra- and extracellular tumor pH, cortical spreading depression in the gyrencephalic feline brain, mapping of the cerebral response to hypoxia by graded symmetric spin-echo EPI, RF coils for combined MR and hyperthermia studies, assessment of myocardial changes and blood velocity and the effects of angiotensin converting enzyme inhibition in diabetic rats, serial MRI, functional recovery, and long-term infarct maturation in a nonhuman primate model of stroke, and MRI velocimetry and quantification of fluid flow through a clinical blood filter and kidney dialyzer.

During the early 1990s, Laurie became interested in the application of MRI to osteoarthritis problems, particularly investigations of articular cartilage, and visualization of subchondral erosion in monoarticular arthritis in the rat. This interest led to a significant collaboration with Dr. J. A. Tyler of the Strangeways Research Institute in Cambridge, an institute that had researched a number of human diseases, but with a specialization in arthritis. The collaboration involved studies of this condition using both animal and human models of the knee joint, especially joint degeneration and three stages of damage in the mouse knee, cartilage swelling and loss in a spontaneous model in the guinea-pig knee, automatic computerized measurement of hyaline cartilage thickness in the distal and proximal interphalangeal joints in fingers, including correlations with histology, as well as visualization of focal cartilage lesions in the excised mini-pig knee, the mass transport of small organic molecules and metal ions through articular cartilage, and of joint flexion, cartilage compression, and solute perfusion. Also studied were detection and monitoring of progressive degeneration of osteoarthritic cartilage, quinolone arthropathy in immature rabbits treated with a fluoroquinolone, and measurement of localized cartilage volume and thickness of human knee joints by computer analysis of 3D MR images.

Following the end of this collaboration, Laurie continued these studies by means of a novel RF coil configuration for *in vivo* and *ex vivo* imaging of arthritic rabbit knee joints, optimization of diffusion measurements using Cramer–Rao lower bound theory and its application to articular cartilage, MRI as a noninvasive means for quantitating the dimensions of articular cartilage in the human knee, and high-resolution MRI relaxation measurements of water in such cartilage of the meniscectomized rat knee.

Laurie was determined to extend the application of MRI into nonmedical areas and embarked on a comprehensive study of food materials. This included food imaging under the skin, packaged foods; quantitative determination of water and lipid in sunflower oil and water–meat/fat emulsions, dairy products in 2D and 3D; visualization of hydration of foods, wheat flake biscuits during baking; NMR imaging of fresh and frozen courgettes and of the migration of liquid triacylglycerol in chocolate. Also studied were the binding of manganese ions and interdiffusion of sodium ions in brine-cured pork; authentication of the freeze thawing of beef, lamb, and pork;

measurement of textural changes in food by MRI relaxometry; 3D imaging of chocolate confectionery and changes during its storage; effects of freeze thawing on the MRI parameters of trout, cod, and mackerel; and 3D mapping of microwave-induced heating patterns and assurance of minimum temperatures achieved in microwave and conventional food processing, including a comparison with fiber optic thermometry. Drying of strawberries, heating of commercial baby foods in glass jars, and quantitation of moisture and structural changes during convection cooking of chicken were also studied.

Laurie's studies of process tomography included 3D MRI of biofilm reactors, for example, metal accumulation and deposition in a *Citrobacter* process, measurement of the flow field through the reactor, and identifying particles in industrial systems using susceptibility artifacts. Nondestructive testing of materials was also thoroughly investigated during this period, including void detection in carbon-fiber composites and comparison with ultrasound methods, Fickian diffusion and inhomogeneities in rubbery polymers, polymerization of methyl methacrylate, simple NMR imaging of solid polymers at elevated temperatures, voids in cement slurries, the origins of contrast in spin-echo images of plant tissues, temperature mapping in solid polymers using the temperature dependence of NMR relaxation times, and noninvasive imaging of fungal colonization, host response, and immobilized, long-lived free radicals at the host–pathogen interface in sycamore wood. A related project concerned initial defense responses in sapwood of *Eucalyptus nitens*, following wounding and fungal inoculation. Studies of the structure of concrete were also performed, including observation of fractures and flexural cracks in loaded concrete beams by MRI. A novel, non-line-of-sight method was reported for coating hydroxyapatite onto the surface of support materials by biomineralization.

Laurie also used MRI extensively for the characterization of physical phenomena, including fluids in porous solids, ingress of solvent into polymers, visualization of fluid motion by tagged MRI, 3D MRI of very slow coherent motion in complex flow systems, dynamic NMR imaging of rapid depth filtration of clay in porous media, visualization of viscous fingering, automatic analysis of tagged images of laminar fluid flow, diffusion coefficients in aqueous gels measured by 1D NMR imaging, spatial and temporal mapping of water in soil, saturation gradients in drainage of porous media, thermal convection, pore geometry, 3D autocorrelation for the determination of large pore sizes, strategies for rapid rheometry by 3D MRI velocimetry, drying in granular beds of nonporous particles, NMR flow imaging of aqueous polysaccharide solutions, Bayesian analysis for quantitative NMR flow and diffusion imaging, and gelation processes. This research topic also included MRI measurements of pulsatile flow, rheometry and detection of apparent wall slip for Poiseuille flow of polymer solutions and particulate dispersions by NMR velocimetry, visualization of the diffusion of metal ions and organic molecules by MRI of water, soil–water transport, flow and diffusion images from Bayesian spectral analysis of motion-encoded NMR data, infiltration into a heterogeneous soil, water distribution in fungal lesions in sycamore wood determined gravimetrically and by NMR imaging, chemically resolved NMR velocimetry, measurement of paramagnetic tracer ion diffusion, comparison of MRI velocimetry with computational fluid dynamics, rapid MRI and velocimetry of cylindrical Couette flow, noninvasive visualization and quantitation of filtration and separation processes, recurrent ponded infiltration into structured soil, measurement of particle separation, visualization of air bubbles, and measurement of the settling and packing of solid particles from aqueous suspensions. Subjects of petrochemical interest were also investigated, including thin-slice chemical shift imaging of oil and water in sandstone rock, spatially resolved T_1 relaxation measurements and diffusion for brine-saturated reservoir cores, and clay filtration in rock cores.

Studies of calcium alginate that had been started at UBC were continued in Cambridge, especially MRI visualization of the geometries of objects fabricated from this material, mapping of the spatial variation of alginate concentration in gels, and the gelation of sodium alginate with calcium ions. Related topics were the NMR flow imaging of aqueous polysaccharide solutions, and a combined magnetization transfer and null point technique for study of gelation processes by quantitative MRI, which led to a better understanding of the properties of alginate solutions and gels.

Laurie's scientific output included more than 570 publications and 14 patents. He was also a prolific and highly sought after science lecturer. The available records indicate that between January, 1985 and September, 2005, he delivered 514 invited lectures and seminars worldwide.

Laurie also had an impact on the annual NMR meeting in the USA, known as the Experimental NMR Conference (ENC). He lectured on some of his important work at the conference, and in 1997, was responsible for keeping it in the alternate years at Asilomar Conference Center, a beautiful site on the Pacific coast near Monterey, California, where it had been held since the early 1970s. Since then, the NMR field had expanded in importance and capability, and by 1997, the number of attendees at the ENC had reached ~ 1600, a number that was thought to be too large for the Asilomar facilities. The organizers of the conference decided to abandon the Asilomar location for a hotel in Houston, Texas. However, Laurie organized a successful grass-roots petition to have the conference remain in Asilomar, and the relocation to the site in Texas was canceled. In recent years, the number of attendees at the ENC has declined to 800–1100, due to various factors, such as economic depression, pharmaceutical mergers, and the splintering of the NMR community into separate conferences for protein NMR, small-molecule NMR, solids NMR, and MRI. However, following Laurie's lead, the ENC remains at Asilomar in the alternate years. He has been described by NMR spectroscopist Dr. Gene Mazzola as "the savior of Asilomar."

Honors and Awards, 1971–2005

Alfred P. Sloan Foundation Fellow, 1971–1973; Jacob Bielly Faculty Research Prize, UBC, 1974; Tate and Lyle Award for Carbohydrate Chemistry, The Chemical Society, 1974; Merck, Sharpe, and Dohme Lecture Award, Chemical Institute of Canada, 1975; Corday Morgan Medal and Prize, Chemical Society, 1976; Barringer Award, Spectroscopy Society of Canada, 1981; Fellow of the Royal Society of Canada, 1982; Canada Council Killam Research Fellow, 1982–1984; Lederle Professor, Royal Society of Medicine, 1984; Fellow of the Chemical Institute of Canada; Fellow of Emmanuel College, Cambridge, 1987; Interdisciplinary Award, Royal Society of Chemistry, 1988; Chemical Analysis and Instrumentation Award, Royal Society of Chemistry, 1990; M.A., Cambridge University, 1990; Fellow, Cambridge Philosophical Society; Honorary D.Sc., University of Bristol, 2000; Life Fellow of Emmanuel College, Cambridge, 2005.

Epilogue

The latter part of Laurie Hall's career in Cambridge was marred by tragedy when his friend Professor Regitze Vold (Gitte) died after falling accidentally from a balcony. Gitte had traveled from the University of California, San Diego to attend an NMR symposium in Cambridge in April 1999, in appreciation of Professor Ray Freeman. This tragedy had an unsettling effect on Laurie, and indeed, on the NMR community in general, as Gitte was well liked and her work widely respected. The events surrounding her death, including the inquest, were extremely stressful for Laurie, as he had designed into the building the balcony from which Gitte fell.

According to the university regulations of the time, Laurie was required to retire at age 67, for which he was quite unprepared. Owing to the unique status of his department, the university was uncertain how to handle it, closing it for a time, a development that was extremely depressing for Laurie, since he had built it from the ground up. However, due to the fact that his department was privately funded, he was able to secure an extension for a while to clear out the laboratory, and to complete work with students. Following his retirement in 2005, Laurie's health began to deteriorate rapidly, but he was reluctant to visit a physician, perhaps fearing what might be revealed. He had suffered from high blood pressure for several years and had had several blackouts. His condition led to some kind of stroke while he was traveling from Cambridge to London by train in September, 2006—he awoke in hospital along the way.

This event caused major short-term memory loss, and difficulty managing his affairs. No longer able to drive or fly, Laurie took the bus, except when visitors driving to Cambridge were able to offer him transportation. His son, D'Arcy, helped him considerably after his stroke, but a three-barn renovation project that Laurie had started some years earlier with his two sons remained unfinished when he died.

About a year before his death, Laurie was allowed to fly again, and visited Canada to attend and greatly enjoy, a family celebration of his 70th birthday. This was held in Whistler, BC, the stunningly beautiful location of the International Carbohydrate Symposium in 2006, and the Winter Olympics in 2010. The visit was a wonderful opportunity for Laurie to meet for one last time with all of his children and grandchildren.

Laurie died of massive heart failure at his home near Cambridge, on August 28, 2009. Survivors include his wife of 47 years, Winifred Margaret, twin boys D'Arcy and Dominic, two daughters Gwendolen and Juliet, and two grandsons and four granddaughters. A celebration of Professor Hall's life was conducted by Vanessa Dennis, Accredited Officiant of the British Humanist Association, at West Suffolk Crematorium on September 7, 2009, with a final tribute from Winifred Hall. Professor Hall was also remembered by colleagues, friends, and family in a memorial service held in Emmanuel College Chapel on May 8, 2010, including a reflection by Juliet Fallowfield, a discourse "Early Years and Beyond" by the Reverend Michael Withers, reminiscences from Ray Freeman (in absentia, but read by Dr. Finian Leeper), Alun Lucas, Keith McLauchlan, Gareth Morris, and Jeremy Sanders, and the reading of a sonnet from John Donne by Winifred Hall.

I count myself fortunate to have known Laurie Hall and to have been inspired by his scientific work. Much to my chagrin, I could never keep up with reading all of his prodigious output. Laurie's science was a bright star that burned fiercely throughout his career. He was a brilliant and exceptionally hard-working man, and truly an entrepreneur in science, being daring and risk-taking by disposition. His scientific legacy includes substantial contributions to the synthesis and NMR of carbohydrates and other molecules, and to the technology and application of NMR imaging.

BRUCE COXON

Acknowledgments

Sincere thanks are due to Professor Laurie Colebrook (retired) for providing very considerable information on Laurance Hall. This chapter has been preceded by at least three obituaries: by L. D. Colebrook, September 25, 2009, for the British NMR Discussion Group website, www.nmrdg.org.uk/History_of_the_NMR-DG/History_of_the_NMRDG.html; by Ray Freeman, The Daily Telegraph, October 1, 2009; and author unknown, The Times, December 1, 2009.

Some Selected Review Articles by L. D. Hall

1. L. D. Hall, Nuclear magnetic resonance, *Adv. Carbohydr. Chem.*, 19 (1964) 51–93.
2. L. D. Hall and J. F. Manville, Studies of carbohydrate derivatives by nuclear magnetic double-resonance. Part I. Studies of deoxy-sugars by proton magnetic resonance spectroscopy, in S. Hanessian, (Ed.), *Deoxy-Sugars, Advances in Chemistry Series*, Vol. 74, American Chemical Society, Washington, DC, 1968, pp. 228–252.
3. L. D. Hall, Solutions to the hidden resonance problem in proton nuclear magnetic resonance spectroscopy, *Adv. Carbohydr. Chem.*, 29 (1974) 11–40.
4. L. D. Hall, Spin lattice relaxation: A fourth dimension for proton NMR spectroscopy. The Tate and Lyle Lecture Award, *Chem. Soc. Rev.*, 4 (1975) 401–420.
5. L. D. Hall, A fourth dimension for proton NMR spectroscopy, The Merck, Sharp, and Dohme Lecture, *Chem. Can.*, 28 (1976) 19–23.
6. L. D. Hall, K. F. Wong, and W. Schittenhelm, High resolution nuclear magnetic resonance spectroscopy. A probe of the structure and solution conformation of sucrochemicals, *ACS Symposium Series: Sucrochemistry*, No. 41, American Chemical Society, Washington, DC, 1977, pp. 22–39.
7. J. M. Berry, L. D. Hall, D. G. Welder, and K. F. Wong, Proton spin lattice relaxation: A new quantitative measure of aglycon sugar interactions, in W. A. Szarek and D. Horton, (Eds.), *Anomeric Effect, Origin and Consequences*, No. 87, American Chemical Society, Washington, DC, 1979, pp. 30–49, (Chapter 3).
8. L. D. Hall, High resolution nuclear magnetic resonance spectroscopy, in W. Pigman and D. Horton, (Eds.), *The Carbohydrates*, Vol. IIB, Academic Press, New York, USA, 1981, pp. 1299–1325.
9. L. D. Hall, *In-vivo* imaging of human biochemistry. Positron emission tomography and nuclear magnetic resonance, *Chem. Can.*, 35 (1983) 23–26.
10. L. D. Hall and P. G. Hogan, Paramagnetic pharmaceuticals for functional studies, in V. R. McReady, M. D. Leach, and P. J. Ell, (Eds.), *Functional Studies Using Nuclear Magnetic Resonance*, Springer-Verlag, Berlin, Germany, 1986, pp. 107–127.
11. L. D. Hall, NMR chemical microscopy, in C. Lunsford, (Ed.), *Natural Products Research: The Impact of Scientific Advances*, The Fifth Philip Morris Science Symposium, pp. 74–105.
12. L. D. Hall and S. C. R. Williams, Nuclear magnetic resonance imaging, in A. Rescigno and A. Boicelli, (Eds.), *Cerebral Blood, Mathematical Models, Instrumentation, and Imaging Techniques, Proceedings of the NATO ASI Series, June 2–13, 1986, L'Aquila, Italy*, Vol. 153, Plenum Press, New York, USA, 1988, pp. 186–199.
13. T. A. Carpenter, and L. D. Hall, (Eds.), Future prospects for NMR-imaging-spectroscopy in studies of human nutrition, Bristol Myers Squibb Symposium, September 12, 1989, New Techniques in Nutritional Research, Academic Press, New York, USA, 1990, pp. 335–359, (Chapter 16).
14. T. A. Carpenter, L. D. Hall, and R. C. Hawkes, NMR imaging: A new method for non-destructive testing, *Spectrosc. World*, 2 (1990) 20–24.
15. J. J. Attard, L. D. Hall, N. Herrod, and S. L. Duce, Materials mapped with NMR, *Phys. World* (1991) 41–45.

16. P. Jackson, N. J. Clayden, J. A. Barnes, T. A. Carpenter, L. D. Hall, and P. Jezzard, New analytical techniques for advanced polymer composites, in A. Kwakernaak and L. van Arkel, (Eds.), *Proceedings of the 12th International European Chapter Conference of the Society for the Advancement of Material and Process Engineering, Maastricht, The Netherlands, May 28–30, 1991,* Advanced Materials: Cost Effectiveness, Quality Control, Health and Environment, Elsevier, Amsterdam, The Netherlands, 1991, pp. 277–288.
17. P. Jezzard, J. J. Attard, T. A. Carpenter, and L. D. Hall, Nuclear magnetic resonance imaging in the solid state, *Progr. NMR Spectrosc.*, 23 (1991) 1–41.
18. R. J. Hodgson, T. A. Carpenter, and L. D. Hall, Magnetic resonance imaging of osteoarthritis, in K. E. Kuettner, R. Schleyerbach, J. G. Peyron, and V. C. Hascall, (Eds.), *Articular Cartilage and Osteoarthritis,* Raven Press, New York, USA, 1992, pp. 629–642.
19. L. D. Hall, A new window into the human body, *Sci. Public Affairs (BAAS)* (1993) 25–29.
20. A. J. Lucas, G. K. Pierens, M. Peyron, T. A. Carpenter, L. D. Hall, R. C. Stewart, D. W. Phelps, and G. F. Potter, Quantitative porosity mapping of reservoir rock cores by physically slice selected NMR, in P. F. Worthington and C. Chardaire-Rivière, (Eds.), *Advances in Core Evaluation III: Reservoir Management,* Reviewed Proceedings of the Society of Core Analysts Third European Core Analysis Symposium, France, September 14–16, 1992, Gordon and Breach Science Publishers, Reading, UK, 1993, pp. 3–24.
21. J. J. Wright and L. D. Hall, NMR imaging of packaged foods, in M. Mathlouthi, (Ed.), *Food Packaging and Preservation,* Chapman and Hall (Blackie Academic and Professional), New York, USA, 1994, pp. 197–208 (Chapter 11).
22. L. D. Hall, J. J. Herrod, T. A. Carpenter, and P. J. McKenna, Magnetic resonance imaging in schizophrenia: A review of clinical and methodological issues, in R. J. Ancill, S. Holliday, and J. Higginbottam, (Eds.), *Schizophrenia Exploring, the Spectrum of Psychosis,* Wiley, Chichester, UK, 1994, pp. 115–135 (Chapter 7).
23. J. J. Attard and L. D. Hall, Magnetic resonance imaging, *Encyclopaedia Appl. Phys.*, 9 (1994) 47–70.
24. M. H. G. Amin, L. D. Hall, R. J. Chorley, K. S. Richards, T. A. Carpenter, and B. W. Bache, Visualisation of static and dynamic water phenomena in soil using magnetic resonance imaging, in V. P. Singh and B. Kumar, (Eds.), *Subsurface Water Hydrology,* Kluwer Academic Publishers, Dordrecht, The Netherlands, 1996, pp. 3–16.
25. L. D. Hall, The image maker, *Chem. Britain* (1996) 43–45.
26. S. L. Duce and L. D. Hall, Foods & grains studied by MRS & MRI, in D. M. Grant and R. K. Harris, (Eds.), *Encyclopaedia of NMR,* Wiley, Chichester, UK, 1996, pp. 2091–2093.
27. T. J. Norwood and L. D. Hall, Multiple quantum coherence imaging, in D. M. Grant and R. K. Harris, (Eds.), *Encyclopaedia of NMR,* Wiley, Chichester, UK, 1996, pp. 3138–3143.
28. E. J. Fordham, D. Xing, J. A. Derbyshire, S. J. Gibbs, T. A. Carpenter, and L. D. Hall, Flow and diffusion images from Bayesian spectral analysis of motion-encoded NMR data, in S. Sibisi and J. Skilling, (Eds.), *Maximum Entropy, and Bayesian Methods,* Proceedings of the Fourteenth International Workshop on Maximum Entropy and Bayesian Methods, Cambridge, UK, 1994, Springer, New York, USA, 1996, pp. 1–12.
29. L. D. Hall, P. J. Watson, and J. A. Tyler, Magnetic resonance imaging and the progression of osteoarthritis, osteoporosis, and ageing, in D. Hamerman, (Ed.), *Osteoarthritis Public Health Implications for an Ageing Population,* The Johns Hopkins University Press, Baltimore, USA, 1997, pp. 215–229 (Chapter 12).
30. J. Rama, L. D. Hall, and N. M. Bleehen, Magnetic resonance, hyperthermia and oncology, *Curr. Sci.*, 76 (1999) 794–799.
31. M. Paterson-Beedle, K. P. Nott, L. E. Macaskie, and L. D. Hall, Study of biofilm within a packed-bed reactor by three-dimensional magnetic resonance imaging, *Methods Enzymol.*, 337 (2001) 285–305.

32. L. D. Hall, M. H. G. Amin, S. Evans, K. P. Nott, and L. Sun, Quantitation of diffusion and mass transfer of water by MRI, in Z. Berk, R. B. Leslie, P. J. Lillford, and S. Mizrahi, (Eds.), *Water Science for Food, Health, Agriculture and Environment ISOPOW 8,* Technomic Publishing Company Inc., Lancaster PA, USA, 2001, pp. 255–273.
33. K. P. Nott and L. D. Hall, New techniques for measuring and validating thermal processes, in P. Richardson, (Ed.), *Improving the Thermal Processing of Foods, Part 4,* Woodhead Publishing Limited, Cambridge, UK, 2004, pp. 385–407.

POTENTIAL TREHALASE INHIBITORS: SYNTHESES OF TREHAZOLIN AND ITS ANALOGUES

AHMED EL NEMR[a] and EL SAYED H. EL ASHRY[b,c]

[a]Environmental Division, National Institute of Oceanography and Fisheries, Kayet Bey, El Anfoushy, Alexandria, Egypt
[b]Chemistry Department, Faculty of Science, Alexandria University, Alexandria, Egypt
[c]Fachbereich Chemie, Universität Konstanz, Fach M 725, Konstanz, Germany

I. Introduction	46
II. Natural Occurrence and Characterization of Trehazolin	47
III. Biosynthesis of Trehazolin	48
IV. Retrosynthetic Analysis of Trehazolin and Its Analogues	49
V. Synthesis of the Sugar Isothiocyanate Fragment	50
VI. Synthesis of the Trehazolamine Fragment	53
1. Synthesis from D-Glucose	54
2. Synthesis from D-Mannose	60
3. Synthesis from D-Ribose	63
4. Synthesis from D-Arabinose	65
5. Synthesis from D-Mannitol	66
6. Synthesis from Sugar 1,5-Lactones	66
7. Synthesis from *myo*-Inositol	68
8. Synthesis from D-Ribonolactone	73
9. Synthesis from Cyclopentadiene	77
10. Synthesis from Acylated Oxazolidinones	81
11. Synthesis from *cis*-2-Butene-1,4-diol	82
12. Synthesis from Norbornyl Derivatives	85
13. Synthesis from 3-Hydroxymethylpyridine	85
VII. Synthesis of Trehazolins	87
1. Trehazolins Having Trehazolamines at the Anomeric Center	87
2. Trehazolines with Trehazolamines on Positions Other than the Anomeric Center	100
3. Synthesis of 1-Thiatrehazolin	104
4. Deoxynojirimycin Analogues	106
Acknowledgments	108
References	108

ABBREVIATIONS

Ac, acetyl; AIBN, azobis(isobutanonitrile); All, allyl; AR, aryl; Bn, benzyl; t-BOC, tert-butoxycarbonyl; Bu, Butyl; Bz, benzoyl; CAN, ceric ammonium nitrate; Cbz, benzyloxycarbonyl; m-CPBA, m-chloroperoxybenzoic acid; DAST, diethylaminosulfur trifluoride; DBU, 1,8-diazabicyclo[5.4.0]undec-7-ene; DCC, N,N'-dicyclohexylcarbodiimide; DCM, dichloromethyl; DCMME, dichloromethyl methyl ether; DDQ, 2,3-dichloro-5,6-dicyano-1,4-benzoquinone; DEAD, diethyl azodicarboxylate; L-(+)-DET, L-(+)-diethyl tartrate; L-DIPT, L-diisopropyl tartrate; D-DIPT, D-diisopropyl tartrate; DMAP, 4-dimethylaminopyridine; DME, 1,2-dimethoxyethane; DMF, N,N-dimethylformamide; DMP, 2,2-dimethoxypropane; Et, ethyl; Im, imidazole; KHMDS, potassium hexamethyldisilazane; Me, methyl; Me_2SO, dimethyl sulfoxide; MOM, methoxymethyl; MOMCl, methoxymethyl chloride; Ms, methylsulfonyl; MS, molecular sieves; NBS, N-bromosuccinimide; NIS, N-iodosuccinimide; NMO, N-methylmorpholine N-oxide; PCC, pyridinium chlorochromate; Ph, phenyl; PMB, p-methoxybenzyl; PPTs, pyridinium p-toluenesulfonate; i-Pr, isopropyl; Py, pyridine; rt, room temperature; TBAF, tetrabutylammonium fluoride; TBS, tert-butyl dimethylsilyl; TBDMSCl, t-butylchlorodimethylsilane; Tf, trifluoromethylsulfonyl; Tf_2O, trifluoromethylsulfonic anhydride; TFA, trifluoroacetic acid; THF, tetrahydrofuran; TMS, trimethylsilyl; TPAP, tetra-n-propylammonium perruthenate; p-TsOH, p-toluenesulfonic acid

I. INTRODUCTION

Enzymes play essential roles in the biological functions of all living organisms. Their regular functions catalyzing metabolic reactions may be modulated by molecules acting as inhibitors, and such inhibitors may be effective in treating some diseases. Such inhibitors are finding increasing uses as chemotherapeutic agents. Many inhibitors[1-3] have been isolated from natural sources,[4,5] and much synthetic effort on them and their analogues is in progress. The subject of enzyme inhibitors is considered to be one of the "hot" areas of research.[1-8]

The biosynthesis of various glycoconjugates is controlled by the glycosyltransferases, and the glycosidases are involved in the degradation of polysaccharides and glycoconjugates. Certain glycoconjugates profoundly affect a variety of biological functions, such as signal transduction and cell recognition.[9,10] Glycosidases are those enzymes that catalyze the bonds. In consequence, glycosidase inhibitors have received a great deal of attention as valuable biochemical tools and as potential chemotherapeutic agents.[1-10] The development of specific inhibitors of glycosidases is an important target in the field of glycobiology.

Trehalase (EC 3.2.1.28) is an enzyme that specifically hydrolyzes α,α-trehalose (**1**),[11] a nonreducing (1 ↔ 1)-linked disaccharide possessing two molecules of α-D-glucopyranose. Hydrolysis by this enzyme is essential in the life functions of various organisms, such as fungi, insects, and nematodes. Trehalase is widely distributed in insects, plants, and animals.[12,13] Trehalose is found in a variety of organisms including yeast, bacteria, fungi, insects, and nematodes, where it has been found to serve as a source of energy and carbon.[11] Moreover, it might have a signaling function and also play a protective role against different stress conditions. Potent inhibitors of trehalase might thus be useful as insecticides.

Several trehalase inhibitors have been isolated from natural sources: these include deoxynojirimycin, validamycins, validoxylamines, salbastatin, calystegin B4, and trehazolin, the last one being the most potent. Trehazolin is used to control blight sheath of rice, caused by the plant pathogenic fungus *Rhizoctonia solani*, and it thus has potential fungicidal activity.[12–15]

Trehazolin is a slow and tight-binding inhibitor of trehalose.[12,16] It is of the reversible, competitive type with respect to trehalose. The hypothesized[17,18] presence of two subsites in the active center of trehalase has been supported by kinetic studies using two competitive inhibitors.[19] The role of a catalytic acid group and a nucleophilic water molecule in the active site of the enzyme has been suggested.[20,21] The inhibition is attributable to the formation of a complex between trehazolin and trehalase via the nitrogen atom of the anomeric center and the carboxyl group in trehalase.[22] The hydroxyl groups in trehazolin are considered to be topologically essential, providing hydrogen bonding with the active sites for tight binding.[23]

Trehalozin specifically inhibits the hydrolysis of α,α-trehalose (**1**), probably through acting as a close mimic of **1**, or more probably to the postulated glycopyranosyl cation intermediate involved in the hydrolytic step of glycosides, or a transition state leading to it (Fig. 1).[13,24] Trehalose (**1**) and the trehalase enzyme have also been reported to participate in germination of ascospores in fungi[25–29] and in glucose transport in mammalian kidney and intestine.[30] Trehazolin (**2**) and its analogues have important implications in immunology, virology, and oncology.[31] Its structure was elucidated to be that of a pseudodisaccharide consisting of an α-D-glucopyranose moiety linked to a unique aminocyclopentitol, trehazolamine (**5**), through a cyclic isourea group.

II. Natural Occurrence and Characterization of Trehazolin

Trehazolin (**2**) was isolated from a culture broth of *Micromonospora* strain SANK 62390.[14,32] Its structure was deduced from degradation and ^{1}H-NMR analysis[14,32] and

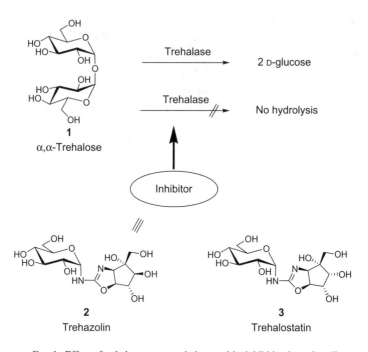

FIG. 1. Effect of trehalase on α,α-trehalose and its inhibition by trehazolin.

confirmed through synthetic studies, which established its absolute configuration. The structure of an inhibitor, isolated from the culture broth of *Amycolalopsis trehalostatica*,[33,34] was first assigned as 5-*epi*-trehazolin, and named trehalostatin (**3**),[33] but later[34–41] was shown to be trehazolin (**2**) itself, through comparison of physical data.

III. BIOSYNTHESIS OF TREHAZOLIN

A speculative biosynthetic pathway for trehazolin (**2**) has been proposed,[42] as shown in Scheme 1. Two molecules of the glucosylamine (**i**) could react with carbon dioxide to give the carbodiimide **ii**, which could be transformed into **iii**. Subsequent regioselective oxidation to **iv**, followed by a stereoselective pinacol-type coupling could afford trehazolin (**2**). The biosynthesis of trehalose has also been the subject of several subsequent chapters.[43–48]

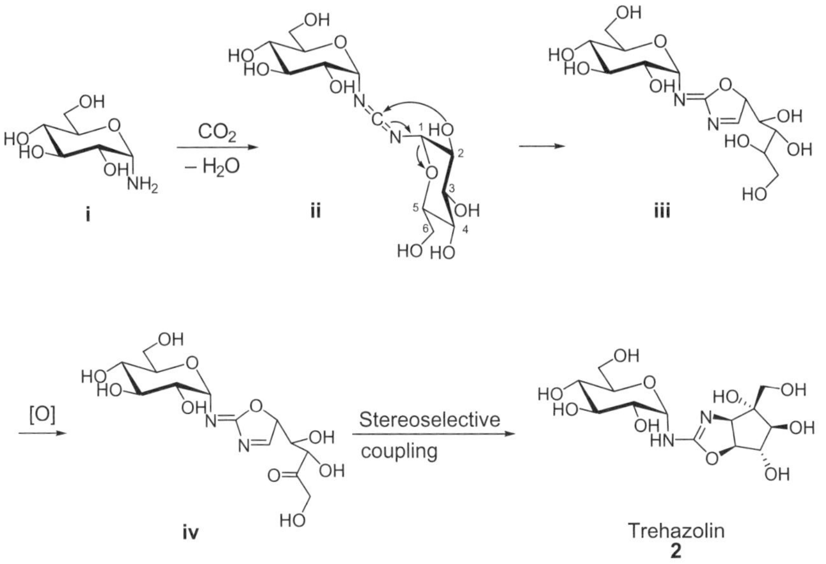

SCHEME 1. Proposed biosynthetic pathway.

IV. RETROSYNTHETIC ANALYSIS OF TREHAZOLIN AND ITS ANALOGUES

Trehazolin (**2**) can be regarded as a glucosylamine derivative, which might be formed by a reaction of D-glucose with trehalamine (**4**); however to the best of our knowledge such an approach has not appeared in the literature. According to the published literature, the precursor for **2** is a thiourea derivative **A**, whose cyclization via a neighboring hydroxyl group should afford **2**. The thiourea **A** can be readily formed from suitable subunits, namely the protected trehazolamine **C**, which can be readily obtained from trehazolamine **5**, and the sugar thiocyanate **B**. Alternatively, **A** can be formed from the reaction of glucosylamine **D** with the isothiocyanate **E**. Consequently, the important point is the generation of trehazolamine **5** or its thiocyanate analogues **E**, that is to say, a functionalized cyclopentitol (Fig. 2).[42]

In order to explore the effect of chemical modification of trehazolin on its biological activity, various analogues of **2** have been prepared. There analogues include effecting a change of the stereochemistry of the stereogenic centers in trehazolamine **4**, a change of the ring size of the cyclopentyl ring, replacement of the glucosyl moiety by other groups and/or a change of the location of the trehalamine component onto other positions of the sugar part (Fig. 2).

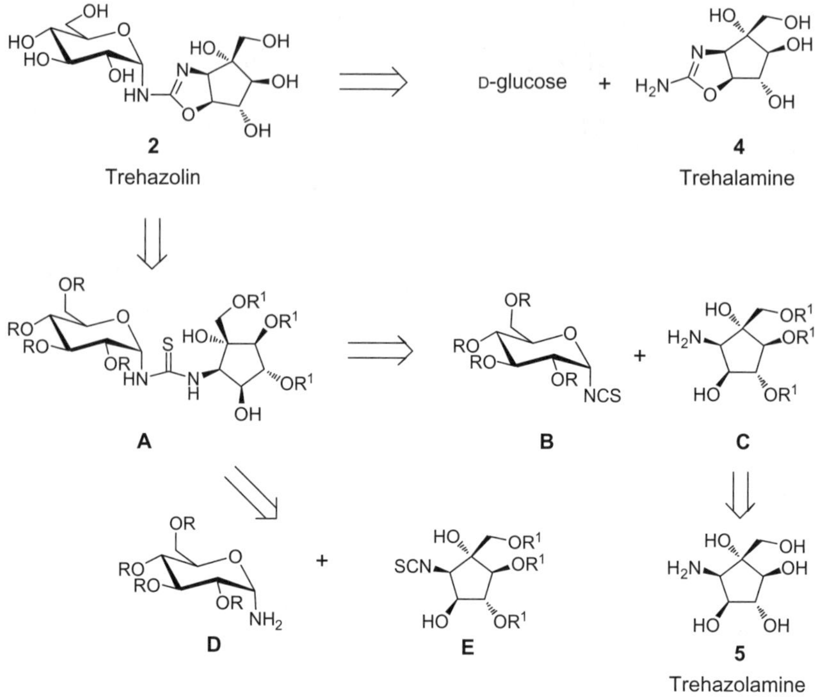

FIG. 2. Retro-synthetic analysis of trehazolin **2**.

Thus, the synthesis of trehazolin and its analogues required two precursors, the glycosyl isothiocyanate or amine analogues thereof, and trehazolamine or isothicyanate analogues, respectively.

V. SYNTHESIS OF THE SUGAR ISOTHIOCYANATE FRAGMENT

Sugar derivatives having isothiocyanate groups on the various carbon atoms were used for the synthesis of trehazoline and its analogues. Thus, the glycosyl isocyanate, where the isothiocyanate group is at the anomeric center, as required for the synthesis of trehazolin **2**, was readily synthesized by reaction of the glycosyl donor **6** with a metal isothiocyanate (Scheme 2).[23]

The isothiocyanate **10**, having an unprotected hydroxyl group at C-6, was prepared from the anhydro sugar **8**. Benzylation of **8** produced **9**, which was treated with (Bu$_4$N)NCS and BF$_3$·OEt$_2$ to give **10** in 50% yield (Scheme 3).[49]

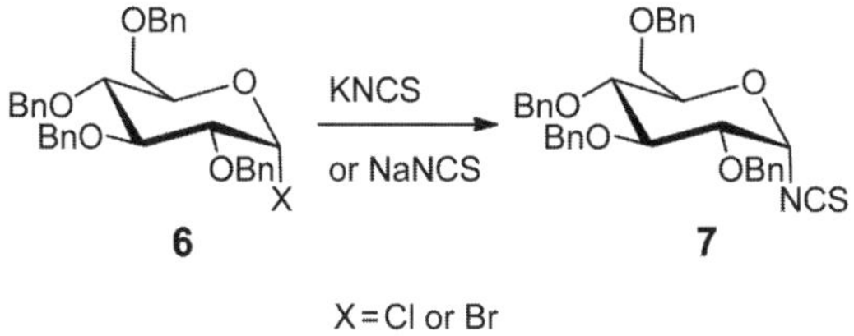

SCHEME 2. Synthesis of the isothiocyanate.

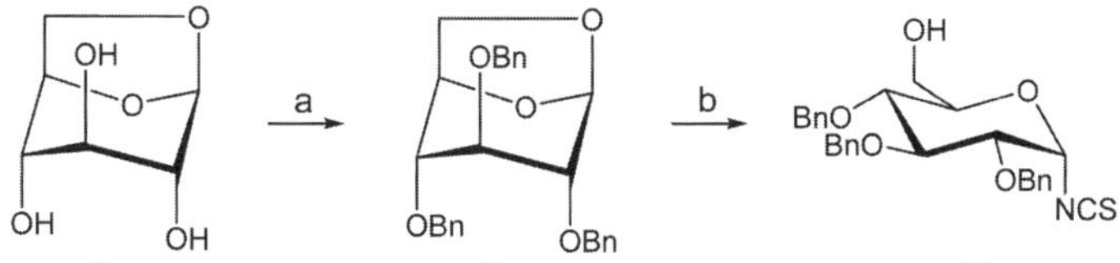

SCHEME 3. (a) BnCl, 65%; (b) (Bu$_4$N)NCS, BF$_3$·OEt$_2$, toluene, 23 °C, 48 h, 50%.

The isothiocyanate linked at C-6 of the sugar part was introduced on the D-glucose framework by conversion into **11**, which gave the corresponding azido derivative **12** under Mitsunobu conditions.[50] Reduction of the azido group of **12** produced the corresponding 6-amino-6-deoxy sugar **13**. Treatment of **13** with carbon disulfide and Et$_3$N, followed by reaction with 2-chloro-1-methylpyridinium iodide and Et$_3$N, led to β-elimination of the corresponding unstable dithiocarbamide triethylamine salt to afford the isothiocyanate derivative **14** (Scheme 4).[51–53]

A galacto analogue having the isothiocyanate group at C-4 was obtained from the D-glucose derivative **15**[54] by its conversion into the 4-mesylate **16**, which upon displacement by the azide ion gave **17**. Reduction of the latter gave the amine **18**, isothiocyanation of which furnished **20** via the intermediate **19**,[51] using a Wittig–Horner–Emmons type of reaction (Scheme 5).

Synthesis of the isomeric (gluco) isothiocyanate **26** started by treatment of D-galactose derivative **21**[55] with methanesulfonyl chloride in pyridine to afford the corresponding mesylate **22**, which underwent S$_N$2-type azidation to furnish the azido D-glucose derivative **23** (Scheme 6).[51] Reduction of the azido group of **23** produced the 4-amino sugar **24**. One-pot isothiocyanation of **24** then afforded the corresponding isothiocyanate **26**, via **25**, in high yield.

By contrast, the D-glucose isothiocyanate derivative **33** was synthesized from D-galactose pentaacetate **27** (Scheme 7).[51]. Reaction of **27** with benzyl alcohol followed

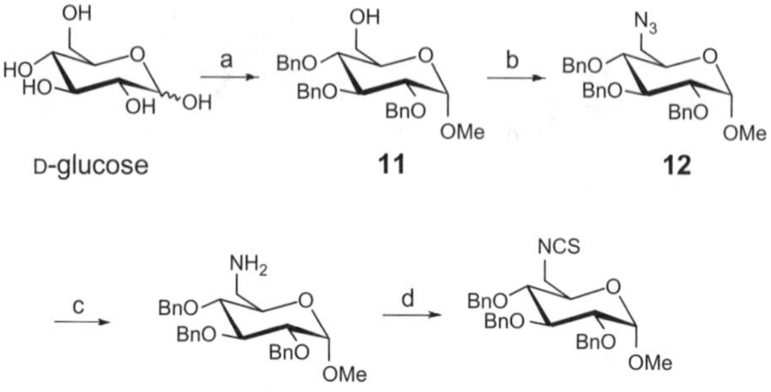

SCHEME 4. (a) Ref. 50; (b) Ph$_3$P, diethylazodicarboxylate, diphenyl phosphorazidate, 0 °C, under N$_2$, 10 min, rt, 1 h, 89%; (c) LiAlH$_4$, Et$_2$O, 0 °C, under N$_2$, 1 h, then 24 °C, 30 min, 83%; (d) CH$_2$Cl$_2$, CS$_2$, Et$_3$N, 24 °C under N$_2$, 3 h then added 2-chloro-1-methylpyridinium iodide, Et$_3$N, 2 h, 96%.

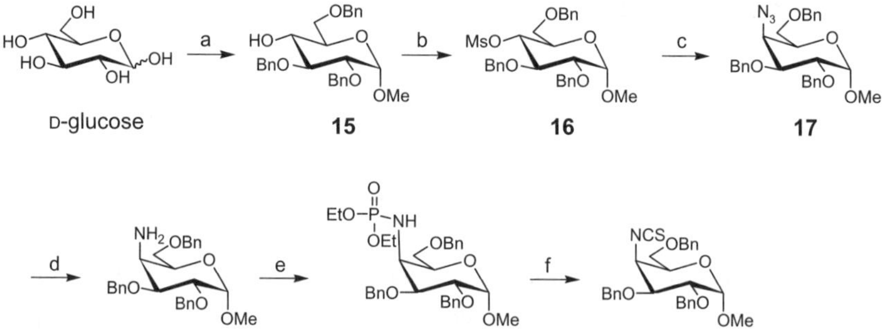

SCHEME 5. (a) Ref. 54; (b) Py, MsCl, 24 °C, 5 h, 95%; (c) NaN$_3$, DMF, 100 °C, 24 h, 76%; (d) LiAlH$_4$, Et$_2$O, 0 °C under N$_2$, 2.5 h, 89%; (e) CH$_2$Cl$_2$, diethyl chlorophosphate, Et$_3$N, 24 °C under N$_2$, then 40 °C, 18 h, 46%; (f) NaH, n-Bu$_4$NBr, benzene, 24 °C under N$_2$, then 70 °C, 2 h, then CS$_2$, 70 °C, 2 h, 34%.

by deacetylation, and subsequent benzylidenation gave **28**. After benzylation of the hydroxyl groups of **28**, regioselective benzylidene opening with borane–trimethylamine complex and aluminum trichloride in tetrahydrofuran (THF) afforded compound **29**. Mesylation of the hydroxyl group at C-4 of **29** yielded mesylate **30**. An S$_N$2-type azidation of **30**, with subsequent reduction of the azido group of **31**, furnished the 4-amino sugar **32**, which was converted into the isothiocyanate **33**.

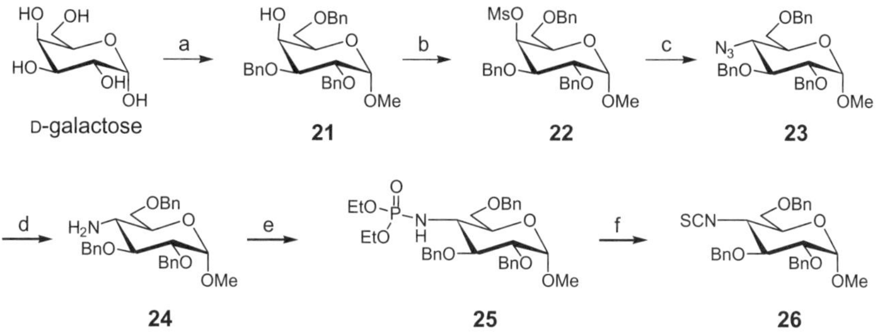

TREHALASE INHIBITORS: SYNTHESIS OF TREHALOZIN AND ANALOGUES 53

SCHEME 6. (a) Ref. 55; (b) MsCl, Py, 24 °C, 6 h, 96%; (c) DMF, NaN$_3$, 120 °C, 2 h, 84%; (d) LiAlH$_4$, Et$_2$O, 0 °C under N$_2$, 2.5 h, 89%; (e) CH$_2$Cl$_2$, diethylchlorophosphate, Et$_3$N, 24 °C, under N$_2$; then 40 °C, 17 h, 56%; (f) NaH, n-Bu$_4$NBr, benzene, 24 °C under N$_2$; then 70 °C, 3 h; then CS$_2$, 70 °C, 1 h, 91%.

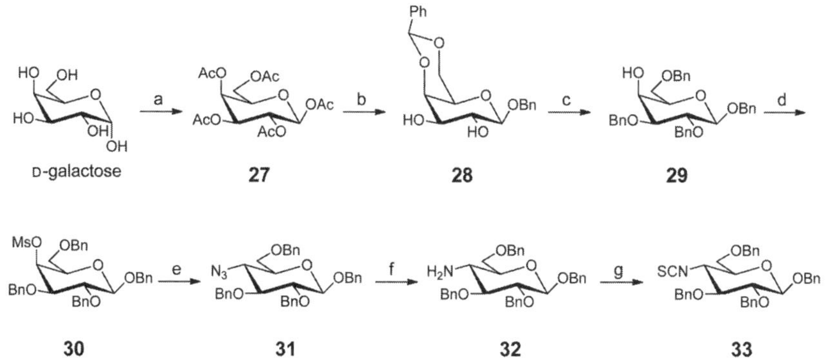

SCHEME 7. (a) 1. Ac$_2$O, Py; (b) CH$_2$Cl$_2$, BF$_3$·Et$_2$O, BnOH, 0 °C, 3.5 h, 72%; 2. MeOH, KOH, rt, 2.5 h; 3. DMF, PhCH(OMe)$_2$, CSA, 24 °C; then 80 °C, 16 h, 81%; (c) 1. NaH, DMF, BnBr, 24 °C, 1 h, 79%; 2. THF, BH$_3$·NMe$_3$, AlCl$_3$, MS 4 Å, 0 °C under N$_2$, 24 °C, 18 h, 91%; (d) Py, MsCl, 24 °C, 15 h, 90%; (e) DMF, NaN$_3$, 100 °C, 20 h, 91%; (f) LiAlH$_4$, Et$_2$O, 0 °C under N$_2$, 2 h, 78%; (g) 1. NaH, n-Bu$_4$NBr, benzene, 24 °C under N$_2$, then 70 °C, 3 h; then CS$_2$, 70 °C, 1.5 h, 87%.

VI. SYNTHESIS OF THE TREHAZOLAMINE FRAGMENT

Trehazolamine (**5**) has five asymmetric centers functionalized with hydroxyl groups and one amine group; one of these stereogenic centers is a tertiary carbon carrying an additional hydroxymethyl group (Fig. 3). Figure 3 shows those trehazolamine analogues (**34–51**) that may be generated from structure of **5** by epimerization, and their syntheses are reported in this chapter. Such a skeleton motivated the use of sugars as precursors for their synthesis, as noted in earlier reviews.[56]

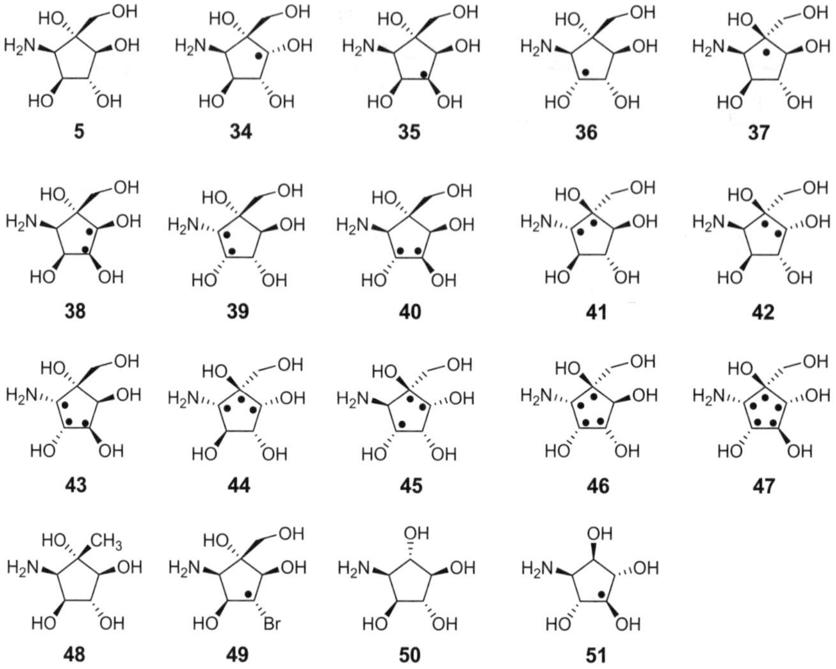

FIG. 3. Structures of trehalozamine and its analogues. The dot within the structure indicates that the carbon at this position(s) is sterically different from natural trehazolamine.

1. Synthesis from D-Glucose

The 4,6-benzylidene acetal **52**,[57] readily prepared from D-glucose, was converted quantitatively into the corresponding open-chain *O*-methyloxime derivative **53** (Scheme 8).[42] Subsequent oxidation[58] of **53** afforded ketone **54**, which underwent intramolecular coupling to give exclusively in 84% yield the diastereoisomer **55** via free-radical cyclization. The only oxidizing agent for conversion of the acetylated *O*-methoxyamine into the oxime ether **56** was found to be lead tetraacetate,[59,60] but the yield was modest (44%). Deacetylation of **56** afforded **57**, which upon reduction afforded **58**, and this underwent complete deprotection in one step to afford trehazolamine (**5**) in 22% overall yield from **52**.

Alternatively, the synthesis of **5** has been effected from 2,3,4,6-tetra-*O*-benzyl-D-glucopyranose (**59**) (Schemes 9 and 10).[61] Reduction of **59** afforded quantitatively the D-glucitol derivative **60**.[62] Swern oxidation of **60** gave **61**, whose cyclization[63]

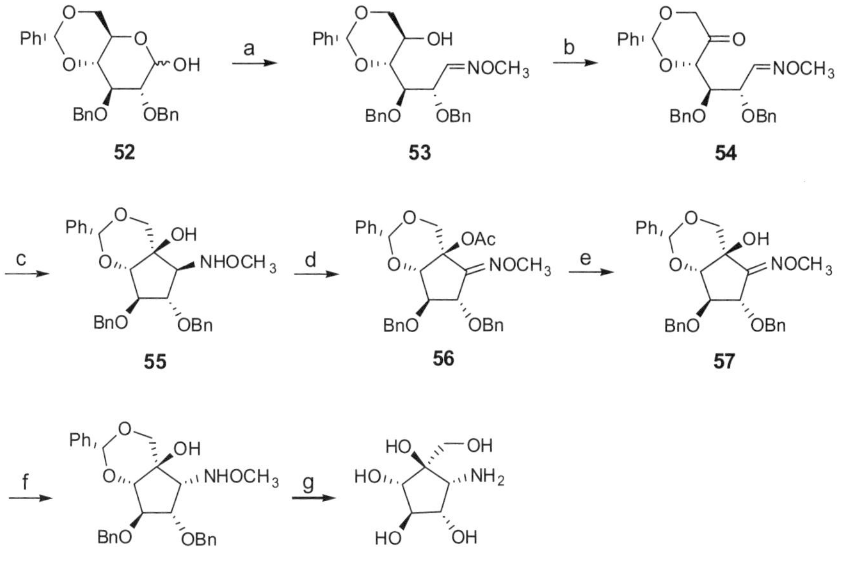

SCHEME 8. (a) CH$_3$ONH$_2$·HCl, Py, 40 °C, quant.; (b) Dess–Martin, periodinane, CH$_2$Cl$_2$, quant.; (c) SmI$_2$ (5 equiv), t-BuOH (2.5 equiv), THF, -78 °C to rt, 84%; (d) 1. Ac$_2$O, Py, DMAP; 2. Pb(OAc)$_4$, benzene, 40 °C, 44% for two steps; (e) 1. K$_2$CO$_3$, CH$_3$OH; (f) LiAlH$_4$, NaOMe, THF, -78 °C, 67% for two steps; (g) Na, NH$_3$ (liquid), -78 °C, 90%.

afforded a 1:1 mixture of **62** and **63** in 90% yield. These compounds were chromatographically inseparable, but they were converted into the separable cyclic thionocarbonates **64** and **65**. However, the mixture of thionocarbonates **66** and **67** gave the same product (**68**) upon heating with triethyl phosphite. Direct epoxidation of **68** afforded an inseparable mixture of epoxides. This problem was solved by performing the epoxidation on the deacetylated derivative **69**. Thus, Sharpless epoxidation of **69** using diisopropyl L-tartrate yielded **70** (93%). Opening the epoxide ring of **70** with LiN$_3$ yielded the azide **71** (89%), which underwent hydrogenolysis to give trehazolamine (**5**) in 39% overall yield from **59**.

Epoxidation of **69** gave **72** as a single diastereoisomer. A sequence of reactions, similar to the foregoing, on **72** produced, via the azide **73**, the trehazolamine diastereoisomer **41** (Scheme 10).[61,64] An alternative direct route to **41** was also developed from the D-glucose derivative **59** (Scheme 10).[61,64] Reductive carbocyclization of the keto-oxime derivative **74**, obtained[65,66] from **59** by using SmI$_2$, took place with

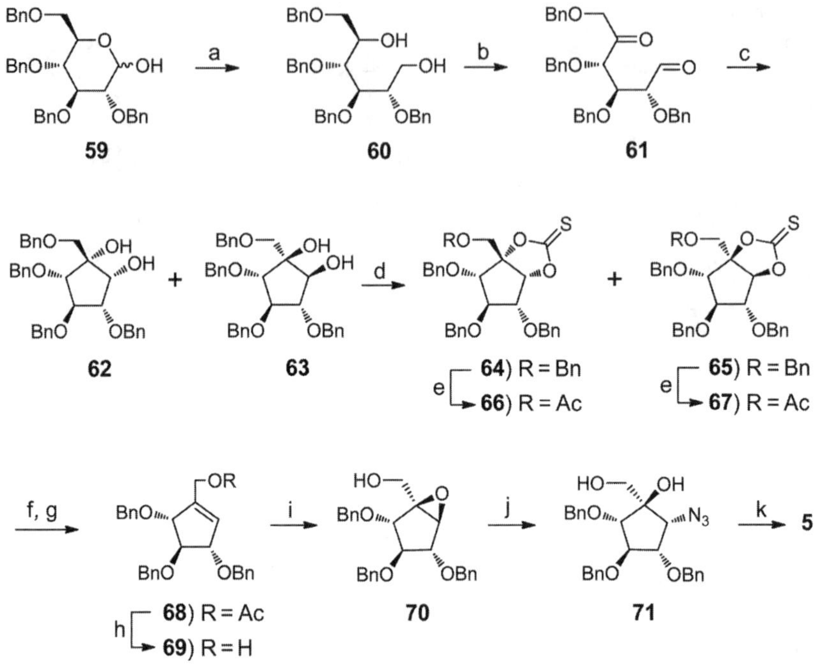

SCHEME 9. (a) NaBH$_4$, EtOH, CH$_2$Cl$_2$ (1:1), rt, 98%; (b) (COCl)$_2$, Me$_2$SO, THF, −65 °C; then Et$_3$N, −65 °C to rt; (c) SmI$_2$, THF, *t*-BuOH, −50 °C to rt, 90% for two steps; (d) 1,1′-thiocarbonyldiimidazole, toluene, 110 °C, 97%; (e) Ac$_2$O, TMSOTf, rt; (f) (EtO)$_3$P, reflux, 97%; (g) Ac$_2$O, TMSOTf, −65 °C, 50%; (h) NaOMe, CH$_2$Cl$_2$, CH$_3$OH, 96%; (i) L-DIPT, Ti(O*i*-Pr)$_4$, *t*-BuO$_2$H, CH$_2$Cl$_2$, −30 °C, 93%; (j) LiN$_3$, NH$_4$Cl, DMF, 125 °C, 89%; (k) H$_2$, Pd(OH)$_2$, EtOH, THF, TFA, 67%.

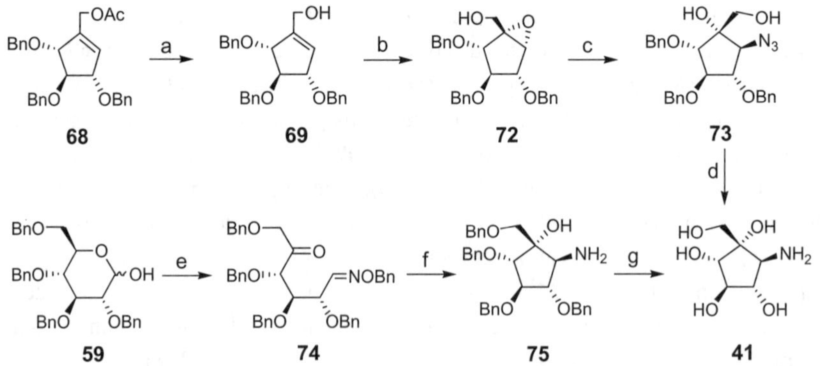

SCHEME 10. (a) NaOMe, CH$_3$OH, CH$_2$Cl$_2$, 96%; (b) D-DIPT, Ti(O*i*-Pr)$_4$, *t*-BuO$_2$H, CH$_2$Cl$_2$, −30 °C, 86%; (c) LiN$_3$, NH$_4$Cl, DMF, 125 °C, 92%; (d) H$_2$, Pd(OH)$_2$, EtOH, THF, TFA, 63%; (e) 1. BnONH$_2$·HCl, Py; 2. oxidation, 81% for two steps; (f) 0.1 M, SmI$_2$, THF, *t*-BuOH, then H$_2$O, −30 °C to rt, 1 h, 88%; (g) H$_2$, Pd(OH)$_2$, on C, EtOH, THF, TFA, 80%.

subsequent N–O reductive cleavage to afford the aminocyclopentitol **75** in 88% yield. Hydrogenation of **75** gave the trehazolamine analogue **41** in 57% overall yield from **59**. Treatment of the benzylated methyl 5-thio-D-glucopyranoside **76**[67] with $ZnCl_2$ in dichloromethyl methyl ether (DCMME) gave the corresponding chloride, which was immediately oxidized to the chlorosulfone **77** in 40% overall yield. Treatment of **77** with *t*-BuOK in THF gave cyclopentene **78** in 40% overall yield (Scheme 11).[62,68]

Trehazolin has been synthesized from D-glucose via conversion[69] into the aldehyde **79**, which upon treatment with hydroxylamine hydrochloride afforded a 4:1 (*anti*:*syn*) mixture of the oximes **80** (Scheme 12).[36,38] Subsequent [2 + 3] cycloaddition of **80** gave the corresponding isoxazoline **81**, which was hydrogenolyzed to give the enone **82**. Silylation of **82** afforded the corresponding silyl ether, whose subsequent reduction afforded a 1:2.5 mixture of **83** and **84** without affecting the double bond. Benzylation of **84**, which possesses the desired configuration, afforded **85**. Removal of the TBS group afforded the the corresponding allyl alcohol **86**. Sharpless epoxidation of **86** gave the epoxide **87** as a single isomer. After benzylation of **87**, the corresponding benzyl ether was treated with NaN_3 and NH_4Cl to afford regiospecifically the azido alcohol **88**. Reduction of **88** and subsequent cleavage of the two methoxymethyl (MOM) groups gave **92**.

Hydrogenation of the azido group in **89**, followed by acetylation, gave **90**. Cleavage of the MOM groups in **90** gave **91**. Complete acetylation of **91** afforded **93**, which upon hydrogenation and subsequent acetylation led to **94**. Hydrolysis of **94**, followed by purification using an ion-exchange resin, afforded the corresponding trehazolamine (**5**) (Scheme 12).[36,38]

A total synthesis of 5-*epi*-trehazolin (trehalostatin, **3**) was accomplished starting with D-glucose (Scheme 13).[71] Thus, benzylation of the allyl alcohol **83**, prepared from D-glucose in 10 steps,[70] gave **95**, and cleavage of the silyl group afforded **96**. Sharpless epoxidation of **96** by L-DIPT or D-DIPT furnished the two stereoisomers of **97** and **98**. Benzylation of the epoxide **98** afforded **99**, which upon azidolysis afforded the

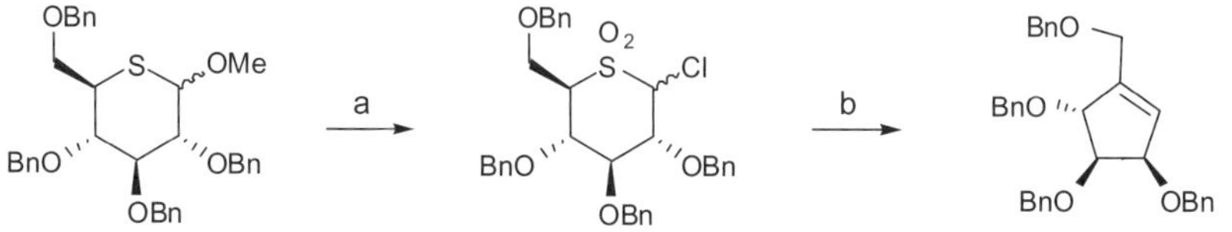

SCHEME 11. (a) 1. $ZnCl_2$ (cat), $CHCl_3$, DCMME, 50 °C; 2. *m*-CPBA, CH_2Cl_2, 40%; (b) *t*-BuOK, THF, −78 °C, 40%.

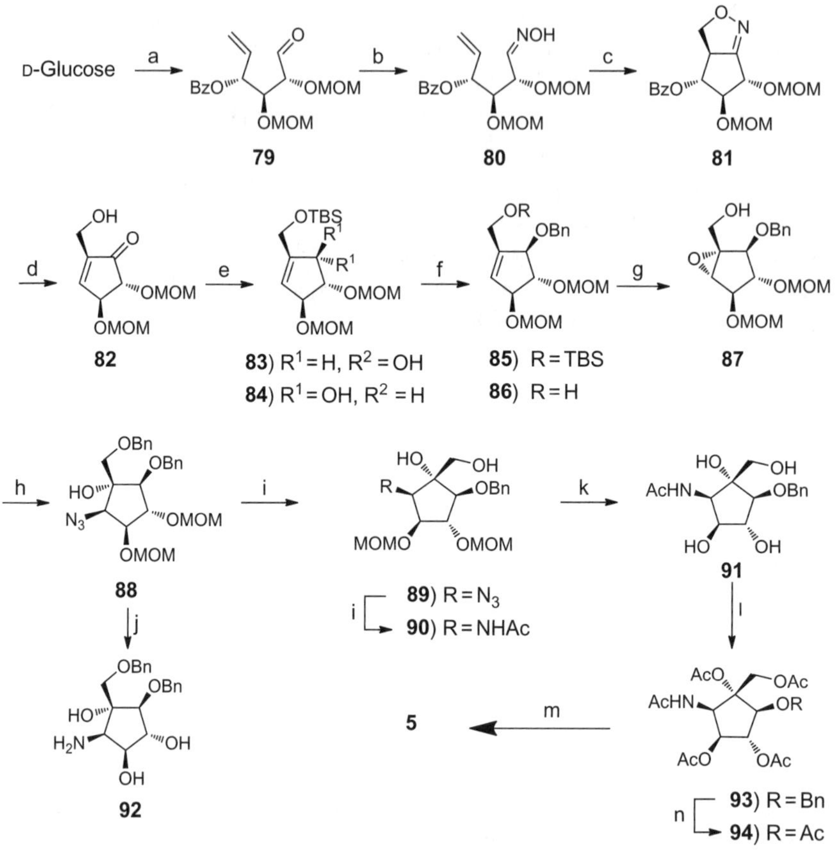

SCHEME 12. (a) Ref. 50; (b) NH$_2$OH·HCl, Na$_2$CO$_3$ (74%); (c) aqueous NaOCl, cat. Et$_3$N, 66%; (d) H$_2$, Raney nickel, B(OH)$_3$, 72%; (e) 1. TBSCl, imidazole, 88%; 2. NaBH$_4$, CeCl$_3$·7H$_2$O; (f) 1. BnBr, NaH; 2. TBAF, 58% for two steps; (g) L-DIPT, Ti(Oi-Pr)$_4$, t-BuOOH, CH$_2$Cl$_2$, −25 °C, 5 h; (h) 1. BnBr, NaH, 98%; 2. NaN$_3$, NH$_4$Cl, ethylene glycol, DMF, 78%; (i) 1. H$_2$, 10% Pd on C; 2. Ac$_2$O, CH$_3$OH, 76%; (j) 1. LiAlH$_4$; 3% HCl in CH$_3$OH; (k) 5% HCl–CH$_3$OH; (l) Ac$_2$O, DMAP, 74%; (m) 1. 2 M HCl; 2. Amberlite, CG-50 (NH$_4^+$) type, 89%; (n) 1. H$_2$, Pd(OH)$_2$ on C, 2. Ac$_2$O, DMAP, 61%.

corresponding azido alcohol **100**. Reduction of the azido group and cleavage of the two MOM groups of the latter compound gave of the corresponding amino alcohol **34**.

The Ferrier reaction was used as a key step for the synthesis of **108** from D-glucose (Scheme 14).[72] Methyl α-D-glucopyranoside (**101**) was converted into benzyl ether **103** via the alkene **102**.[73,74] Regioselective 2,6-ditosylation of **101**, followed by methoxymethylation with dimethoxymethane and P$_2$O$_5$, afforded alkene **102**, which

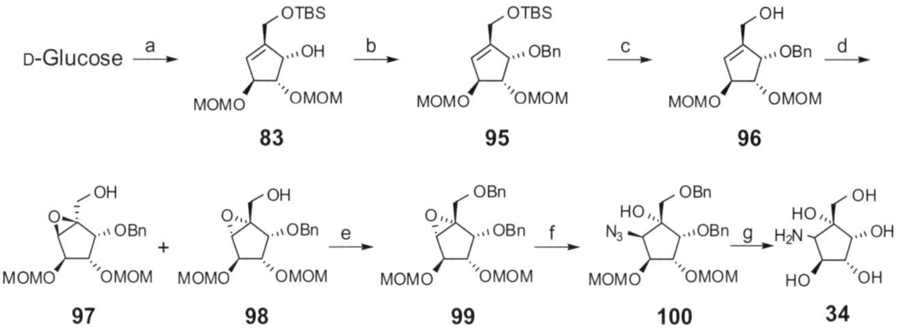

SCHEME 13. (a) Ref. 70; (b) 1.5 equiv of BnBr, 1.5 equiv of NaH, DMF, 0 °C, 2 h, 94%; (c) 1.5 equiv of n-Bu₄NF, THF, 0 °C, 1 h, 91%; (d) 1.5 equiv of L-DIPT (or D-DIPT), 1.4 equiv of Ti(Oi-Pr)₄, 2 equiv of t-BuO₂H, CH₂Cl₂, −25 °C to rt, 5 h; (e) 1.9 equiv of BnBr, 1.7 equiv of NaH, DMF, 0 °C, 2 h, 86%; (f) 13 equiv of NaN₃, 14 equiv of NH₄Cl, DMF–ethylene glycol (5:1), 125 °C, 48 h, 62%; (g) 5.5 equiv of LiAlH₄, Et₂O, 0 °C, 3 h.

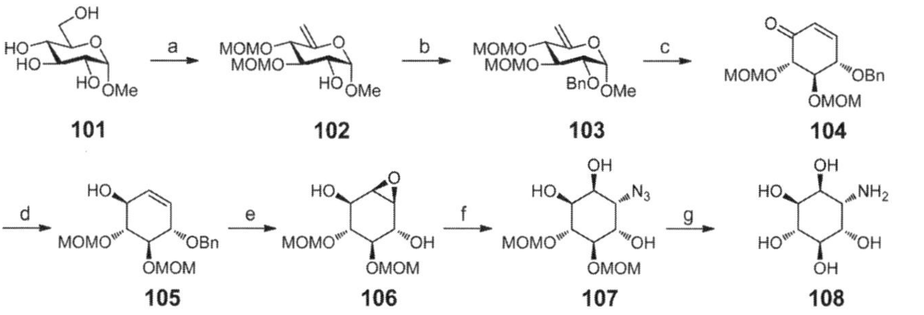

SCHEME 14. (a) 1. TsCl, Py, 0 °C, then rt, 21 h; 2. dimethoxymethane, P₂O₅, 0 °C, 1 h, 60% two steps; 3. NaI, n-Bu₄NI, DBU, 90 °C, 2 h, 53%; 4. NaOMe, MeOH, reflux, 60 h, 90%; (b) NaH, DMF, 0 °C, 30 min, then BnBr, rt, 1 h, 86%; (c) 2:1 acetone–H₂O, Hg(OCOCF₃)₂, rt, 15 h; then to the residue added Py, Ac₂O, rt, 21 h, 76%; (d) MeOH, CeCl₃·7H₂O, NaBH₄, 0 °C, 30 min, 86%; (e) 1. CH₂Cl₂, m-CPBA, rt, 88 h, 91%; 2. MeOH, 20% Pd(OH)₂/C, H₂, rt, 1 h, 86%; (f) DMF, NaN₃, NH₄Cl, rt, 100 °C, 13 h, 86%; (g) MeOH, 20% Pd(OH)₂/C, H₂, rt, 1 h.

was benzylated to give alkene **103**. The Ferrier reaction[55,75–78] converted alkene **103** into the enone **104**, for which a catalytic amount of Hg(OCOCF₃)₂ was used. Reduction of **104** with NaBH₄ and CeCl₃·7H₂O gave allylic alcohol **105**. Epoxidation of **105** was conducted stereospecifically by the effect of the free hydroxyl group, which exerts a hydrogen-bonding effect in directing the m-chloroperoxybenzoic acid (m-CPBA) to yield, after cleavage of the benzyl groups, the epoxy alcohol **106**. Azido opening of the epoxide ring in **106** produced the azido alcohol **107**. The regiospecificity of this opening can be rationalized on the basis of a preferential diaxial attack of the nucleophile on the most stable conformation of **106**. Reduction of the azido group of **107**, followed by deprotection of the two MOM groups of the resulting aminotriol, gave **108**.

2. Synthesis from D-Mannose

Efficient syntheses of trehazolamine (**5**) starting from D-mannose (Scheme 15)[79] have been readily achieved by deacetylation of phenyl 1-thio-α-D-mannopyranoside (**109**),[80] followed by monoacetonation under kinetic conditions to give **110**. Regioselective benzylation of the equatorial hydroxyl group of **110** gave **111**, whose

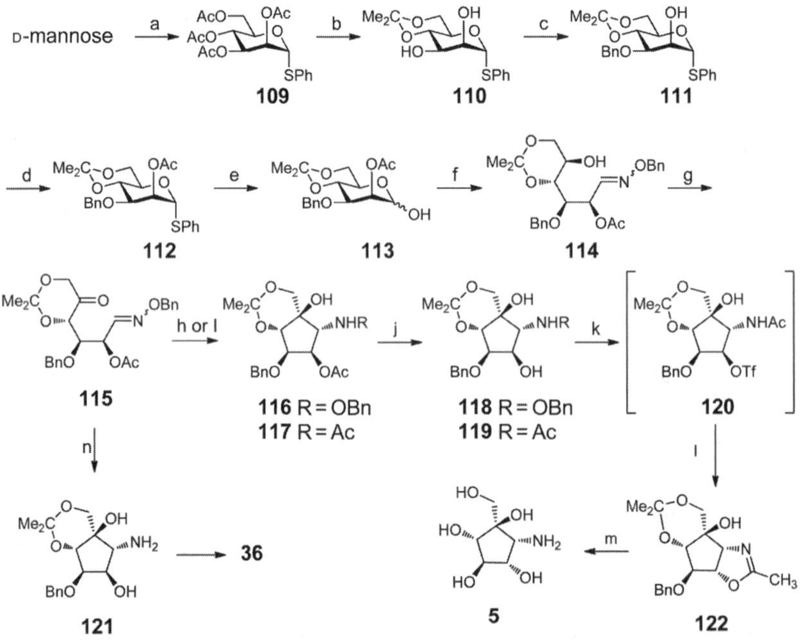

SCHEME 15. (a) 1. Ac$_2$O, Py; 2. PhSH, BF$_3$OEt$_2$, 85%; (b) 1. NaOMe, MeOH; 2. CH$_2$=C(Me)OMe, p-TSOH, DMF, 73%; (c) BnBr, Bu$_4$NBr, Bu$_2$SnO, Ms, MeCN, 60 °C, 98%; (d) Ac$_2$O, Py, 98%; (e) NBS, H$_2$O, acetone, −15 °C, 95%; (f) BnONH$_2$, Py, MeOH, 97%; (g) Dess–Martin periodinane, CH$_2$Cl$_2$, 85%; (h) SmI$_2$ (3 equiv), THF, t-BuOH, −30 °C, 86%; (i) 1. SmI$_2$ (6 equiv), THF, t-BuOH, −30 °C; 2. H$_2$O (25 equiv), rt; 3. Ac$_2$O, Py, 99%; (j) NH$_3$, MeOH, THF, **118** (99%), **119** (90%); (k) Tf$_2$O, Py, CH$_2$Cl$_2$, −20 °C; (l) TfOH, 82%; (m) 1. Pd(OH)$_2$/C, H$_2$, EtOH, TFA; 2. 3 N HCl, 100 °C, 83%. (n) 1. SmI$_2$ (6 equiv), THF, t-BuOH, −30 °C; 2. H$_2$O (25 equiv), rt; 3. LiOH, H$_2$O, 98%.

acetylation produced the fully protected 1-thiomannoside **112**. Mild hydrolysis of the phenylthio glycoside under oxidative conditions provided hemiacetal **113**, which was subsequently condensed with *O*-benzylhydroxylamine to give **114** as a mixture of isomeric oxime ethers. Oxidation of **114** using the Dess–Martin periodinane afforded keto-oxime **115**, whose treatment with SmI_2 led to a smooth cyclization to give the *O*-benzylhydroxylamine **116** as a single diastereoisomer. Using a larger excess of SmI_2 (six equivalents) in the presence of water also effected the subsequent reduction of the hydroxylamino group to give the corresponding free amine, and subsequent *in situ* treatment with excess Ac_2O and pyridine provided **117** in quantitative overall yield. Alternatively, addition of an aqueous solution of LiOH to the crude reaction mixture hydrolyzed the ester to afford aminodiol **121** in quantitative overall yield. Compounds **116** and **121** have the correct stereochemistry of trehazolamine except for the C-2 stereocenter. The O-deacetylation of **116** and **117** gave **118** and **119**. When **119** was treated with triflic anhydride in the presence of pyridine at low temperature, a smooth cyclization took place by intramolecular S_N2 displacement of the transient triflate ester in intermediate **120** by the carbonyl oxygen atom to give oxazoline **122**. Compound **122** was fully deprotected by catalytic hydrogenolysis of the *O*-benzyl group followed by acidic hydrolysis of the oxazoline and the isopropylidene groups to yield trehazolamine (**5**). This route afforded **5** in 14 steps and 29% overall yield from D-mannose.

A total synthesis of a trehazolamine diastereoisomeric analogue from D-mannose, using reductive carbocyclization of a ketone-oxime ether by samarium diiodide, has been reported (Scheme 16).[81] Thus, the precursor **125** was prepared in three steps from D-mannose. Condensation of the readily available hemiacetal **123**[82] with *O*-benzylhydroxylamine afforded oxime **124** as a 1:1 mixture of *E* and *Z* isomers. Oxidation of the free hydroxyl group of **124** took place readily with the Dess–Martin periodinane to give **125**. Treatment of **125** with SmI_2 gave cyclopentitol **127** as a single isomer in high yield. Conversely, treatment of **125** with a larger excess of SmI_2 afforded **126** directly, and this underwent removal of the acetal protecting groups to produce compound **36**.

An asymmetric synthesis of aminocyclopentitols **134–137** has been used in the synthesis of trehazolin via free-radical cycloisomerization of enantiomerically pure, alkyne-tethered oxime ethers derived from D-mannose (Scheme 17).[84] Treatment of 2,3:5,6-di-*O*-isopropylidene-D-mannofuranose (**128**)[85] with ethynylmagnesium bromide gave compound **129**, which underwent sequential "one-pot" acid hydrolysis plus diol cleavage to give **130**, oximation of which afforded the radical precursor **131**, in 41% overall yield from **129**. The free hydroxyl group of **131** was protected as acetate **132** and *tert*-butyldimethylsilyl ether **133**, which were isolated as inseparable

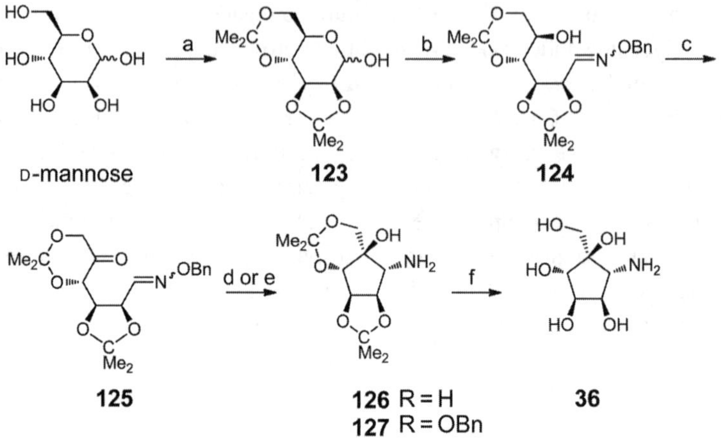

SCHEME 16. (a) MeOC(Me)=CH$_2$, p-TsOH, Drierite, DMF, -10 °C; (b) BnONH$_2$·HCl, Py, MeOH, rt, 65% for two steps; (c) Dess–Martin periodinane, CH$_2$Cl$_2$, rt, 98%; (d) 1. SmI$_2$ (6 equiv), t-BuOH, THF, -30 °C; 2. H$_2$O (30 equiv), -30 °C to rt, 86%; (e) SmI$_2$ (3 equiv), t-BuOH, THF, -30 °C, 85%; (f) aqueous 95% CF$_3$CO$_2$H, rt, 88%.

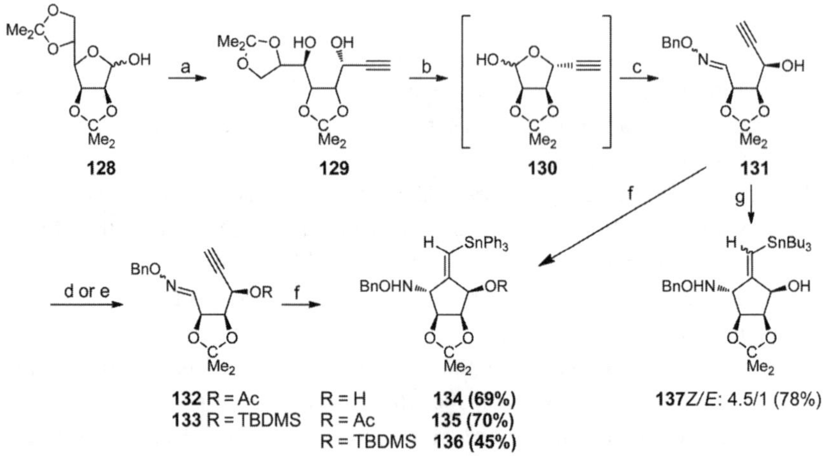

SCHEME 17. (a) HC≡CMg, 80–90%[83]; (b) dry Et$_2$O, H$_5$IO$_6$, rt, 4.5 h, then added H$_5$IO$_6$ and overnight; (c) CH$_2$Cl$_2$, BnONH$_2$·HCl, Py, reflux, 8 h, 41%; (d) Py, Ac$_2$O, rt, overnight, 63%; (e) TBDMSCl, imidazole, CH$_2$Cl$_2$, 0 °C under argon, rt, 21 h, then reflux for 18 h, 48%; (f) Et$_3$N, Ph$_3$SnH, toluene under argon, rt, 4 h; then triethylborane, Ph$_3$SnH, overnight, **134Z** (69%), **134E** (3%), **135** (70%), **136Z** (45%); (g) Bu$_3$SnH, triethylborane, toluene, 41 h, **137Z/E**, 78%

mixtures of *E* and *Z* isomers in 2:1 ratio. Cyclization, using the triethylborane plus triphenyltin hydride-mediated carbocyclization of enyne precursors, gave the vinyltin derivatives **134–136** in good yields. Compounds **134** and **136** were obtained as mixtures of *Z* and *E* isomers, whereas compound **135** was obtained as the single *Z* isomer. Treating **131** with triethylborane and tributyltin hydride gave **137** in good yield as an inseparable 4.5:1 mixture of *Z* and *E* isomers.

3. Synthesis from D-Ribose

Synthesis via free-radical cyclization has been achieved from the D-ribo radical precursors **140** and **141**, prepared from 2,3-*O*-isopropylidene-D-ribose, as 2:1 mixtures of *E* and *Z* isomers (Scheme 18),[84] to give high yields of the corresponding aminocyclopentitol derivatives **145** and **142**, as a mixture of *E* and *Z* isomers that could be separated, and **143** as the exclusive *Z* isomer. The high degree of stereochemical control observed in the cyclization of **140** and **141** can be explained by assuming that, in the early transition state, the favored vinyl radical species is in a chair-like conformation having most of the substituents in the favored quasi-equatorial orientation; the same argument is also valid for compounds **131–133**.

Reaction of compound **142** (*Z/E*) with anhydrous ethanol saturated with hydrogen chloride resulted in simultaneous protodestannylation and hydrolysis of the

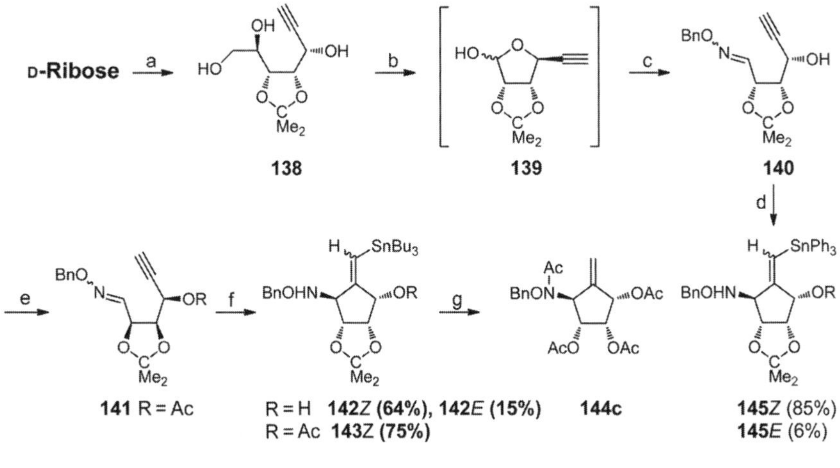

SCHEME 18. (a) Ref. 86; (b) H_2O, $NaIO_4$, rt, 1.5 h; (c) CH_2Cl_2, $BnONH_2 \cdot HCl$, Py, reflux, 4.5 h, 86%; (d) Ph_3SnH, triethylborane, toluene, **145Z** (85%), **145E** (6%); (e) Py, Ac_2O, rt, 2 h, 96%; (f) triethylborane, Bu_3SnH, toluene under argon, 60 °C, 1.3 h, **142Z** (64%), **142E** (15%); (g) HCl, EtOH, rt, 2 h, 92%.

isopropylidene acetal, leading to the triol **146** in high yield (Scheme 19).[84] However, it was obtained in only 20% yield when **143Z** was treated with methanol in acetic acid (catalytic). When **143Z** was treated with anhydrous ethanol saturated with hydrogen chloride, followed by acetylation, the corresponding peracetylated aminocyclitol **144c** (Scheme 18) was obtained in good overall yield (73%). When **145Z** was subjected to the same conditions for protodestannylation, compound **147** was formed in low yield (16%), while trifluoromethanosulfonic acid or cerium ammonium nitrate

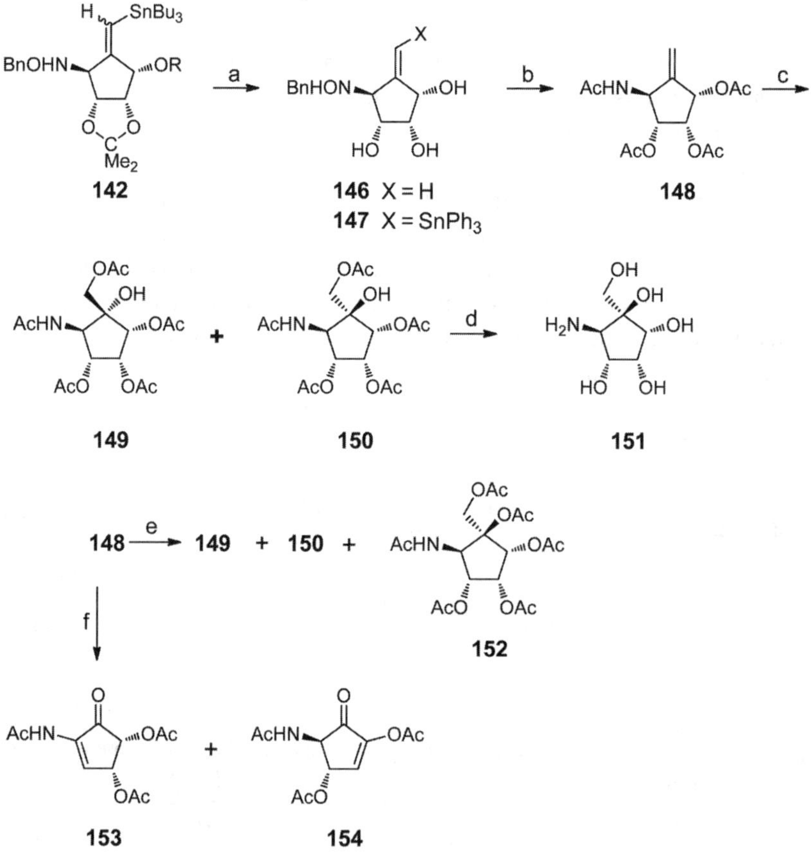

SCHEME 19. (a) HCl, EtOH, rt, 2 h; then EtOH, propylene oxide, 50 °C, overnight, **146** (92%), **147**; (b) 1. SmI$_2$; 2. Ac$_2$O, 79%[87]; (c) OsO$_4$, t-BuOH, NMO, acetone, 40 °C, dark, overnight, sodium hydrogen sulfite, rt, 1.3 h; then Py, Ac$_2$O, 0 °C, then rt, 2 h, **149** (3%), **150** (80%); (d) 2 M HCl, 80 °C, 17 h, 97%; (e) m-CPBA, CH$_2$Cl$_2$, pH 8, rt in dark, 20 h, **152** (14%), then Ac$_2$O, Py; (f) Ozone, MeOH, −78 °C, 3 h, dimethyl sulfide, rt, overnight, **153** (42%), **154** (23%).

in methanol did not work at all. This methodology was also demonstrated by transforming compound **142** into **150** by acid hydrolysis of compound **142** to give **146**, followed by samarium diiodide-mediated cleavage of the nitrogen–oxygen bond and subsequent acetylation, in a "one-flask" operation, to afford peracetate **148** in 79% yield from **146**. Treatment of this allylic acetamide with osmium tetraoxide and *N*-methylmorpholine *N*-oxide (NMO) in 80% aqueous acetone gave almost exclusively, after partial acetylation, the aminocyclopentitol **150** in excellent yield; only traces of the minor isomer **149** were isolated. A similar high *syn*-stereoselectivity has also been observed in the osmylation of some allylic substituted cyclopentanes. Acid hydrolysis of **150** gave in good yield the fully deprotected aminocyclopentitol **151**.

When compound **148** was epoxidized and the resulting mixture treated with sodium acetate in *N,N*-dimethylformamide (DMF) followed by acetylation, compounds **149**, **150**, and **152** were detected (Scheme 19). After purification by column chromatography, compound **150** (37%), a mixture of **149** and **152** (in 1.7:1 ratio; 16% yield), and peracetate **152** (14%) were isolated. In contrast, ozonolysis of **148** gave the ketones **153** and **154** in 42% and 23% yields, respectively.

4. Synthesis from D-Arabinose

The tri-*O*-benzyl trehazolamine **161** has been synthesized from D-arabinose (Scheme 20).[89] The precursors **156** and **157** were first prepared from 2,3,5-tri-*O*-benzyl D-arabinose **155** in 47% overall yield.[88] Removal of the *p*-methoxybenzyl

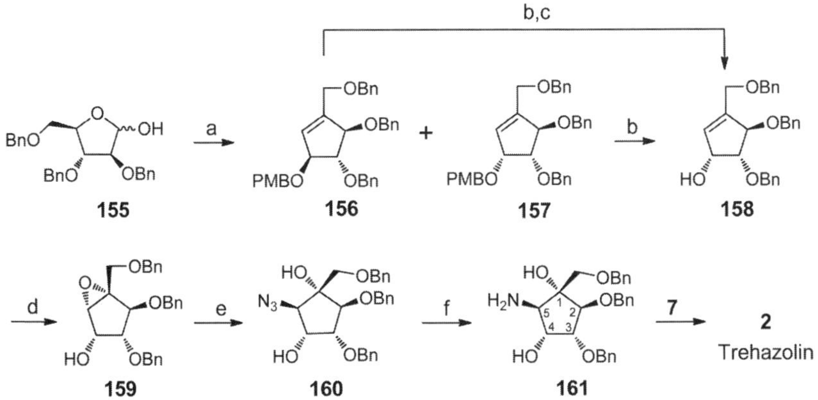

SCHEME 20. (a) Ref. 88; (b) DDQ, CH_2Cl_2, H_2O, 84%; (c) PPh_3, DEAD, $PhCO_2H$, NaOMe, 95%; (d) *m*-CPBA, CH_2Cl_2, 89%; (e) NaN_3, DMF, 97%; (f) PPh_3, THF, 98%.

(PMB) group from **156** followed by inversion of configuration under Mitsunobu conditions afforded **158**; it was also obtained from **157** by removal of the PMB group. The combined **158** was then treated with *m*-CPBA to afford **159**, whose epoxide ring was opened with NaN$_3$ to give **160**. The azido group in **160** was reduced[90] using Ph$_3$P to give the protected aminocyclopentitol **161**, which could be converted[79] into trehazolin **2**. The pseudoanomeric center at C-4 was inverted using triflic anhydride in the presence of pyridine at low temperature to give the corresponding aminooxazoline, which was then subjected to hydrogenolysis to afford trehazolin (**2**).

5. Synthesis from D-Mannitol

D-Mannitol was also used as a precursor for trehazolamine via its conversion into the (*R*)-(−)-epichlorohydrin (**162**),[91,92] which gave the optically active 1-(hydroxymethyl)spiro[2,4]cyclohepta-4,6-diene (**163**) in 60% yield upon treatment with lithium cyclopentadienide (Scheme 21).[49] Conversion of **163** into the corresponding trichloroacetimidate **164**,[93] followed by reaction with I(*sym*-collidine)$_2$ClO$_4$, afforded **165** in 61% yield, which underwent silylation of the secondary carbinol to produce **166** in 95% yield. Treatment of **166** with Li$_2$NiBr$_4$, followed by treatment of the resulting cyclopropylcarbinyl bromide with a solution of dimethyldioxirane in acetone, afforded the epoxide **167**. Epoxide-ring opening via the vicinal trichloroacetamido group by treatment with BF$_3$·OEt$_2$ in toluene, followed by free-radical reduction, produced the oxazoline **168**. Treatment with aqueous PPTs, followed by acetylation and subsequent hydroboration of the terminal alkene and then oxidation of the resulting primary alcohol, provided the aldehyde **169**. Conversion of **169** into the corresponding phenylketone **170** (60% yield) took place by reaction with PhMgBr and LiBr. Norrish type-II cleavage, upon irradiation in benzene, gave the alkene **171**, which without purification underwent reaction with catalytic OsO$_4$ to yield **172** as a single diastereomer.

6. Synthesis from Sugar 1,5-Lactones

A synthesis of precursors for trehalostatin analogues from 2-iodo-5-formyl-1,5-lactone has been reported (Scheme 22).[94] Reaction of the iodoaldehyde **173**[95] with potassium fluoride in acetonitrile in the presence of 18-crown-6 gave **174** in 81% yield, together with two other isomers in low yield. Reaction of **174** with potassium carbonate in methanol at −20 °C gave the iododiol **175** in quantitative yield.

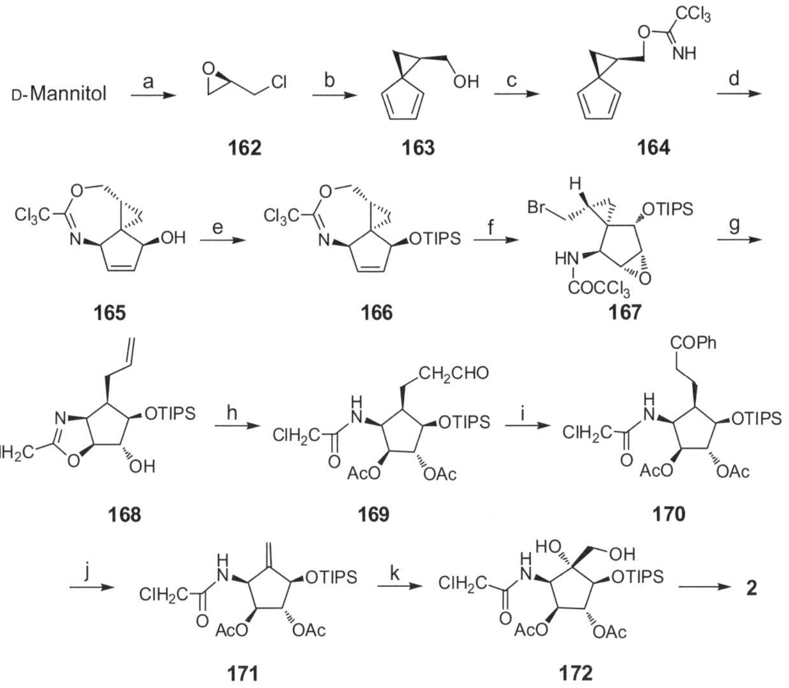

SCHEME 21. (a) Refs. 91 and 92; (b) LiC$_5$H$_5$, NaH, THF, 60%; (c) NaH, Cl$_3$CCN, THF, 95%; (d) I(sym-collidine)$_2$ClO$_4$, NaHCO$_3$, aqueous, CH$_3$CN, 61%; (e) i-Pr$_3$SiOTf, 2,6-lutidine, CH$_2$Cl$_2$, 95%; (f) 1. Li$_2$NiBr$_4$, THF, 80%; 2. (CH$_3$)$_2$CO$_2$, acetone, 65%; (g) 1. BF$_3$·OEt$_2$, 87%; 2. Bu$_3$SnH, Et$_3$B, NaBH$_4$, EtOH, 75%; (h) 1. PPTs, aqueous CH$_3$CN, then Ac$_2$O, DMAP, 77%; 2. CHX$_2$BH, H$_2$O$_2$, 83%; 3. (COCl)$_2$, Me$_2$SO, CH$_2$Cl$_2$, then Et$_3$N, 83%; (i) PhMgBr, LiBr, THF, 60%; (j) 1. hv and then OsO$_4$, NMO; 2. (COCl)$_2$, Me$_2$SO, CH$_2$Cl$_2$, and then Et$_3$N, 100% for two steps; (k) hv and then OsO$_4$, NMO, 75%.

Treatment of iodolactone **174** at room temperature with potassium carbonate in methanol afforded epoxide **176** in quantitative yield; the same epoxide **176** was also obtained by reacting the iododiol **175** under the same conditions. Treatment of **176** with sodium azide gave the azidodiol **177** in 83% yield; this compound is a potential intermediate for a range of highly functionalized targets. Removal of the isopropylidene protecting group by aqueous trifluoroacetic acid (TFA) afforded the tetrol **178** in 96% yield, which on further reaction with aqueous sodium hydroxide and processing with ion-exchange chromatography gave the azidocarboxylic acid **179** in quantitative yield. Hydrogenation of **179** produced the highly functionalized amino acid **180** in 98% yield. In contrast, reduction of **177** with lithium borohydride in THF gave the azidotriol **181** in 94% yield, from which the ketal protecting group was removed by

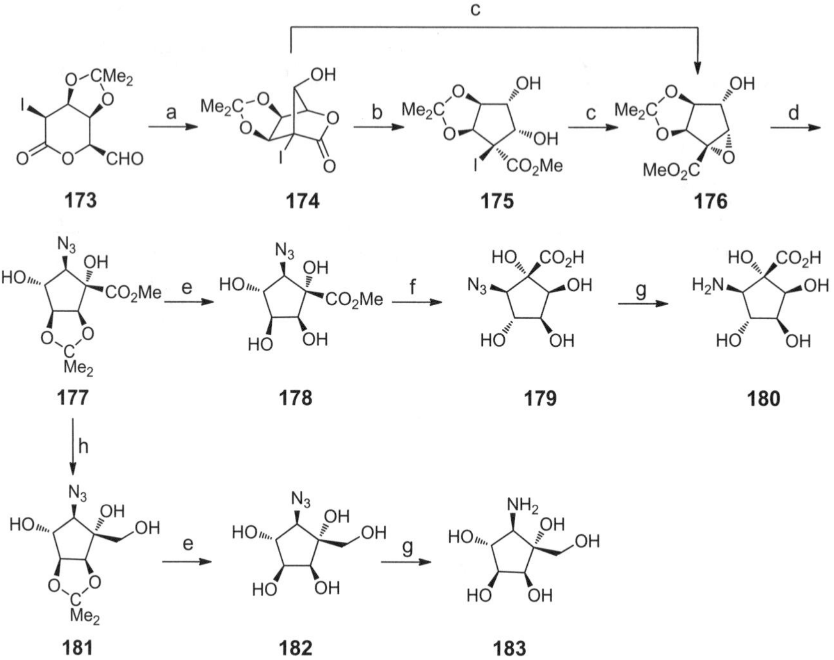

SCHEME 22. (a) KF, 16-crown-6, MeCN, −6 °C; (b) K_2CO_3, MeOH, −20 °C; (c) K_2CO_3, MeOH, rt; (d) NaN_3, NH_4Cl, aq. MeOH; (e) 40% aq. CF_3CO_2H; (f) aq. NaOH, then ion-exchange chromatography; (g) H_2, Pd black, H_2O; (h) $LiBH_4$, THF, 0 °C.

aqueous TFA to furnish the pentahydroxyazide **182** in 80% yield. Hydrogenation of **182** formed the amine **183** in quantitative yield. The aminopentol **183**, which is an analogue of the trehazolin fragment **5**, was tested for its ability to inhibit human liver glycosidases. No strong inhibition of any glycosidases was found, although **183** was a weak inhibitor of α-L-fucosidase (44% at 1 mM) and a weak activator of α-D-glucosidase (33% at 1 mM); the presence of the amino group did not lead to significant inhibition of β-hexosaminidase (only 20% at 1 mM).

7. Synthesis from *myo*-Inositol

(±)-1,2-*O*-Cyclohexylidene-*myo*-inositol (**184**)[96] has been used for the synthesis of trehazolamine (**5**) and the isomers **34**, **37**, and **42** (Fig. 3). Periodate oxidation of **184**, followed by base-catalyzed nitromethane condensation of the generated dialdehyde, gave a mixture of nitro-diols, which was hydrogenated in the presence of Raney nickel, followed by acetylation to afford the three diastereoisomeric 2,3-*O*-cyclohexylidene derivatives **185** (40%), (±)-**186** (5%), and **187** (5%) of 5-acetamido-1,4-*O*-acetylcyclopentane-1,2,3,4-tetrol (Scheme 23).[97] The minor racemic mixture **186** was *O*-deacetylated, *N,O*-isopropylidenated, and then resolved by chromatographic separation of its (*S*)-acetylmandelates to give **188** and **189**. Deacylation of **188** gave **190**, which upon PCC oxidation furnished **191** (Scheme 23). Compound **191** was synthesized from **189** by a similar sequence of reactions.

Compound **191** was transformed into the exo-alkene **193** via the respective spiro epoxide; the enone **192** (11%) was obtained as a side product (Scheme 24).[97] Compound **193** was deprotected, and the triol obtained was selectively mesylated at the allylic position to give, after acetylation, compound **194** (68%). Treatment of **194** with sodium acetate resulted in the inversion of configuration at C-1 to give the tetra-*N,O*-acetyl derivative **195**. Oxidation of **195** with osmium tetraoxide in aqueous acetone, followed by acetylation, afforded **196** (87%) and **197** (13%), whose acid hydrolysis provided the free bases **5** and **37**, respectively.

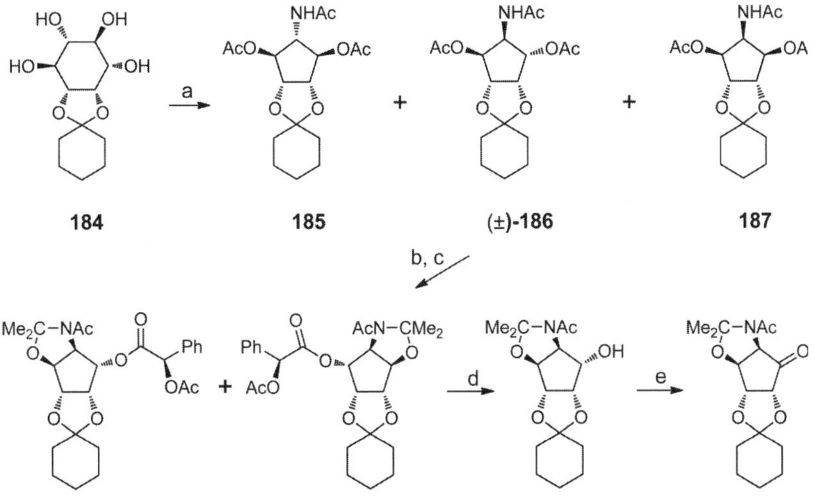

SCHEME 23. (a) 1. NaIO$_4$; 2. CH$_3$NO$_2$, base, H$_2$, Raney nickel and then Ac$_2$O, Py, **185** (40%), (±)-**186** (5%), **187** (5%); (b) NaOMe, CH$_3$OH; 2,2-dimethoxypropane, *p*-TsOH, DMF, 4 h at 50 °C; and then AcOH, CH$_3$OH, 48 h, 75%; (c) (*S*)-*O*-acetylmandelic acid, DMAP, CH$_2$Cl$_2$, DCC, CH$_2$Cl$_2$, 0 °C, 0.5 h, **188** (50%), **189** (48%); (d) NaOMe, CH$_3$OH, rt, quant; (e) PCC, MS 4 Å, rt, 2 h, 98%.

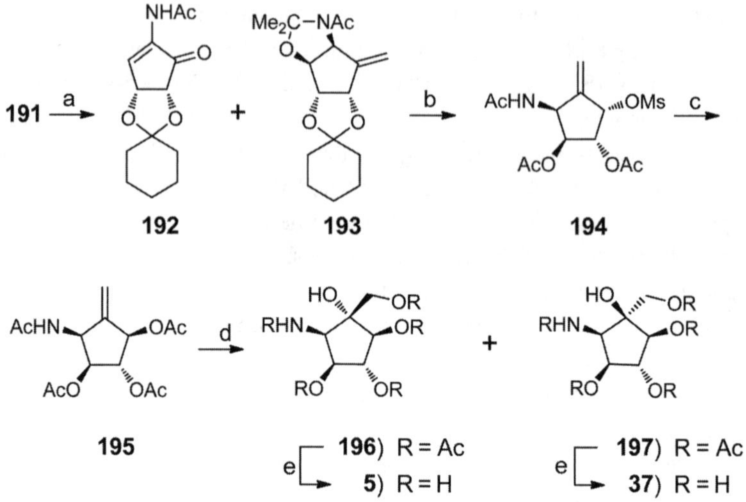

SCHEME 24. (a) CH$_2$N$_2$, Me$_2$SO, ether, P(OCH$_3$)$_3$, 130 °C, 83% (45%), 82% (11%); (b) ref. 95; (c) NaOAc, DMF, 88%; (d) 1. OsO$_4$, acetylation, **196** (87%), **197** (13%); (e) 2 M HCl, 4.5 h at 80 °C; Dowex 50WX2(H$^+$) eluted with aqueous 5% NH$_3$, 94%.

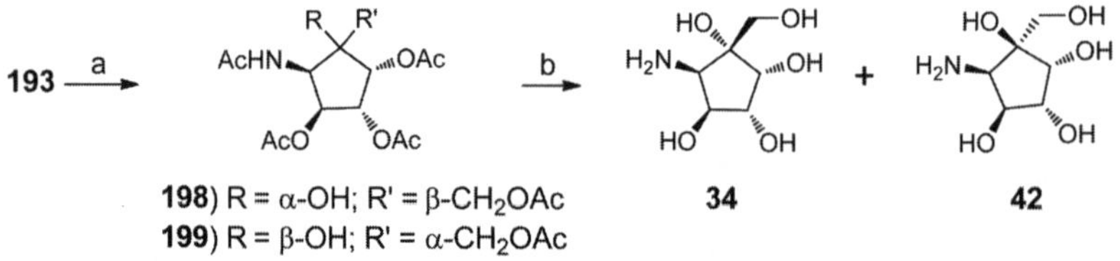

198) R = α-OH; R' = β-CH$_2$OAc
199) R = β-OH; R' = α-CH$_2$OAc

SCHEME 25. (a) 1. OsO$_4$, 2. acetylation; (b) 2 M HCl, 4.5 h at 80 °C; Dowex 50WX2(H$^+$) eluted with aqueous 5% NH$_3$.

Treatment of **193** with osmium tetraoxide, followed by conventional decyclohexylidenation, deisopropylidenation, and acetylation, gave the two branched aminocyclitols, **198** (49%) and **199** (51%), which afforded almost quantitatively the respective free amino alcohols **34** and **42**, upon acid hydrolysis followed by purification over Dowex 50W-X2 (H$^+$) resin (Scheme 25).[97,98]

Unambiguous syntheses of the penta-N,O-acetyl derivatives **205** and **208** have been successfully performed by starting from the DL-(1,4,5/2,3)-5-amino-1,2,3,4-cyclopentanetetrol derivative **187**[96,99–101] (Schemes 26 and 27).[102] O-Deacetylation of **187** gave **200** in 100% yield, and subsequent treatment with 2,2-dimethoxypropane in

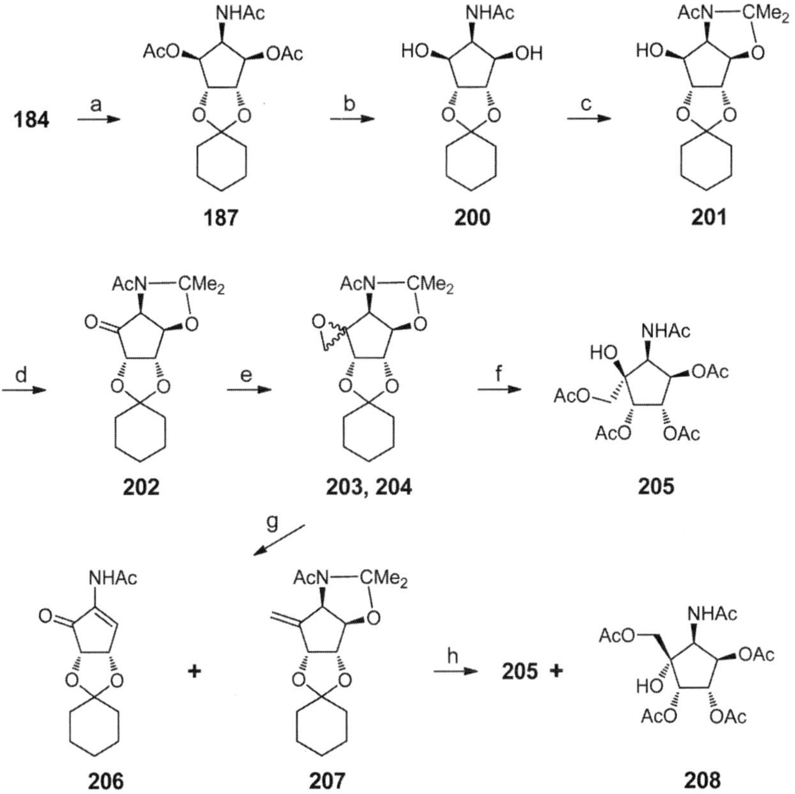

SCHEME 26. (a) Refs. 96,99–101; (b) MeOH, NaOMe, rt, 1 h; then added Amberlite IR 120B (H$^+$) resin, 100%; (c) DMF, DMP, p-TsOH·H$_2$O, 50 °C, 2 h, 100%; (d) 4 Å MS, CH$_2$Cl$_2$, PCC, rt, 1 h, 100%; (e) Me$_2$SO, diazomethane, Et$_2$O, rt, 19 h, 87%; (f) DMF, NaOAc, 120 °C, 19 h and then HCl, 80 °C, 5 h, acetylation, 100%; (g) trimethylphosphite, 130 °C, 51 h, sealed tube, **206**, 11% and **207**, 53%; (h) aq. 80% acetone, OsO$_4$, t-BuOH, NMO, 50 °C, 48 h, rt, 1 h, HCl, 80 °C, 3 h, acetylation, **205**, 45%, **208**, 53%.

DMF gave the *N,O*-isopropylidene derivative **201** (100%). The free hydroxyl group of **201** was readily oxidized to give the ketone **202** (100%), whose reaction with diazomethane afforded a 2:3 mixture (87%) of the two isomeric spiro epoxides **203** and **204**. Without separation, the mixture was directly subjected to epoxide cleavage with sodium acetate, followed by acetylation, giving a sole penta-*N,O*-acetyl derivative **205** (97%). Selective ring opening of **204** may be explained by assuming the assistance of the neighboring *N*-acetyl group at C-1, suggesting that **205** possesses the (1,4,5/2,3)-configuration. Therefore, an attempt was made to convert the spiro epoxide into the exo-alkene **207**, which was expected to be oxidized with osmium

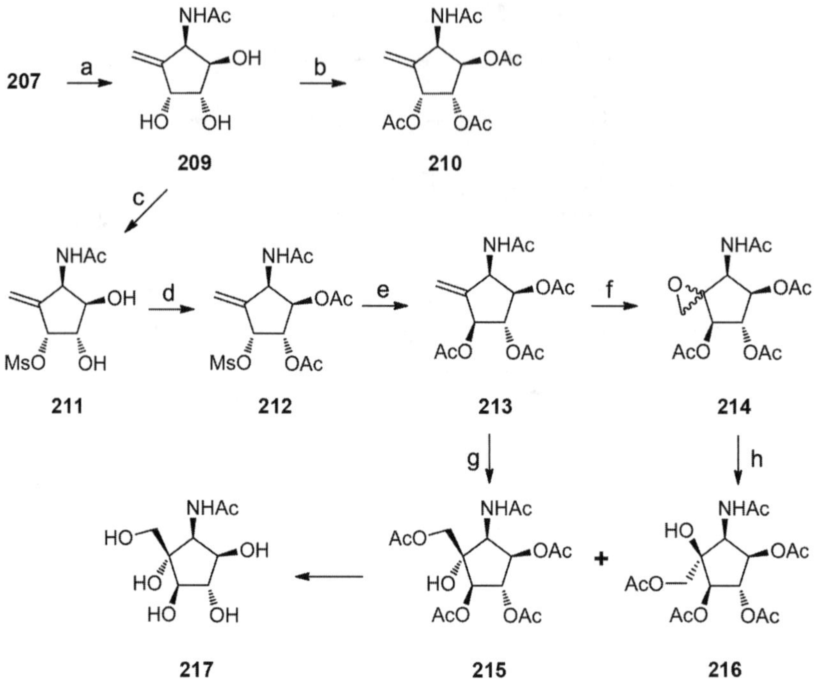

SCHEME 27. (a) Aq. 80% AcOH, 80 °C, 24 h; (b) Py, Ac$_2$O, rt, 2 h, 83%; (c) MsCl, Py, 49%; (d) Py, MsCl, 0 °C, 2 h; Ac$_2$O, rt, 1.5 h, 65%; (e) NaOAc, DMF, 80 °C, 19 h, 60%; (f) ClCH$_2$CH$_2$Cl, m-CPBA, phosphate buffer (pH 8), rt, 10.5 h, 64%; (g) OsO$_4$, t-BuOH, NMO, rt, 19.5 h, Ac$_2$O, Py, **216**, 94%, **217**, 9%; (h) aq. 80% DMF, NaOAc, 120 °C, 20 h, **216** (83%).

tetraoxide to afford the two isomers with no selectivity. Thus, treatment of the mixture of **203** and **204** with trimethylphosphite gave the alkene **207** (53%), together with the unsaturated ketone **206** (11%). cis-Hydroxylation of **207** with osmium tetraoxide in the presence of NMO gave, after acid hydrolysis and acetylation, **208** (53%) and **205** (45%).

Alternatively, synthesis of compound **215** (4-epimer of **208**) started by initial inversion of the OH group at C-1 of **207** (Scheme 27).[35,96,99–101] Acid hydrolysis of **207** gave the triol **209** (100%), which was identified as its tetraacetate **210**, whose allylic hydroxyl group was selectively sulfonylated with mesyl chloride to afford **211**, which was then converted into the acetate **212** (65%). On treatment with an excess of sodium acetate in DMF, **212** afforded **213** (60%). Oxidation of **213** with osmium tetraoxide gave, after acetylation, **214** and **216**. Furthermore, epoxidation of **213** gave a single spiro epoxide **214** (64%), which was transformed exclusively into **216** (83%)

by treatment with sodium acetate followed by acetylation. The amide function seems to facilitate the sterically hindered top-side attack of the epoxide.

The aminocyclitol D-**48** was synthesized from L-**218** (Scheme 28).[103] Since direct peroxyacid oxidation of the exo-methylene group of compound L-**218** had been shown to give selectively the undesired β-spiro epoxide,[35,102] it was first converted into L-**219**, followed by oxidation with *m*-CPBA to give the desired α-spiro epoxide D-**220** (80%). The alcohol D-**221** obtained in 73% yield by reductive cleavage of epoxide D-**220** with lithium triethylborohydride (LiBHEt₃) in THF was formed as a 2:3 mixture of two alcohols. However, hydrolysis of D-**221** with hydrochloric acid gave D-**48** (94%), which was further characterized by conversion into the pentaacetyl derivative **222** (82%).

Compound (±)-**223** was *O*-deacetylated and again protected with MOM groups to give (±)-**224** (94%) (Scheme 29).[103] *O*-Deisopropylidenation of (±)-**224** with aqueous acetic acid, followed by esterification with (*S*)-*O*-acetylmandelic acid gave a mixture of the (*S*)-*O*-acetylmandelates, which were separable by column chromatography to give the enantiomers D-**225** (48%) and L-**226** (49%), and these were deprotected to give quantitatively the respective amino alcohols D-**50** and L-**51**.

8. Synthesis from D-Ribonolactone

D-Ribonolactone was converted into the allylic alcohol **227**,[104] whose condensation with PMB isothiocyanate followed by anti-Markovnikov iodocyclization with iodine afforded the iodo oxazolidinone **228** (82%) (Scheme 30).[105] The latter was treated with a mixture of acetic anhydride and sulfuric acid, followed by activated zinc to

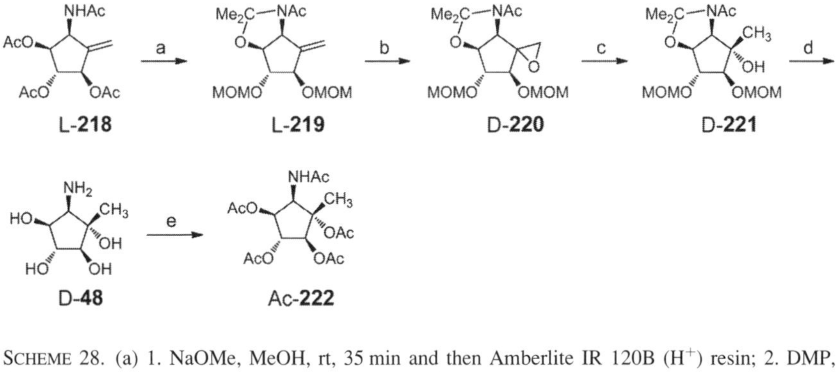

SCHEME 28. (a) 1. NaOMe, MeOH, rt, 35 min and then Amberlite IR 120B (H⁺) resin; 2. DMP, *p*-TsOH, DMF, rt, 6 h; 3. MOMCl, CH₂Cl₂, reflux, 5 h, 61%; (b) *m*-CPBA, ClCH₂CH₂Cl, phosphate buffer solution, rt, 10.5, 80%; (c) THF, lithium triethylborohydride, 0 °C under Ar, 3.3 h; then 28% aq. H₂O₂, 73%; (d) 2 M HCl, 80 °C, 12 h, Dowex 50 W-X2 (H⁺) resin, 5% aq. NH₃, 94%; (e) Ac₂O, Py, rt, 3 h, 82%.

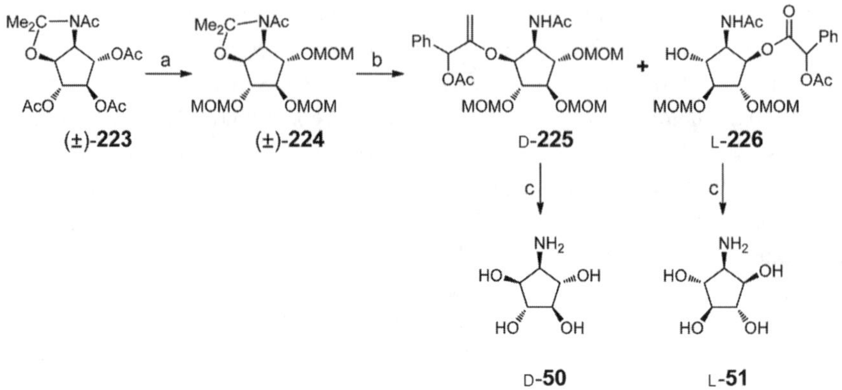

SCHEME 29. (a) CH_2Cl_2, diisopropylethylamine, MOMCl, reflux, 4 h, 93%; (b) 1. 60% aq. AcOH, 60 °C, 2 h, 64%; 2. DMAP, (S)-(+)-O-acetylmandelic acid, CH_2Cl_2, DCC, 0 °C, 10 min, L-**226** (49%), D-**225** (48%); (c) 2 M HCl, 80 °C, 5 h, 100%.

furnish the allylic acetate **228** (90%). This compound underwent inversion at C-2′ under Mitsunobu condition, and the resulting alcohol was epoxidized to produce **230**. Hydrolysis of epoxide **230**, followed by acetylation of the resulting triol **231**, afforded **232**, which was treated with ceric ammonium nitrate (CAN) to furnish the triacetate **233**. Finally, **233** was converted into hexaacetate **234** in two steps. Concurrently, benzylation of **231**, followed by treatment with CAN in aqueous CH_3CN of intermediate **235**, afforded **236**.

A protected trehazolamine was synthesized in a stereocontrolled manner from D-ribonolactone via 2,3-O-isopropylidene-D-ribonolactone (**237**) (Schemes 31),[106,107] whose oxidation under Pfitzner–Moffat or Dess–Martin conditions gave aldehyde **238**, which was treated with TBDMSCl and 1,4-diazabicyclo[2.2.2]octane (DABCO) in DMF to give the (Z)-silyl enol ether **239**. An aldol–Wittig type of reaction of **239** gave **240**, whose selective reduction gave alcohol **241**. Treatment of **241** with benzyloxycarbonyl chloride and 4-(dimethylamino)pyridine gave **242** (84%). Epoxidation of **242** with MCPBA gave **243** (87%), which has the epoxide ring in the trans disposition to all substituents on the cyclopentane ring. Hydrogenolysis of **243** quantitatively gave alcohol **244**, azidolysis of which gave triol **245** (92%). The triol was protected as the bis(isopropylidene) derivative (80%), and protection the remaining alcohol with a PMB group gave **246** in 90% yield. The terminal acetal of **246** was removed regioselectively, and subsequent dibenzylation gave **247** in 89% yield. Removal of the PMB group from **247** with DDQ-H_2O in CH_2Cl_2 gave alcohol **248** in 92% yield. Reduction of the azide group of **248** gave the amine **249**.

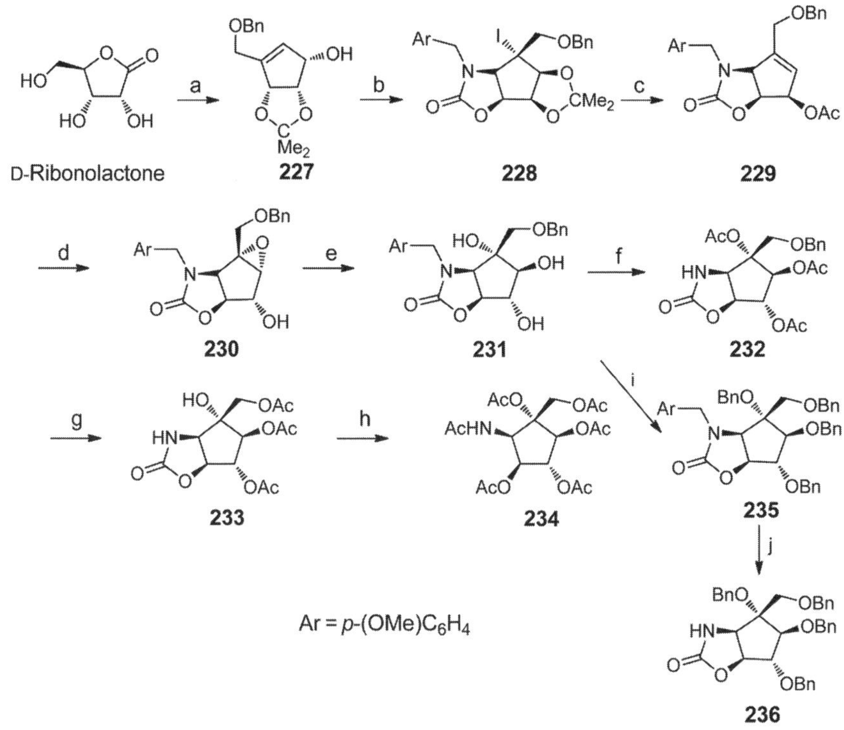

TREHALASE INHIBITORS: SYNTHESIS OF TREHALOZIN AND ANALOGUES 75

Ar = p-(OMe)C$_6$H$_4$

SCHEME 30. (a) Six steps, ref. 104; (b) 1. NaH, p-(OCH$_3$)C$_6$H$_4$CH$_2$NCS, CH$_3$I; 2. I$_2$, THF, Na$_2$CO$_3$, Na$_2$SO$_3$, 82% overall; (c) Ac$_2$O, H$_2$SO$_4$, Zn, THF, 90%; (d) 1. K$_2$CO$_3$, aqueous CH$_3$OH; 2. PhCO$_2$H, DEAD, Ph$_3$P, toluene; 3. Na$_2$CO$_3$, aqueous CH$_3$OH, 83% for three steps; 4. CF$_3$CO$_3$H, CH$_2$Cl$_2$, Na$_2$CO$_3$, −20 °C, 90%; (e) NaOBz, aqueous DMF, 100 °C, 12 h, 89%; (f) 1. Ac$_2$O, Py, CH$_2$Cl$_2$, DMAP; 2. CAN, aqueous CH$_3$CN, 87% for two steps; (g) H$_2$, 10% Pd on C; CH$_3$OH, 98%; (h) 1. 2 M aqueous KOH, EtOH, reflux, 12 h; 2. Ac$_2$O, Py, DMAP, 70%; (i) NaH, BnBr, THF, n-Bu$_4$NI; (j) CAN, aq. CH$_3$CN, 86% two steps.

Methoxybenzylation of **241** gave **250** in 69% yield (Scheme 32).[106,107] Deprotection of the silyl ether of **250** with n-Bu$_4$NF, followed by reprotection of the liberated alcohol with a benzyl group, gave **251**. Epoxidation of **251** afforded stereospecifically epoxide **252** in 55% yield by attack on the double bond from the least-crowded face of the cyclopentene ring. Removal of the PMB group from **252** by treatment with 2,3-dichloro-5,6-dicyano-1,4-benzoquinone (DDQ) and H$_2$O afforded the corresponding alcohol **253** in 99% yield, and subsequent azidolysis gave the azido diol **254** in 99% yield. Reduction of the azide group of **254** with triphenylphosphine gave the amine **255** (97%).

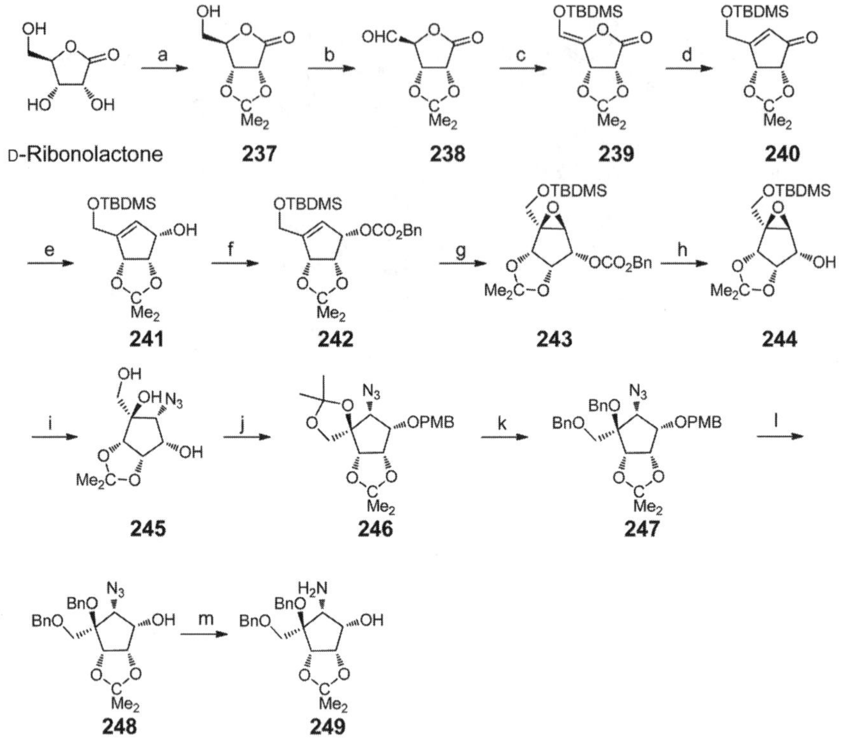

SCHEME 31. (a) 2,2-Dimethoxypropane, PPTS, 50 °C, 1 h; then 1 M HCl, 24–25 °C, 1 h, 82%; (b) Me$_2$SO, CH$_2$Cl$_2$, DCC, H$_3$PO$_4$, rt, 5 h, 80%; or Dess–Martin reagent, CH$_2$Cl$_2$, rt under N$_2$, 30 min, 95%; (c) DMF, TBDMSCl, DABCO, 0–5 °C, 1 h, 1 M HCl, 43%; (d) (bromomethyl)-triphenylphosphonium bromide, t-BuOK, THF, −78 °C under N$_2$, 1.5 h, then t-BuLi, 5 h, then 17 h at 24 °C, 46%; (e) CeCl$_3$·7H$_2$O, EtOH, NaBH$_4$, rt, 30 min, 99%; (f) DMAP, CH$_2$Cl$_2$, benzyl chloroformate, 0 °C, 1 h; then rt, 30 min, 84%; (g) m-CPBA, CHCl$_3$, rt, 4 days, 89%; (h) EtOAc, 10% Pd on C, rt, 30 min, quantitative; (i) NaN$_3$, NH$_4$Cl, DMF, 100 °C, 16 h, 92%; (j) 1. DMF, 2,2-dimethoxypropane, p-TsOH.H$_2$O, rt, 16 h, 82%; 2. DMF, THF, NaH, 4-methoxybenzyl chloride, rt, 90%; (k) 1. 85% AcOH, 40–45 °C, 24 h, 64%; 2. BnBr, NaH, DMF, 10 °C, 16 h, 89%; (l) H$_2$O, DDQ, CH$_2$Cl$_2$, rt, 4 h.

Treatment of the azido diacetal **256** with benzyl bromide and sodium hydride yielded the ether **257** quantitatively (Scheme 33).[106,107] And this was subjected to azide reduction to give amine **258**, and subsequent deacetonation gave **259**.

Reaction of **255** first with 2% HCl in MeOH and then with Et$_3$N and benzyl isothiocyanate afforded **260** in 48% yield (Scheme 34).[106,107] The reaction of **260** with 2-chloro-3-ethylbenzoxazolium tetrafluoroborate gave oxazoine **261**(45%), hydrogenolysis of which gave oxazoline **262** in 45% yield.

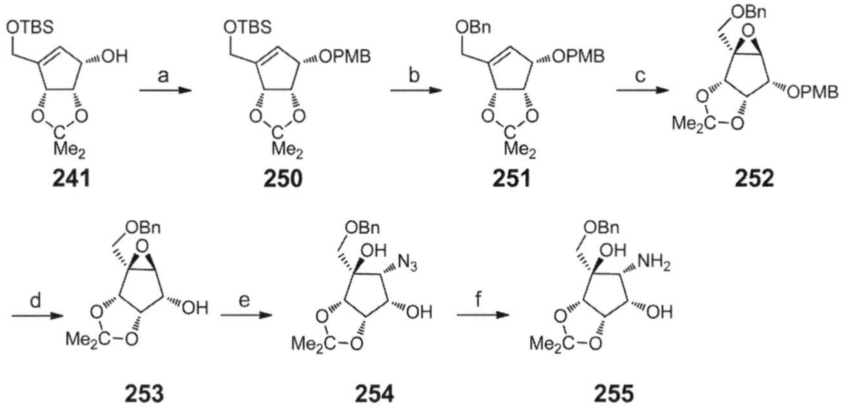

SCHEME 32. (a) DMF, 4-methoxybenzyl chloride, NaH, rt, 3 h, 69%; (b) 1. THF, n-Bu$_4$NF, rt, 3 h, 93%; 2. DMF, NaH, BnBr, rt, 2 h; (c) m-CPBA, CHCl$_3$, rt in dark, 24 h, 55%; (d) H$_2$O, DDQ, CH$_2$Cl$_2$, 24 °C, 3 h, 99%; (e) NaN$_3$, NH$_4$Cl, DMF, 16 h, 100 °C, 99%; (f) Ph$_3$P, THF, rt, 9 days, 97%.

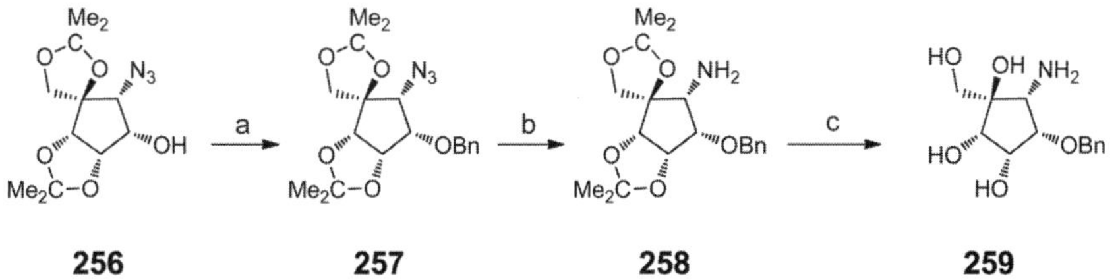

SCHEME 33. (a) DMF, BnBr, NaH, rt, 1 h, quantitative; (b) Ph$_3$P, THF, rt, 7 days, then added H$_2$O, 15 h; (c) 2% HCl in MeOH, 50 °C, 4 h; then add Et$_3$N, rt, 2 h, 72%.

9. Synthesis from Cyclopentadiene

Enantioselective total syntheses of (−)-6-epitrehazolin and (+)-trehazolin were achieved by the synthesis of **275**, which began with an asymmetric heterocycloaddition between [(benzyloxy)methyl]cyclopentadiene (**263**),[108] prepared from thallous cyclopentadienide, and the acylnitroso compound arising from *in situ* oxidation of (*S*)-mandelohydroxamic acid (**264**) with tetrabutylammonium periodate. Cycloaddition led to a mixture of **265** and its diastereomer (Scheme 35).[109] The inseparable mixture was reduced to afford cyclopentenes **266** and **268** in 40% and 11% overall yields, respectively, from thallous cyclopentadienide. Catalytic osmylation of **266** favored *syn* addition, while the osmylation of diacetate **267** was more selective and nearly quantitative, affording, after acetylation, compounds **270** and **269** in >5:1 ratio.

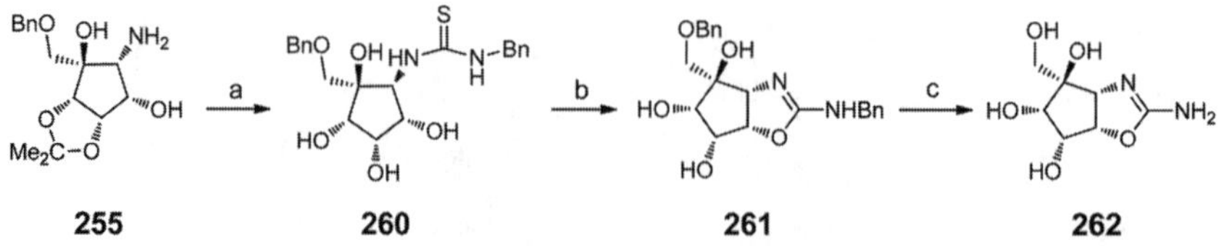

SCHEME 34. (a) THF–H$_2$O (5:1), Et$_3$N, benzyl isothiocyanate, 22 °C, 2 h, 48%; (b) CH$_2$Cl$_2$, Et$_3$N, 2-chloro-3-ethylbenzoxazolium tetrafluroborate, 0–5 °C under N$_2$, 1.5 h, 45%; (c) MeOH, Pd(OH)$_2$ on carbon, H$_2$, 60 °C, 30 min, 31%.

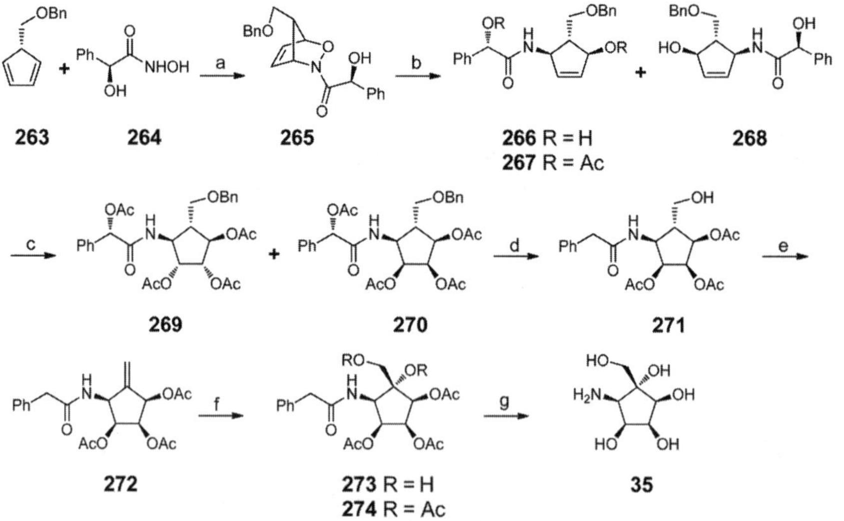

SCHEME 35. (a) PhCH$_2$OCH$_2$Cl, cyclopententadienylthallium, Et$_2$O, −30 °C under Ar, then −20 °C, 30 h; then n-Bu$_4$NIO$_4$, MeOH, −20 °C, 24 h; then rt, 20 h; (b) Na$_2$HPO$_4$, Na amalgam, THF, MeOH, 0 °C, 8 h, **266** (40%); or Py, DMAP, Ac$_2$O under Ar, 20 h, **267** (100%); (c) NMO, acetone, H$_2$O, 0 °C, OsO$_4$, rt, 48 h; then Py, DMAP, Ac$_2$O, under Ar, rt, 24 h, **270** (82%), **269** (16%); (d) EtOH, Pd(OH)$_2$/C, H$_2$, rt, 8 h, 100%; (e) o-nitrophenyl selenocyanate, THF, rt, n-Bu$_3$P under Ar, 4 h; then THF, H$_2$O$_2$, rt, 5 h, 84%; (f) 1. aq. OsO$_4$, NMO, 0 °C, 36 h, **273** (100%); 2. Py, DMAP, Ac$_2$O under Ar, rt, 24 h, **274** (96%); (g) HCl–MeOH, sealed, 90 °C, 24 h, 96%.

Catalytic hydrogenolysis of **270** furnished the debenzylated product **271**, whose nitrophenylselenation followed by *in situ* oxidative elimination gave alkene **272** in 83% yield. The exocyclic alkene group in **272** underwent stereoselective vicinal hydroxylation upon using OsO$_4$ and NMO, to afford diol **273**. Further evidence for the assigned structure came from the observation that **273** underwent slow intramolecular 1,5-acetyl migration to the corresponding primary acetate. The resulting mixture was converted into a single pentaacetate (**274**) in 96% overall yield from **272**. Exhaustive hydrolysis of **274** furnished the aminocyclopentitol **35** in 96% yield.

In an alternative approach, synthesis of the bromo derivative **280** was achieved by reaction of N-bromosuccinimide–H_2O with amide **266**, utilizing the anchimeric participation of the mandelamide carbonyl group, as depicted in **275**, to afford **276** in 85% yield (Scheme 36).[109] Compound **276** was acetylated to give triester **277**, which was subjected to benzylic deoxygenation by hydrogenolysis over Pd–C with no trace of dehalogenation, to give **278**. Subsequent selenation, followed by oxidative elimination, furnished alkene **279** in 80% yield. Stereoselective vicinal dihydroxylation of **279** followed by acetylation of the product afforded the hexasubstituted cyclopentane **280** in 83% yield after chromatography. Hydrolysis of **280** in dilute HCl afforded the corresponding aminobromocyclopentitol, which without purification, was used in further steps.

Amide **266** was also used for the synthesis of **5** (Scheme 37),[109] employing *syn*-epoxidation to afford **284** in 96% yield. Alternatively, alkaline hydrolysis of **266** furnished **281**, which could be trichloroacetylated to afford alkene **282**. Treatment of **282** with potassium peroxymonosulfate ("oxone") formed the corresponding *syn*-epoxide **283** in 60% yield. Epoxide **284** proved resistant to acid but did undergo slow hydrolysis upon using 2:1 H_2O–TFA to afford in 1:1.6 ratio the desired product **285** (32%) and its isomer **286** (52%), which for further characterization were transformed into the corresponding peracetates **289** (93%) and **290** (90%). Under similar conditions,

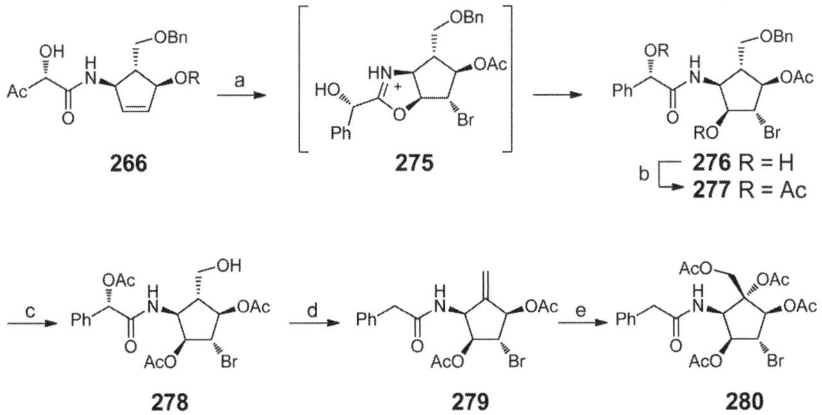

SCHEME 36. (a) NBS, 1,4-dioxane, H_2O, rt, 18 h, **276** (85%); (b) Ac_2O, DMAP, rt, overnight, **277** (94%); (c) EtOH, Ph(OH)$_2$/C, rt, H_2, 4 h, **278** (99%); (d) THF, MS 4 Å, 2-nitrophenyl selenocyanate, under argon, *n*-Bu$_3$P, rt, overnight, then H_2O_2, 50 h, **279** (80%); (e) NMO, 2% aqueous OsO_4, acetone, H_2O, rt, 40 h, then excess Ac_2O, DMAP, rt, overnight, **280** (65%).

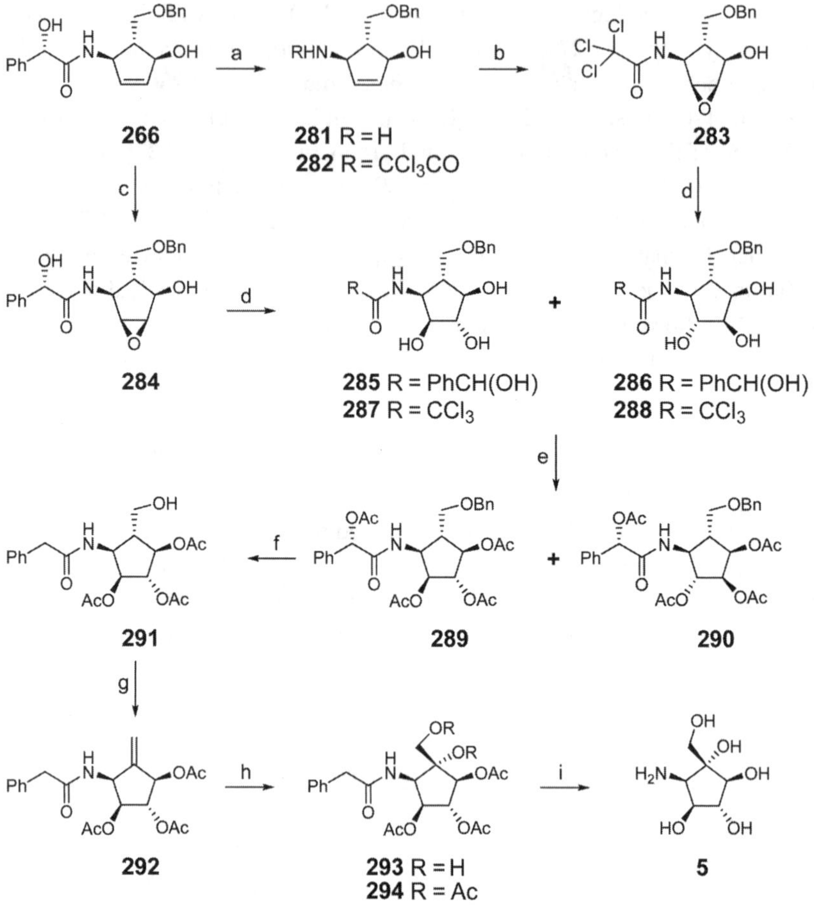

SCHEME 37. (a) 1. MeOH, NaOMe; 2. Hexachloroacetyl chloride, **282** (36%); (b) "Oxone," 60%; (c) 1,4-dioxane, *m*-CPBA, rt, overnight, 96%; (d) H$_2$O, TFA, 0 °C, 1 h; then rt, 48 h, **285** (32%), **286** (52%); (e) Ac$_2$O, DMAP, Py, **289** (93%), **290** (90%); (f) EtOH, Pd(OH)$_2$/C, H$_2$, rt, 4 h, quantitative; (g) 2-nitrophenyl selenocyanate, THF, MS 4 Å, *n*-Bu$_3$P, rt, 4 h, 30% H$_2$O$_2$, 5 h, rt, 95%; (h) 1, acetone, H$_2$O, NMO, 2% aqueous OsO$_4$, rt, 24 h; 2. Ac$_2$O, Py, DMAP, rt, overnight, **294** (65%); (i) pentacetate, 0.5 M HCl, MeOH, 4 days, 87%.

the combined halogen electron-withdrawing effects in epoxytrichloroacetamide **283**, which were expected to influence the regiochemistry of epoxide opening, had only a modest effect on the product distribution, leading to **287** and **288** in 1:1 ratio.

Reductive cleavage by hydrogenolysis of both the benzylic ether and ester substituents in **289** gave the hydroxy phenylacetamide **291** in quantitative yield.

Nitrophenylselenation of the primary alcohol in **291**, followed by *in situ* oxidative elimination, furnished alkene **292** in 95% yield. Stereoselective osmylation of the exocyclic alkene in **292**, using OsO$_4$–NMO gave diol **293**, which was additionally characterized as its pentaacetate **294**, formed in 65% yield. Acid hydrolysis of **294** furnished the target aminocyclopentitol (+)-**5** in 87% yield.

10. Synthesis from Acylated Oxazolidinones

An asymmetric synthesis of the aminocyclopentitol has been achieved from an acylated oxazolidinone (Scheme 38).[110] Thus, the acylated oxazolidinone **295** was subjected to boron triflate-catalyzed condensation with 3-butenal to yield the *syn* aldol product **297** in 63% yield. Similarly, the *N*-acyloxazolidinethione **296** delivered the aldol adduct **298** in 75% yield when enolized with TiCl$_4$-(−)-sparteine and then

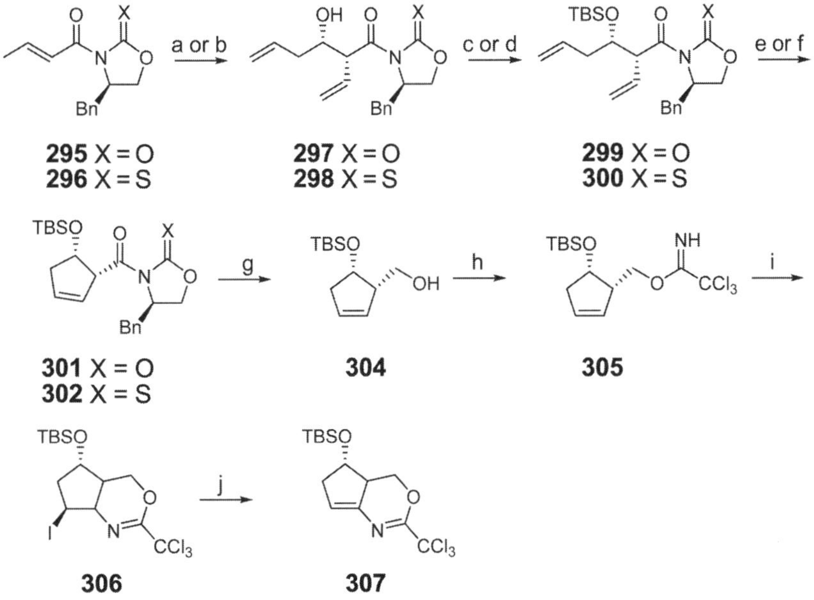

295 X = O
296 X = S

297 X = O
298 X = S

299 X = O
300 X = S

301 X = O
302 X = S

304

305

306

307

SCHEME 38. (a) CH$_2$Cl$_2$, −78 °C, Bu$_2$BOTF, Et$_3$N, 1 h; then added 3-butenal, 1 h; 0 °C, 1 h; then H$_2$O$_2$, 0 °C, 1 h, **297** (90%); (b) CH$_2$Cl$_2$, TiCl$_4$, −78 °C, 5 min, (−)-sparteine, 2 h; then added 3-butanal, 1 h, **298** (75%); (c) CH$_2$Cl$_2$, 2,6-lutidine, TBDMSOTf, 2 h, **299** (92%); (d) CH$_2$Cl$_2$, 2,6-lutidine, TBDMSOTf, 1.5 h, **300** (70%); (e) CH$_2$Cl$_2$, Grubbs' metathesis catalyst [Cl$_2$(Cy$_3$P)$_2$Ru=CHPh], reflux, 2 h, **301** (92%); (f) CH$_2$Cl$_2$, [Cl$_2$(Cy$_3$P)$_2$Ru=CHPh], reflux, 2.5 h, **302** (91%); (g) MeOH, Et$_2$O, lithium borohydride, 0 °C, 2 h; then 25 °C, 1 h, **304** from **301** (83%), **304** from **302** (75%); (h) Cl$_3$CCN, NaH, 82%; (i) NIS, CH$_2$Cl$_2$, 92%; (j) DBU, 80 °C, benzene.

treated with 3-butenal. Attempted ring-closing metathesis on diene **297** or **298** led to incomplete conversion into the desired cyclopentene (56:44 ratio of starting material to product), presumably attributable to coordination of the homoallylic alcohol to the metal center in the intermediate alkylidene. To circumvent the problem, the secondary alcohol was protected as its TBDMS ether to provide **299** or **300** in high yield, and exposure of the dienes **299** and **300** to the Grubbs catalyst generated cyclopentene **301** or **302** in 85% yield over two steps. Reductive removal of the chiral auxiliary, followed by treatment of the resultant alcohol **304** with trichloroacetonitrile, afforded **305**. *N*-Iodosuccinimide-promoted cyclization produced the iodide **306**. Elimination using DBU resulted in the undesired cyclopentene regioisomer **307**. However, the aminocyclopentane **319** was prepared from **304** as shown in Scheme 39.[110]

Treatment of the alcohol **304** with *p*-toluenesulfonyl isocyanate followed by cyclization with iodine afforded the iodide **308** (77%), which underwent elimination with DBU to yield 91% of the cyclopentene **309**. Reductive removal of the sulfonamide of alkene **309** using sodium naphthalide, followed by acylation of the nitrogen with di-*tert*-butyl dicarbonate furnished carbamate **310** in 88% yield (Scheme 39).[110] Epoxidation was attempted after hydrolysis of the carbamate. Cesium carbonate-mediated hydrolysis of the carbamate, and protection of the alcohol with acetic anhydride gave ester **312** in 95% overall yield. Epoxidation of the less-hindered face of the alkene using dimethyldioxirane gave 74% of epoxide **313**. Acid-promoted nucleophilic opening of the epoxide by the carbonyl oxygen of the neighboring carbamate generated oxazolidinone **314** in 83% yield. Deprotection of the acetate under basic conditions provided the primary alcohol **315**. Treatment of alcohol **315** with 2-nitrophenyl selenocyanate and tributylphosphine gave the intermediate selenide **316**, which was immediately oxidized to give the exocyclic alkene **317**. Osmylation of the alkene gave 75% of triol **318**, which was subjected to reaction with aqueous base followed by peracetylation with acetic anhydride to give ester **319**.

11. Synthesis from *cis*-2-Butene-1,4-diol

The aminocyclitol moiety was synthesized in a stereocontrolled manner from *cis*-2-butene-1,4-diol (Scheme 40)[112] by conversion into epoxide **321** via Sharpless asymmetric epoxidation in 88% yield.[111] Oxidation of **321** with IBX, followed by a Wittig reaction with methyl-triphenylphosphonium bromide and KHMDS, produced alkene **322**. Dihydroxylation of the double bond of **322** with OsO_4 gave the diol **323**, which underwent protection of the primary hydroxyl group as the TBDMS ether to furnish **324**. The secondary alcohol of **324** was oxidized with Dess–Martin periodinane to

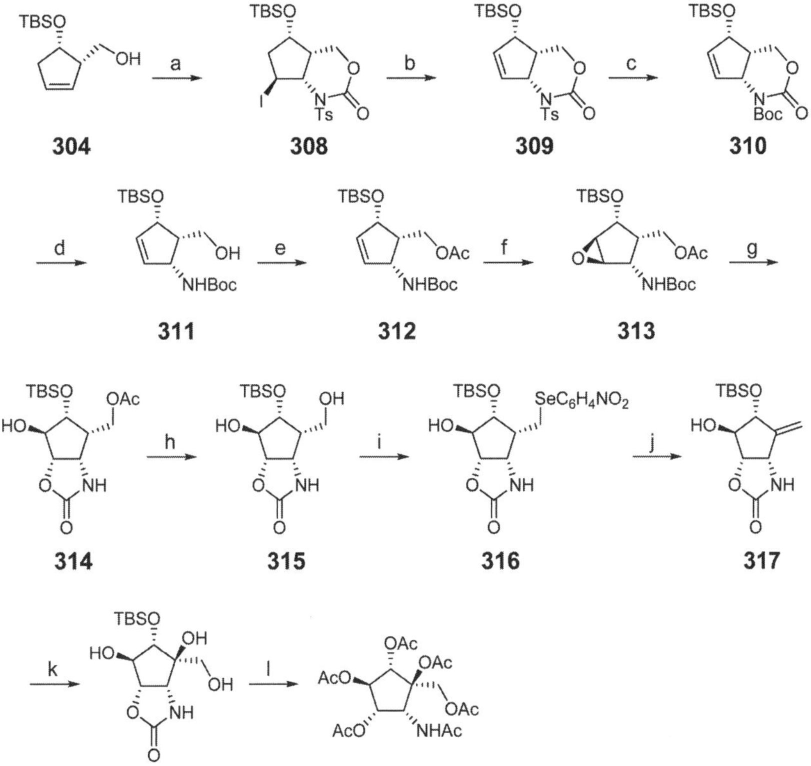

SCHEME 39. (a) THF, *p*-toluenesulfonyl isocyanate, 12 h, 95%; then alkene, ether, K_2CO_3, 0 °C, I_2, 25 °C, 36 h, 82%; (b) benzene, DBU, reflux, 1 h, 91%; (c) naphthalene, 1,2-dimethoxyethane, sodium, sonicated, 2 h; then added alkene 15, THF, −78 °C, 92%; THF, Et_3N, 0 °C; and then Boc_2O, DMAP, 25 °C, 12 h, 96%; (d) MeOH, 0 °C, $CsCO_3$, 25 °C, 48 h, 94%; (e) CH_2Cl_2, Et_3N, Ac_2O, DMAP, 25 °C, 6 h, 95%; (f) dimethyldioxirane, acetone, 5 h, 74%; (g) CH_2Cl_2, CSA, 12 h, 83%; (h) MeOH, K_2CO_3, rt, 2 h, 98%; (i) 2-nitrophenyl selenocyanate, THF, *n*-Bu_3P, rt, 12 h; (j) THF, H_2O_2, 25 °C, 12 h, 79%; (k) NMO, acetone, H_2O, OsO_4, rt, 12 h, 75%; (l) EtOH, 2 N KOH, reflux, 12 h; then Ac_2O, Py, DMAP, rt, 12 h, 80%.

produce the epoxy ketone **325**, which was subjected to reaction with lithium (trimethylsilyl)diazomethane (TMSCLiN$_2$) in DME at 0 °C to produce the cyclized product **326** as a single diastereomer in 51% yield.

Opening of the epoxide ring of **326** was performed with methanol or allyl alcohol under acidic conditions to provide the corresponding ethers **327** or **328**. Epoxidation of **327** with *m*-CPBA gave the epoxide **329** and its isomer in 2.5:1 ratio with an 88% combined yield, while epoxidation of **328** gave epoxide **330** with a 3.5:1 ratio in a

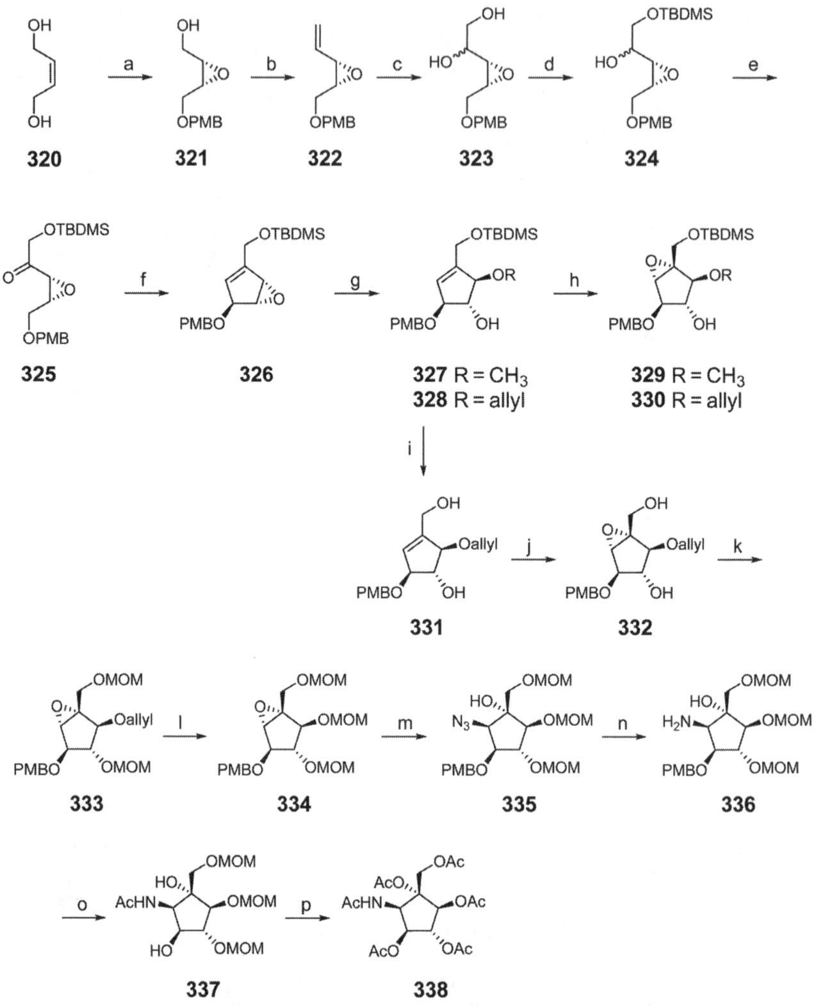

SCHEME 40. (a) Ref. 111; (b) 1. IBX, Py, Me$_2$SO, 45 °C; 2. Ph$_3$PCH$_3$Br, KHMDS, toluene, THF, −20 °C to rt (83% in two steps); (c) OsO$_4$, NMO, THF–H$_2$O, rt (95%); (d) TBSCl, imidazole, DMF, rt (88%); (e) Dess–Martin periodinane, CH$_2$Cl$_2$, rt (85%); (f) TMSCLiN$_2$, DME, 0 °C to rt (51%); (g) MeOH or allyl alcohol, p-TsOH, −78 to 0 °C; (h) epoxidation OsO$_4$, NMO, THF–H$_2$O, rt; (i) TBAF, THF, rt (69% in two steps); (j) L-(+)-DET, Ti(Oi-Pr)$_4$, TBHP, CH$_2$Cl$_2$, −20 °C (73%); (k) MOMCl, iPr$_2$NEt, CH$_2$Cl$_2$, rt (87%); (l) 1. PdCl$_2$, NaOAc, AcOH, H$_2$O, rt (85%); 2. MOMCl, iPr$_2$NEt, CH$_2$Cl$_2$, rt (86%); (m) NaN$_3$, NH$_4$Cl, DMF–ethylene glycol, 120 °C (78%); (n) 1. DDQ, CH$_2$Cl$_2$–H$_2$O, rt (87%); 2. 10% Pd/C, H$_2$, EtOAc, rt (97%); (o) Ac$_2$O, MeOH, rt (quant.); (p) 1. AcCl, MeOH, rt; 2. Ac$_2$O, DMAP, Py, rt (40% in two steps).

lower yield (10%) because of competitive reaction with the allyl group. Attempted demethylation of methyl ether **329** was unsuccessful. Removal of the TBDMS group of **328** produced the diol **331**, which was subjected to Sharpless epoxidation and purification by silica gel chromatography to produce the epoxide **332** as a single isomer. After protecting the two hydroxyl groups of **332** as the MOM ethers, the allyl group of **333** was removed with $PdCl_2$ in acetic acid and the resulting hydroxyl group was protected as the MOM ether to give **334**. Epoxide opening of **334** with sodium azide followed by removal of the PMB group gave the azide **335**, reduction of which provided the amino diol **336**. Treatment of **336** with methanolic hydrogen chloride followed by acetylation of the crude product resulted in a complex mixture containing hexaacetate **338**. Therefore, the amino group of **336** was first acetylated to give acetamide **337**, which was then subjected to deprotection followed by acetylation to provide the peracetate **338**.

12. Synthesis from Norbornyl Derivatives

Synthesis of aminocyclopentane analogues was also achieved from norbornyl derivative **339**[113] (Scheme 41).[114] For elaboration to the aminocyclopentitol structures, the free hydroxyl group in **340** was activated as its mesylate **341**. On exposure to base, **341** readily underwent elimination to the desired exocyclic methylene compound **342**, which was the key precursor for generation of the 1,2-diol functionality present in trehazolamine and related compounds. Catalytic dihydroxylation of **342** furnished a readily separable mixture (55:45) of the diastereomeric diols **344** and **345**. Introduction of the amino group into **344** and **343** was achieved through an azidation–reduction sequence. Thus, displacement of the mesylate group in **344** with sodium azide furnished the azide **345**, which was reduced to the corresponding amine; removal of the isopropylidene group with dilute HCl led directly to the amine hydrochloride **346**. An identical sequence from **343** led to the salt **348** via the intermediate azido compound **347**.

13. Synthesis from 3-Hydroxymethylpyridine

A strategy for the synthesis of trehazolamine has been developed from 1-methoxyethoxymethyl-3-pivaloyloxymethylpyridinium perchlorate via amino cyclopentane derivatives (Schemes 42 and 43).[115] Pyridinium salt **351** was prepared by a two-step sequence from 3-(hydrooxymethyl)pyridine **349** via **350** (Scheme 42). Irradiation of **351** in aqueous $NaHCO_3$ led to production of the separable bicyclic aziridines **352** (20%) and **353** (16%). Ring opening of **352** by acetic acid, followed by removal of the

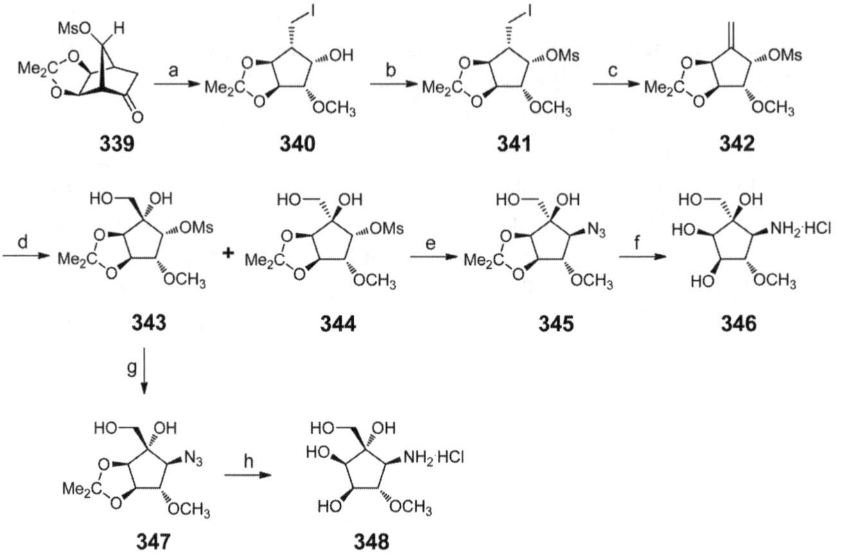

SCHEME 41. (a) Ref. 113; (b) MsCl, Py, DCM, rt, overnight, 93%; (c) DBU, CH₃CN, rt, overnight, 90%; (d) OsO₄, NMMO, 4:1 Me₂CO–H₂O, 96 h, 84%; (e) NaN₃, 1:1 DMF–HMPA, 110 °C, 10 h, 89%; (f) H₂, Lindlar's catalyst, EtOH, 2 h; HCl (5%), 1:4 ether–H₂O, 36 h, ∼95%; (g) NaN₃, 1:1 DMF–HMPA, 110 °C, 10 h, 96%; (h) H₂, Lindlar's catalyst, EtOH, 2 h; HCl (5%), 1:4 ether–H₂O, 36 h, 90%.

MEM group and peracetylation, provided the Acetamidocyclopentenetriol derivative **354** (80%, three steps). Inversion of the C-3 hydroxyl group was set up by transformation of **354** into the isopropylidene acetal **355**. Treatment of **355** with the Burgess salt, followed by ring opening of the bicyclic-oxazolidine intermediate with NaH₂PO₄ in aqueous THF, led to formation of the epimeric alcohol **356**. Hydroxyl-directed epoxidation of **356** produced epoxide **357**, which was then transformed into the hexacetylated (±)-trehazolamine **358** by regioselective epoxide ring opening, acetal cleavage, and acetylation.

Other trehazolamines were prepared by the photocyclization of 3-pivaloyloxymethyl-*N*-(tetra-*O*-acetyl-α-D-glucopyranosyl)-pyridinium perchlorate **359** (Scheme 43).[115] Irradiation of **359** in aqueous NaHCO₃ led to a mixture of isomeric *N*-glucosyl bicyclic aziridines, which could be partially separated by chromatography on silica gel to yield pure **362** (15%) and a mixture of **363** and **360** or **361** (30%). Aziridine ring opening of **362**, followed by hydrolytic cleavage of the glycosylic C–N bond and peracetylation gave the (−)-enantiomer of the acetamidocyclopententriol derivative (−)-**364**. Treatment of the mixture containing **363** and **360** or **361** with acetic acid generated a separable mixture of the *N*-glucosylaminocyclopentenes **365** (46%) and **366** (47%). The former

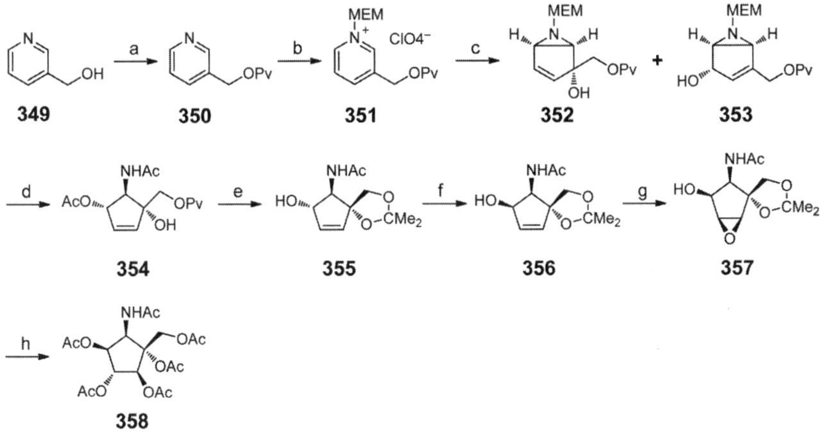

SCHEME 42. (a) Et$_2$O, PvCl, TEA, 0 °C, 3 h, 98%; (b) 1. MEMCl, MeCN, 0 °C, 20 min; 2. AgClO$_4$, MeCN, 0 °C, 1 h, 100%; (c) $h\nu$, aq. NaHCO$_3$; (d) 1. HOAc, MeCN, 25 °C, 5 h; 2. 0.1 M HCl, Ac$_2$O, Py, DMAP, 25 °C, 12 h, 80%; (e) 1. NaOMe, MeOH, 25 °C, 3 h, 83%; 2. Me$_2$C(OMe)$_2$, (−)-CSA, MeCN, 25 °C, 30 min, 87%; (f) THF, Burgess salt, aq. NaH$_2$PO$_4$, 70 °C, 24 h, 82%; (g) m-CPBA, CH$_2$Cl$_2$, 25 °C, 48 h, 87%; (h) 1. aq. NaOBz, H$_2$O, 110 °C, 32 h; 2. 80% aq. HOAc, 80 °C, 24 h; 3. Ac$_2$O, Py, DMAP, 25 °C, 24 h, 90%.

365 served as the precursor for acetamidocyclopentenes (+)-354 formed by sequential glucosyl cleavage and acetylation. The absolute stereochemistry of (+)-354 was determined by X-ray crystallography on the ester 367, derived by conversion of (+)-354 into the acetal (+)-355 and acylation.

The trehazolin aminocyclitol moiety (trehazolamine) 5 (Scheme 44)[116,117] was treated with benzyl chloroformate in THF–H$_2$O containing pyridine at 0–5 °C, and the resulting N-benzyloxycarbonyl derivative was converted into the trisilylated (368) and tetrasilylated (369) ethers with TBSCl and 4-dimethylaminopyridine in DMF. Compound 368 was also silylated at 20–25 °C for 4 days to give 369. Hydrogenolysis of 369 gave 370.

VII. Synthesis of Trehazolins

1. Trehazolins Having Trehazolamines at the Anomeric Center

The synthesis of trehalamines has generally involved the thiourea derivative of trehazolamines by intramolecular cyclization with one of the neighboring hydroxyl groups. Thus, treatment of 88 with benzyl isothiocyanate afforded the thiourea derivative 371. Compound 371 was hydrogenolyzed to cleave the two MOM groups,

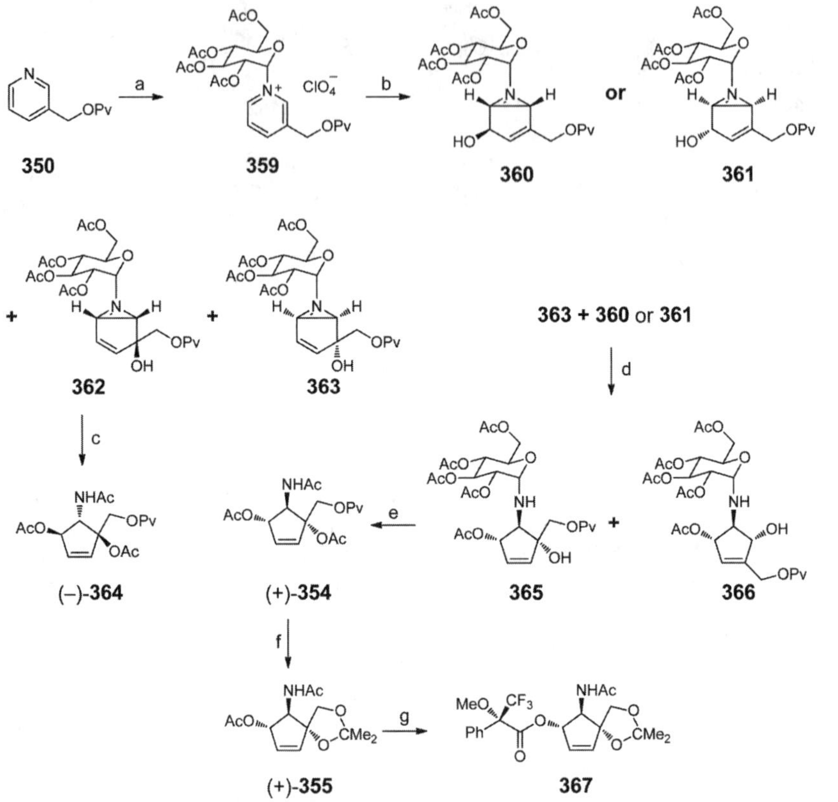

SCHEME 43. (a) 1. Tetra-O-acetyl-α-D-glucopyranosyl bromide, CH₃CN, 60 °C, 24 h; 2. AgClO₄, MeCN, 0 °C, 1 h, 42%; (b) hv, aq. NaHCO₃; (c) 1. HOAc, MeCN, 25 °C, 12 h; 2. 0.1 M HCl, 25 °C, 12 h; 3. Ac₂O, Py, DMAP, 25 °C, 12 h, 89%; (d) HOAc, MeCN, 25 °C, 12 h, **362** (46%), **360**, **361**, **363** (47%); (e) 1. 0.1 M HCl, MeOH, 25 °C, 12 h; 2. Ac₂O, Py, DMAP, 25 °C, 12 h, 95%; (f) 1. NaOMe, MeOH, 25 °C, 3 h, 85%; 2. Me₂C(OMe)₂, (−)-CSA, 25 °C, 30 min, 83%; (g) DMAP, Py, R-(−)-α-methoxy-α-trifluoromethylphenylacetyl chloride, CH₂Cl₂, 25 °C, 12 h, 97%.

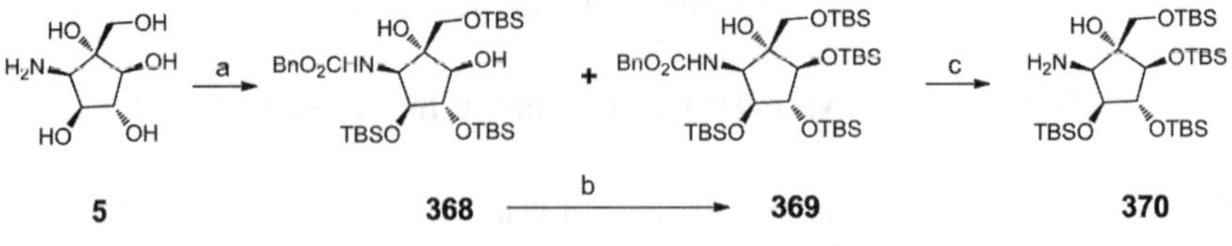

SCHEME 44. (a) ClCO₂Bn, Py, H₂O–THF (2:1), 0–5 °C, 30 min, concentrated; then TBDMSCl, DMAP, DMF, 20–25 °C, 16 h, **368** (34%), **369** (11%); (b) TBDMSCl, DMAP, DMF, 20–25 °C, 4 days, ∼62% (recovery **368**, 38%); (c) H₂, Pd/C, THF, 24 °C, 8 h, 92%.

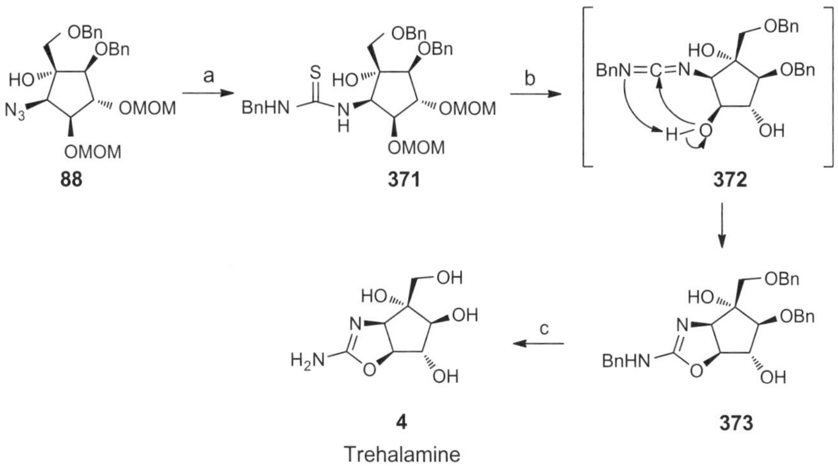

SCHEME 45. (a) BnNCS, 83%; (b) 0.5 M aqueous HCl, 2-chloro-3-ethylbenzoxazolium tetrafluoroborate, Et₃N, 74%; (c) H₂, Pd(OH)₂ on C, 71%.

and the resulting product was treated with 2-chloro-3-ethylbenzoxazolium tetrafluoroborate and triethylamine to afford the corresponding aminooxazoline **373** via **372**. Hydrogenolysis of **373** gave the trehalamine (**4**) (Scheme 45).[38]

Toward a complete synthesis of trehazolin (**2**), 2,3,4,6-tetra-*O*-benzyl-1-deoxy-α-D-glucopyranosyl isothiocyanate (**7**)[52] was brought into reaction with the amines **5** and **374** in the presence of triethylamine to afford the α-D-glucopyranosyl thiourea derivatives **375** or **376**, respectively (Scheme 46).[38] Subsequent treatment with 2-chloro-3-ethylbenzoxazolium tetrafluoroborate and triethylamine afforded the corresponding amino oxazoline derivatives **377** and **378**. Finally, hydrogenation over Pd(OH)₂ on carbon afforded trehazolin (**2**).

Treatment of **5** with 4 M HCl, followed by treatment with isothiocyanate **379** in pyridine–DMF, produced **380**. This was treated with yellow HgO to afford the desired glycosyl oxazoline **381**. Finally, deprotection of the trehazolin tribenzyl ether **381** afforded trehazolin (**2**) (Scheme 47).[49]

The *N*-alkyl and *N*-aryl ureas **382** and **383** were prepared in high yield by treatment of **121** with the corresponding isocyanates (Scheme 48).[79] The reaction of **382** and **383** with triflic anhydride in pyridine–CH₂Cl₂ at low temperature produced the respective 2-aminooxazolines **384** and **385** in high yield, with concomitant inversion of stereochemistry at the required position. Complete deprotection of **384** by hydrogenolysis under acidic conditions afforded trehalamine (**4**). Deprotection of **385** under

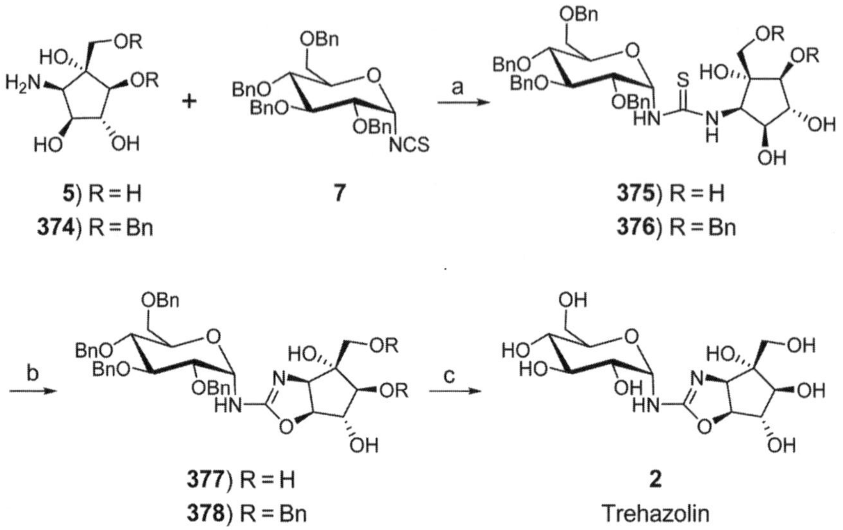

SCHEME 46. (a) Et$_3$N, 69%; (b) EtBF$_4$, Et$_3$N, 68%; (c) H$_2$, Pd(OH)$_2$ on C, 44%.

similar conditions gave **386**, a product that was reported to be a moderate α-glucosidase inhibitor.[79] Condensation of the aminocyclitol **121** with the benzylated α-D-glucosyl isothiocyanate **7**[52] gave the thiourea derivative **387**. Treatment of **387** with aqueous acetonitrile in the presence of yellow HgO gave the urea derivative **388** in quantitative yield. The reaction of **388** with triflic anhydride and pyridine effected cyclization to the 2-aminooxazoline **389** in 92% yield. Hydrogenolysis of the benzyl groups and acid cleavage of the isopropylidene acetal afforded trehazolin (**2**) in 34% overall yield.

A total synthesis of 5-*epi*-trehazolin (trehalostatin, **3**) was accomplished via the azido alcohol **100** (Scheme 49).[71] Reduction of the azido group and cleavage of the two MOM groups of the latter compound, and coupling of the corresponding amine hydrochloride with 2,3,4,6-tetra-*O*-benzyl-α-D-glucopyranosyl isothiocyanate (**7**) in the presence of triethylamine afforded thiourea derivative **390**. Cyclization using 2-chloro-3-ethylbenzoxazolium tetrafluoroborate gave an aminooxazoline **391**. Hydrogenolysis of compound **391** and purification on Amberlite CG-50 (NH$_4^+$ type/H$^+$ type, 3/2 resin) afforded 5-*epi*-trehazolin (**3**). This compound displays much weaker inhibitory activities toward trehalases than trehazolin, indicating that the stereochemistry at the C-5 position of the trehazolin aglycon has a significant influence upon the inhibitory activities of trehalase analogues.

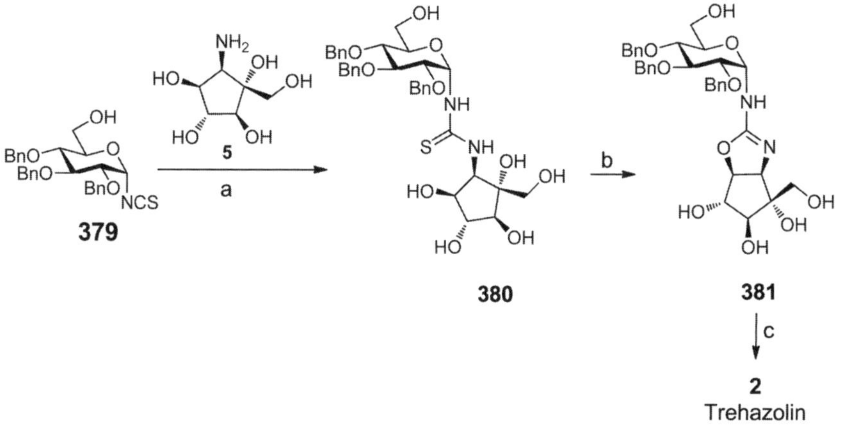

TREHALASE INHIBITORS: SYNTHESIS OF TREHALOZIN AND ANALOGUES 91

SCHEME 47. (a) 4 M HCl, 60 °C, 8 h, then added aminocyclitol, Py, DMF, DMAP, MeOH, 77%; (b) yellow HgO, Et$_2$O, Me$_2$CO; (c) Pd(OH)$_2$/C, H$_2$, MeOH, 40%.

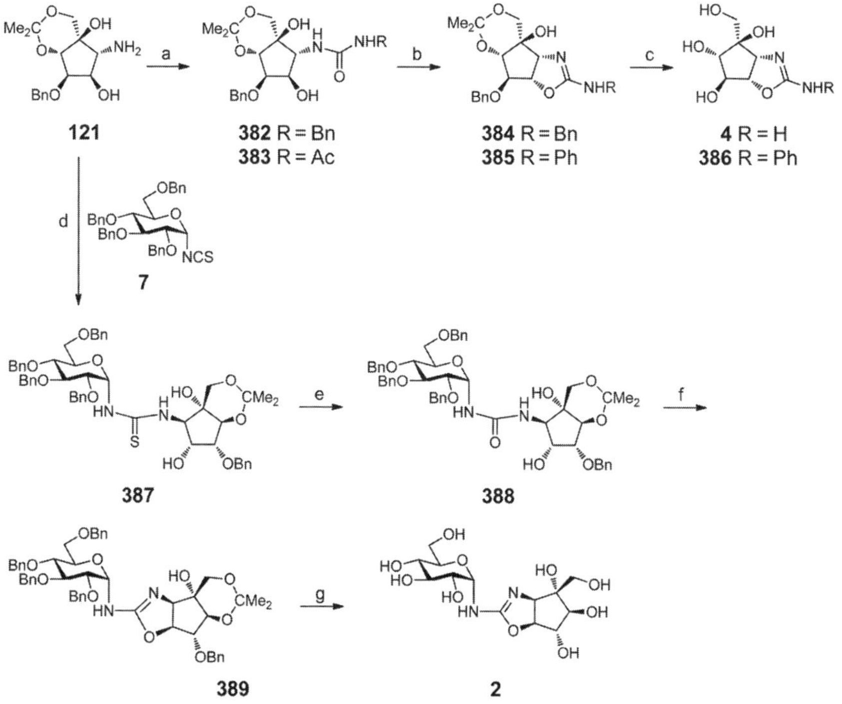

SCHEME 48. (a) BnNCO or PhNCO, CH$_2$Cl$_2$, rt, **382** (93%), **383** (99%); b. Tf$_2$O, Py, CH$_2$Cl$_2$, −20 °C, **384** (85%), **385** (65%); (c) Pd(OH)$_2$/C, H$_2$, EtOH, TFA, **4** (85%), **386** (65%); (d) THF, rt, 96%; (e) MeCN, H$_2$O, HgO, 98%; (f) Tf$_2$O, Py, CH$_2$Cl$_2$, −20 °C to rt, 92%; (g) 1. Pd(OH)$_2$/C, H$_2$, EtOH, TFA; 2. 2 M HCl, rt, 83%.

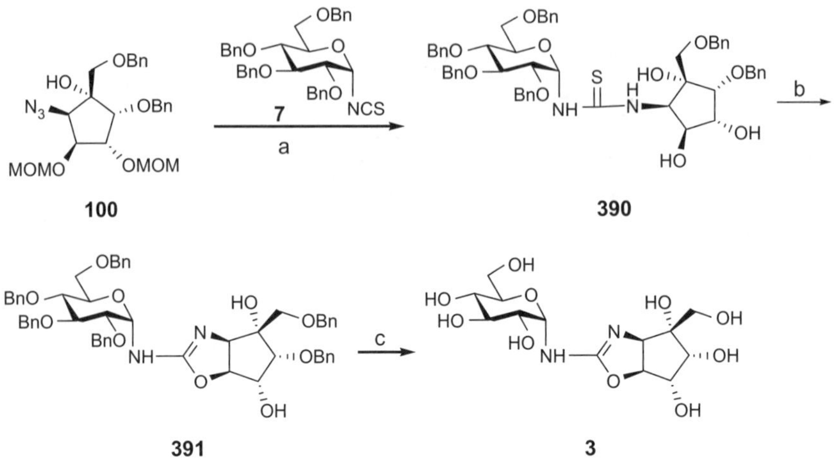

SCHEME 49. (a) 5.5 equiv of LiAlH$_4$, Et$_2$O, 0 °C, 3 h, 5% HCl in MeOH, 60 °C, 7 h, and then 1.2 equiv of 2,3,4,6-tetra-*O*-benzyl-α-D-glucopyranosyl isothiocyanate, 1.2 equiv of Et$_3$N, THF, rt, 21 h, 54%; (b) 2.6 equiv of 2-chloro-3-ethylbenzoxazolium tetrafluoroborate, MeCN, 0 °C, 1 h, then 5.2 equiv of Et$_3$N, 0 °C, 3 h, 71%; (c) H$_2$, Pd(OH)$_2$/C, MeOH, 60 °C, 30 min, 20%.

Coupling of **34** with isothiocyanate **7** afforded trehalostatin **3** via compound **392**. Likewise, the coupling of epimer **42** with **7** afforded **393**, which was converted into the trehalostatin analogues **396** and **397** via the intermediates **394** and **395**, respectively (Scheme 50).[97,98]

A complete synthesis of trehazolin (**2**) and its diastereoisomer involved coupling of **5**[35,102] with **7**[52] in a mixture of DMF and methanol, affording a 93% yield of a diastereoisomeric mixture of the thiourea derivatives **398** (Scheme 51).[118] Without separation, the mixture was treated directly with mercuric oxide in diethyl ether, resulting in simultaneous ring-closure through attack of the neighboring *cis*-hydroxyl group and giving rise to an inseparable mixture (96%) of **399**. This was subjected to debenzylation followed by acetylation to give the acetylated product **400**. Deacetylation of **400** afforded trehazolin (**2**).

Coupling of the amino alcohol D-**48** and the isothiocyanate **7** in 75% aq. DMF gave the thiourea derivative D-**401** (100%), which was treated with an excess of yellow mercury(II) oxide in diethyl ether to afford D-**402** (93%). Removal of the benzyl ether groups by sodium in liquid ammonia was followed by purification by a column of Dowex 50W-X2 (H$^+$) resin to give the trehazolin analogue D-**403** (90%). The hepta-*N*,*O*-acetyl derivative D-**404** fully supported the structures assigned for D-**403** (Scheme 52).[118]

TREHALASE INHIBITORS: SYNTHESIS OF TREHALOZIN AND ANALOGUES 93

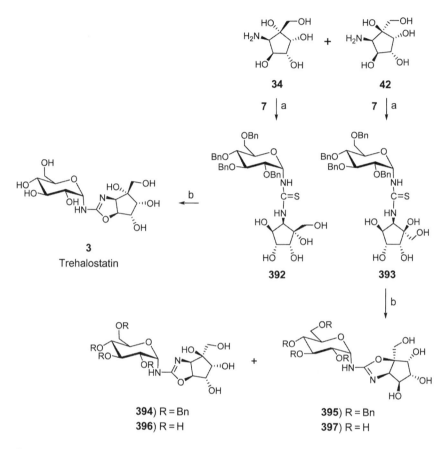

SCHEME 50. (a) aqueous 75% DMF, 4 h, rt, 92%; (b) 1. Ether, HgO, 3 h, 100%; 2. Na, liquid NH$_3$, 94%.

The amino alcohols D-**50** and L-**50** were converted into D-**408** and L-**408** via coupling with isothiocyanate **7** to give the thiourea D-**405** (93%), which was similarly converted into the cyclic isourea D-**406** 100%) (Scheme 53).[98] Conventional O-debenzylation afforded the trehazolin analogue D-**407** (96%), which was further characterized as the octa-N,O-acetyl derivative D-**408**. Likewise, the diastereoisomer L-**407** was synthesized in 91% overall yield by similar reaction sequence, starting from the L-**405** (97%) obtained from isothiocyanate **7** and aminotetrol L-**50**.

Replacement of the aminocyclitol part of trehazolin (**2**) by validamine was performed to to give the analogue **412** (Scheme 54).[103] Coupling of isothiocyanate **7** with

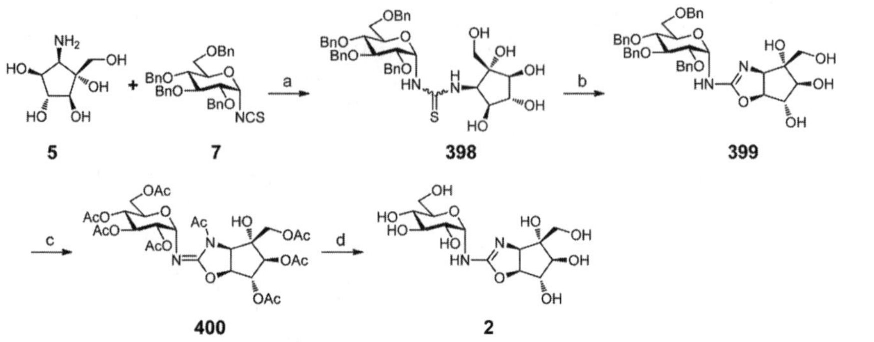

SCHEME 51. (a) DMF, MeOH, 93%; (b) HgO, Et$_2$O, 96%; (c) 1. Liquid NH$_3$, Na, −78 °C, 2. Ac$_2$O, Py, DMAP, 92%; (d) NaOMe, MeOH, rt, 71%.

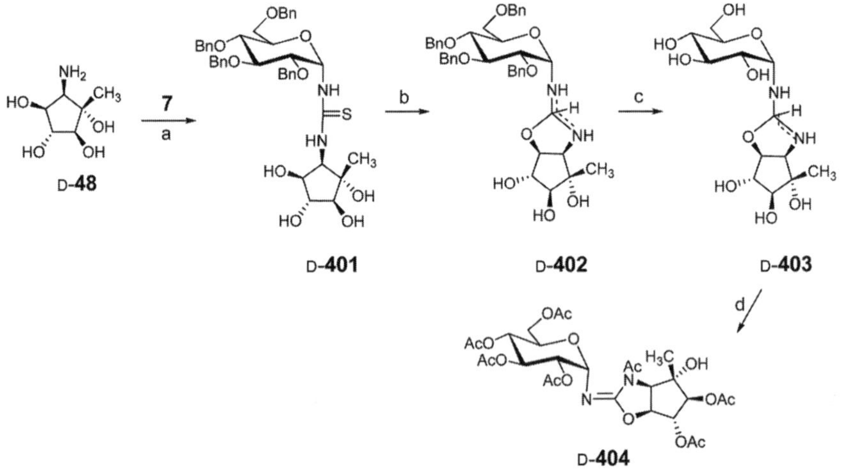

SCHEME 52. (a) 75% aq. DMF, rt, 22.5, 100%; (b) yellow HgO, Et$_2$O, rt, 24 h, 93%; (c) Na/liquid NH$_3$, THF, −78 °C, 15 min; then added NH$_4$Cl, 90%; (d) Ac$_2$O, Py, 72%.

validamine **409**[98] gave the thiourea **410** (93%), which upon treatment with yellow mercury(II) oxide afforded the cyclic isourea **411** (95%), whose O-debenzylation with sodium in liquid ammonia gave the free base **412** (86%), which was further characterized as the octa-*N,O*-acetyl derivative **413** (90%).

The trehazolin analogues D-**403**, D-**407**, L-**407**, and **412** were subjected to bioassay for trehalase-inhibitory activity. They were found to completely lack activity against α-amylase, β-D-glucosidase, α-D-mannosidase, maltase, and sucrase.[103]

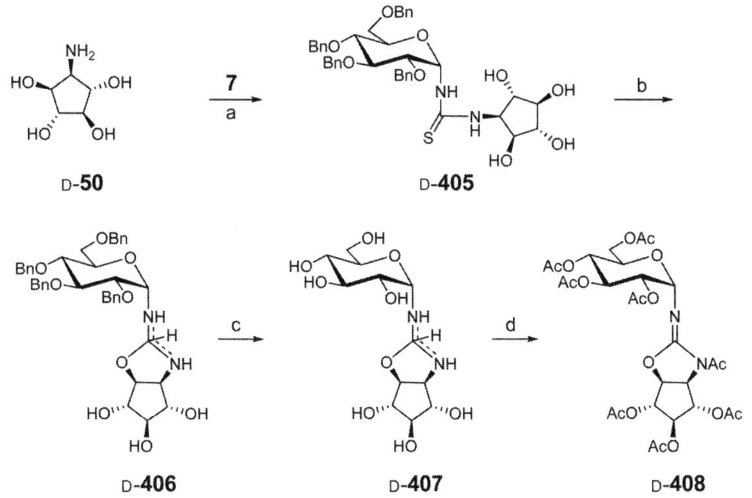

SCHEME 53. (a) 75% aq. DMF, rt, 2 h, 93%; (b) yellow HgO, acetone–Et$_2$O, rt, 30 h, 100%; (c) Na/liquid NH$_3$, THF, −78 °C, 10 min; then added NH$_4$Cl, 96%; (d) Ac$_2$O, Py, rt, 3 h, 91%.

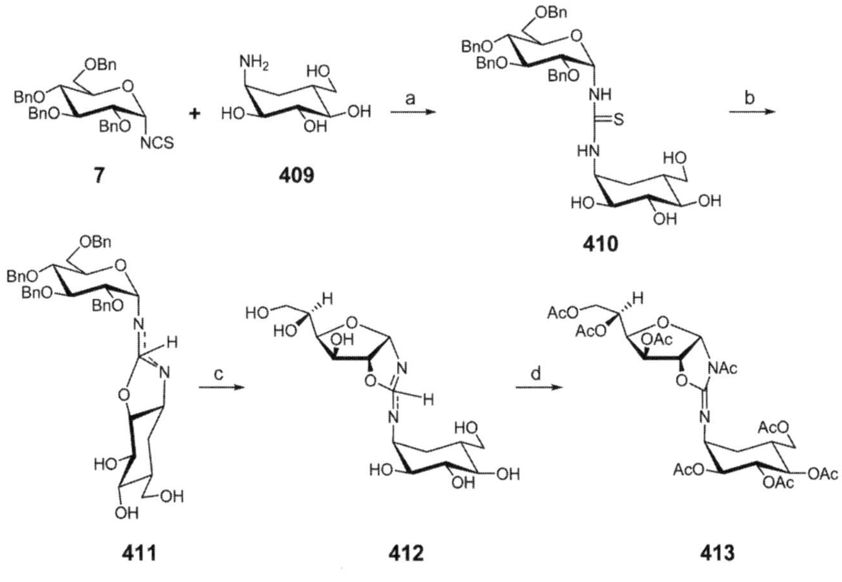

SCHEME 54. (a) 75% aq. DMF, rt, 40 min, 93%; (b) Et$_2$O, yellow HgO, rt, 13.5 h, 95%; (c) Na/liquid NH$_3$, THF, −78 °C, 10 min; then added NH$_4$Cl, 86%; (d) Ac$_2$O, Py, rt, 3 h, 90%.

Amine **249** was treated with 2% HCl–MeOH followed by reaction with Et$_3$N and isothiocyanate **7** to give **414**. Treatment of **414** with 2-chloro-3-ethylbenzoxazolium tetrafluoroborate gave the protected oxazoline **415** (90%), which upon hydrogenolysis with Pd(OH)$_2$ on carbon gave the trehalostatin diepimer **416** in 46% yield (Scheme 55).[106,107]

Condensation of **258** with isothiocyanate **7** gave thiourea **417** (80%), which could be cyclized in 70% yield to aminooxazoline **418** using yellow mercuric oxide. Deprotection of **418** by hydrogenolysis afforded (+)-**416** (82%) (Scheme 56).[106,107]

Treatment of **255** with 2% HCl in MeOH removed the acetal group and then reaction with **7** in presence of Et$_3$N gave benzylated product **419** in 71% yield. Removal of the benzyl groups from the glucosylamine derivative **420** was accomplished by hydrogenolysis with Pd(OH)$_2$ on carbon to give the trehalostatin analogue **416** in 37% yield (Scheme 57).[106,107]

Treatment of trehazolamine analogue **38** with isothiocyanate **7** in the presence of Et$_3$N afforded **421** in 84% yield. Oxazoline formation from **421** with 2-chloro-3-ethylbenzoxazolium tetrafluoroborate gave **422** (84%), hydrogenolysis of which gave diepimer **416** in 43% yield (Scheme 58).[106,107]

The amine **258** was transformed into the thiourea **423**, and then to 2-aminooxazine **424** in 51% yield. Removal of the benzyl groups from **424** by hydrogenolysis gave **416** in 42% yield. It seems that the unstable **425** underwent transformation to intermediate **426**, which underwent conversion into the more stable isomer **416** (Scheme 59).[106,107]

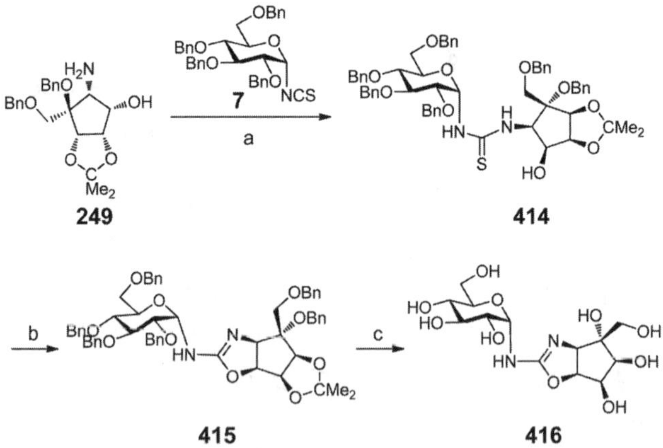

SCHEME 55. (a) Triphenylphosphine oxide, 2% HCl, MeOH, 50 °C, 4 h, then THF, H$_2$O, Et$_3$N, rt, 2 h, 22%; (b) MeCN, 2-chloro-3-ethylbenzoxazolium tetrafluoroborate, 0 °C, under N$_2$, 1 h, then Et$_3$N, 15 min, then rt 30 min, 90%; (c) MeOH, Pd(OH)$_2$ on C, H$_2$, 60 °C, 30 min.

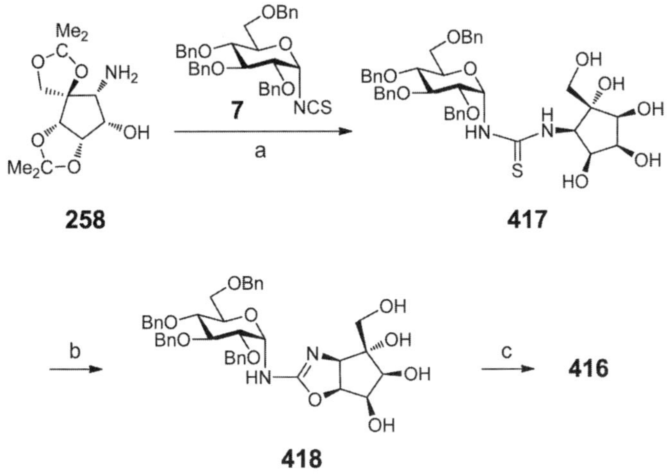

258 → **417**

418 → **416**

SCHEME 56. (a) 2% HCl in MeOH, 50 °C, 2 h; then added Et$_3$N, rt, 2 h, 89%; (b) MeCN, 2-chloro-3-ethylbenzoxazolium tetrafluoroborate, 0 °C under N$_2$, 1 h; then added Et$_3$N, 30 min, 84%; (c) MeOH, Pd(OH)$_2$ on carbon, H$_2$, 60 °C, 30 min, 46%.

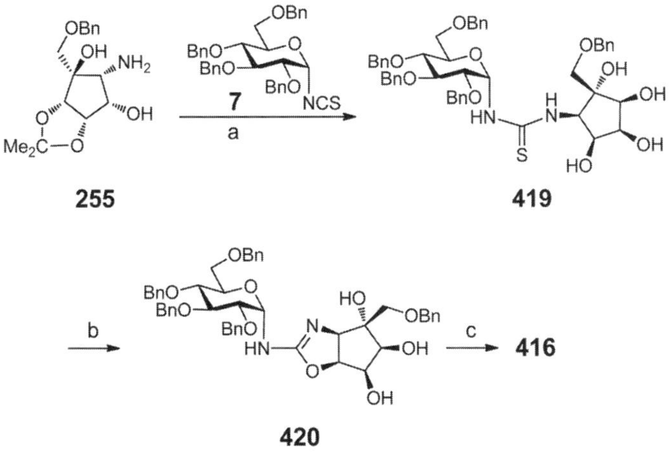

255 → **419**

420 → **416**

SCHEME 57. (a) 2% HCl in MeOH, 50 °C, 3 h; then added Et$_3$N, rt, 2 h, 71%; (b) MeCN, 2-chloro-3-ethylbenzoxazolium tetrafluoroborate, 0 °C under N$_2$, 2 h; then added Et$_3$N, 30 min, 85%; (c). MeOH, Pd(OH)$_2$ on carbon, H$_2$, 60 °C, 30 min, 37%.

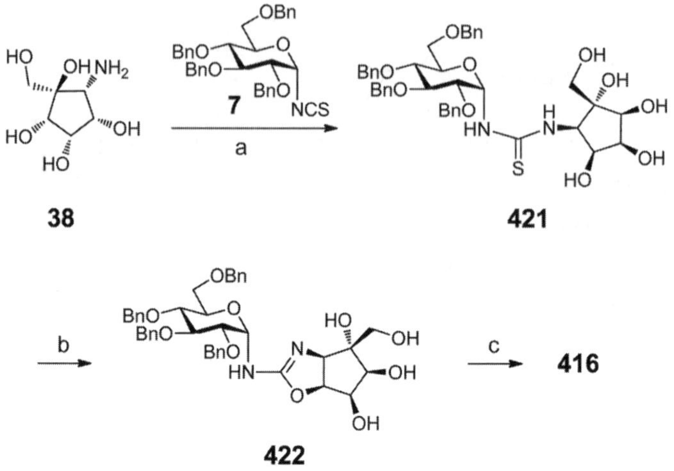

SCHEME 58. (a) 2% HCl in MeOH, 50 °C, 2 h; then added Et$_3$N, rt, 2 h, 89%; (b) MeCN, 2-chloro-3-ethylbenzoxazolium tetrafluoroborate, 0 °C under N$_2$, 1 h; then added Et$_3$N, 30 min, 84%; (c) MeOH, Pd(OH)$_2$ on carbon, H$_2$, 60 °C, 30 min, 46%.

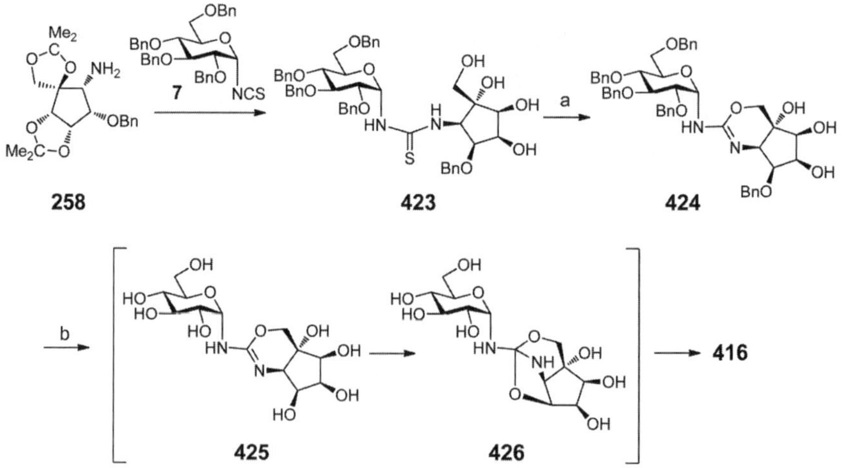

SCHEME 59. (a) MeCN, 2-chloro-3-ethylbenzoxazolium tetrafluoroborate, 0 °C under N$_2$, 1 h; then added Et$_3$N, 15 min, 71%; (b) MeOH, Pd(OH)$_2$ on carbon, H$_2$, 60 °C, 30 min.

The synthesis of the bromodeoxy and epoxy analogues of trehazoline were achieved by reaction of the aminobromocyclopentitol **280** with glucosyl isothiocyanate **7** to give the thiourea **427** in 50% yield, which underwent cyclization by the action of yellow HgO to afford aminooxazoline **428** in 60% yield. The *trans*-bromohydrin **428** was stirred in a suspension of potassium carbonate in methanol to form the epoxide **429** in 81% yield (Scheme 60).[109]

Total synthesis of the β anomer **433** of trehazolin, utilized the corresponding aminotriol hydrochloride **88**, which upon coupling with 2,3,4,6-tetra-*O*-benzyl-β-D-glucopyranosyl isothiocyanate in the presence of triethylamine[52], afforded the thiourea **431** (Scheme 61).[119] Cyclization of compound **431** with 2-chloro-1-methylpyridinium iodide and triethylamine yielded **432** in 70% yield. However, cyclization using 2-chloro-3-ethylbenzoxazolium tetrafluoroborate instead of 2-chloro-1-methylpyridinium iodide caused partial anomerization, with the result that compound **432** was contaminated with the α anomer. This is presumably due to the tetrafluoroborate species acting as a Lewis acid. Cyclization conditions using 2-chloro-1-methylpyridinium iodide are applicable for the synthesis of acid-sensitive aminooxazolines. Hydrogenolysis of compound **432** over palladium hydroxide on carbon cleaved the benzyl groups to afforded the anomer, compound **433**. The activities of this trehazolin β anomer (**433**) toward two trehalases were found to be much weaker than that of natural trehazolin (**2**).[119]

The cyclohexyl analogue was prepared by addition of isothiocyanate **7** to the six-membered aminocyclitol hydrochloride **434** to give the adduct **435**, which was treated

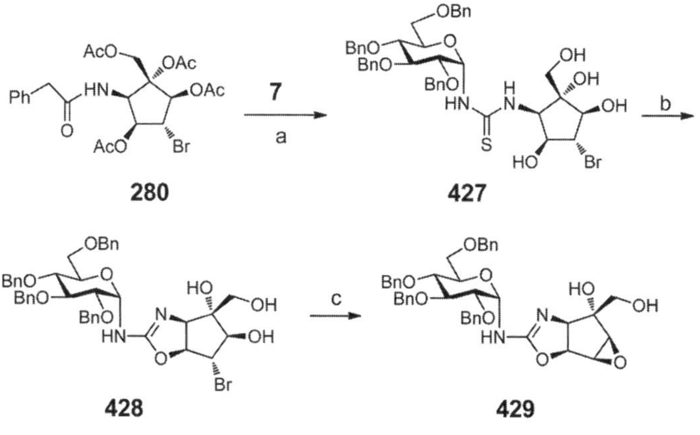

SCHEME 60. (a) 0.5 M HCl, MeOH, reflux, 60 h, concentrate to residue; then H$_2$O, THF, rt, 5 days, 55%; (b) THF, MS 4 Å, yellow HgO, rt in dark, 16 h, 86%; (c) K$_2$CO$_3$, bromohydrin, MeOH, rt, 60 h, 81%.

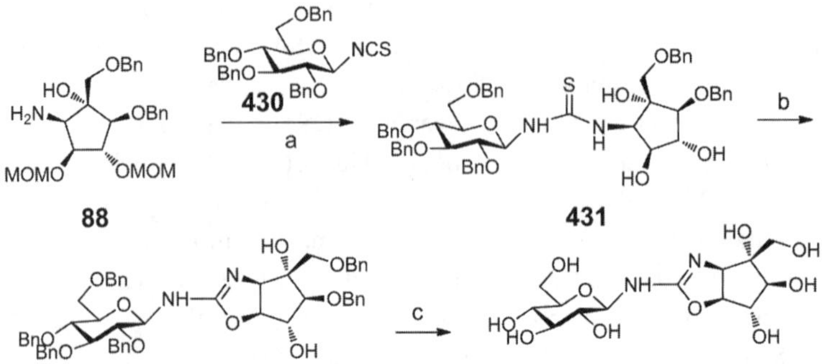

SCHEME 61. (a) 4.0 equiv of LiAlH$_4$, Et$_2$O, 0 °C, 4 h, 5% HCl in MeOH, 60 °C, 4 h, and then 1.2 equiv of isothiocyanate **430**.[52] 1.6 equiv of Et$_3$N, THF, rt, 13 h, 79%; (b) (Method A) 1.7 equiv of 2-chloro-3-ethylbenzoxazolium tetrafluoroborate, MeCN, 0 °C, 1 h, then quenched with 3.3 equiv of Et$_3$N, 0 °C, 1 h, 43% isolated. (Method B) 2.0 equiv of 2-chloro-1-methylpyridinium iodide, MeCN, 0 °C, 1 h; then quenched with 4.0 equiv of Et$_3$N, 0 °C, 1 h, 70%; (c) H$_2$, Pd(OH)$_2$/C, MeOH, 60 °C, 30 min, 51%.

with 2-chloro-3-ethylbenzoxazolium tetrafluoroborate and triethylanilne to afford the oxazoline **436** (Scheme 62).[72] Hydrogenolysis of **436** gave a mixture that was inseparable by chromatography. This mixture consists of three structural isomers **437**, **438**, and **439**.

2. Trehazolines with Trehazolamines on Positions Other than the Anomeric Center

Trehazolamines analogues having the trehazolamine moiety on position-6 of the sugar component were prepared by coupling of the 6-isothiocyanoglucose derivative **14** with **5** to yield the corresponding thiourea derivative **440** (Scheme 63).[51] Treatment of **440** with 2-chloro-3-ethylbenzoxazolium tetrafluoroborate and Et$_3$N gave **441**, which upon hydrogenolysis afforded the 6-substituted analogue **442**.

Coupling of the 4-isothiocyanate **20** with **5** gave thiourea **443**, which upon treatment with 2-chloro-3-ethylbenzoxazolium tetrafluoroborate and Et$_3$N, and subsequent hydrogenolysis of the resulting aminooxazoline **444** yielded the 4-substituted product **445** (Scheme 64).[51] The inhibitory activities of pseudodisaccharides **442** and **445** toward such rat intestinal α-glucosidases as trehalase, maltase, isomaltase, and sucrase were assayed by a one-step method in a microtiter plate, and they showed only weak or no inhibitory activities. Compound **442** inhibits maltase and

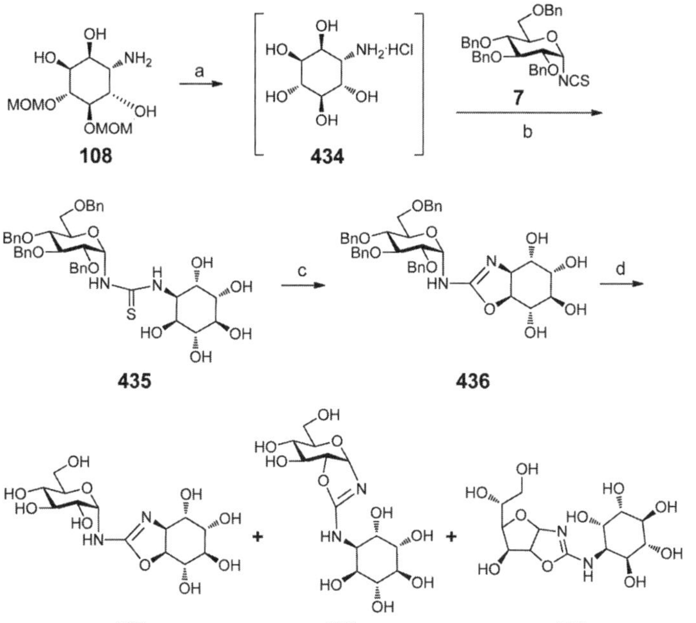

SCHEME 62. (a) 10% HCl in MeOH, 50 °C, 30 min; (b) Et$_3$N, THF, H$_2$O, rt, 16 h, 40% for three steps; (c) MeCN, 0 °C under N$_2$, 1 h, 60%; (d) MeOH, 20% Pd(OH)$_2$/C, H$_2$, 60 °C, 360 min, 44%.

sucrase more potently than does trehazolin, while **445** displayed no inhibitory activities toward maltase, isomaltase, and sucrase, as compared with trehazolin.

Syntheses of the maltose-type trehazolin derivatives **448** and **451** from D-galactose have been reported (Schemes 65 and 66).[51] These were expected to manifest specific inhibitory activity toward maltases. Coupling of **26** with **5** gave the thiourea derivative **446**, whose treatment with 2-chloro-3-ethylbenzoxazolium tetrafluoroborate and Et$_3$N, afforded the aminooxazoline derivative **447**. Subsequent hydrogenolysis of **447** furnished **448**.

Similarly, coupling of the 4-isothiocyanate **33** with **5** gave the thiourea derivative **449**, which was converted into aminooxazoline **450**. Subsequent hydrogenolysis of **450** furnished the 4-linked analogue **451** (Scheme 66).[51]

Hydrolysis of the oxazolidinone **236** gave the corresponding amino alcohol, which without isolation was coupled with the sugar 4-isothiocyanate **26**[53] to provide the thiourea **452** (Scheme 67).[105] Cyclization of **452** to the isourea **453** was carried out with yellow mercuric oxide, and the resulting oxazoline **453** was debenzylated to give

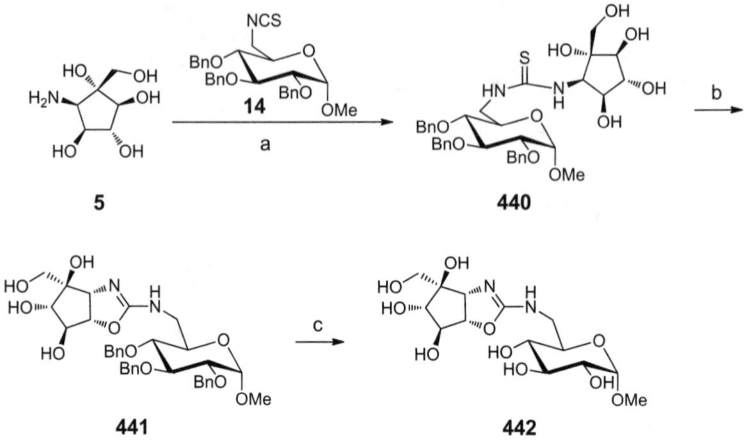

SCHEME 63. (a) THF, 24 °C, 4 days, 88%; (b) 2-chloro-3-ethylbenzoxazolium tetrafluorobrorate, MeCN, 0 °C, under N$_2$, 1 h, Et$_3$N, 0 °C, 2 h, 47%; (c) MeOH, 20% Pd(OH)$_2$, H$_2$, 60 °C, 30 min, 55%.

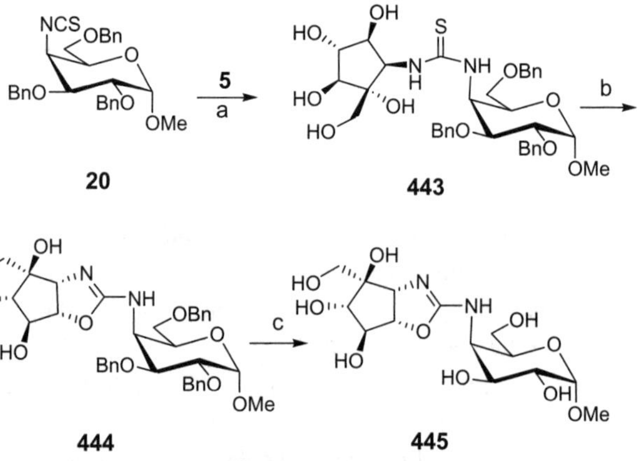

SCHEME 64. (a) H$_2$O, THF, 24 °C, 7 days, 87%; (b) 2-chloro-3-ethylbenzoxazolium tetrafluoroborate, MeCN, 0 °C under N$_2$, 1 h, then added Et$_3$N, 2 h, 81%; (c) MeOH, 20% Pd(OH)$_2$/C, 24 °C, H$_2$, 60 °C, 30 min, 46%.

454, acetylated to give **455** in 75% yield, and the latter converted into the 4-linked analogue **448** by treatment with sodium methoxide.

The 6'-deoxy-6'-fluoro trehazolin **460** has been synthesized (Scheme 68).[120] 2,3,4-Tri-O-benzyl-6-deoxy-6-fluoro-α-D-glucopyranosyl isothiocyanate (**457**) was

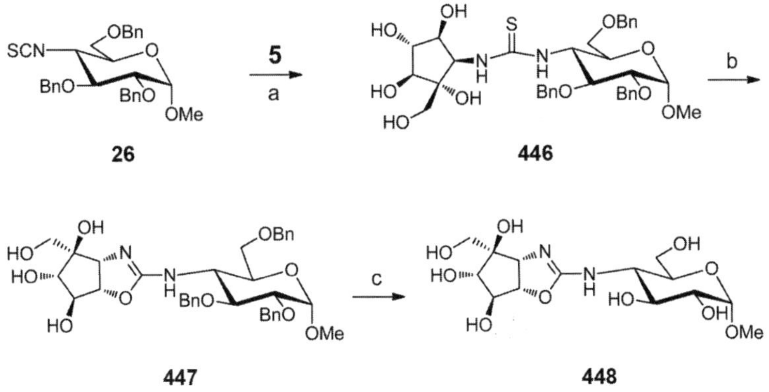

TREHALASE INHIBITORS: SYNTHESIS OF TREHALOZIN AND ANALOGUES 103

SCHEME 65. (a) THF, 24 °C, 6 days, 94%; (b) 2-chloro-3-ethylbenzoxazolium tetrafluoroborate, MeCN, 0 °C under N_2, 1 h; then Et_3N, 0 °C, 2 h, 72%; (c) MeOH, 20% $Pd(OH)_2$/C, 24 °C, then H_2 at 60 °C, 1 h, 57%.

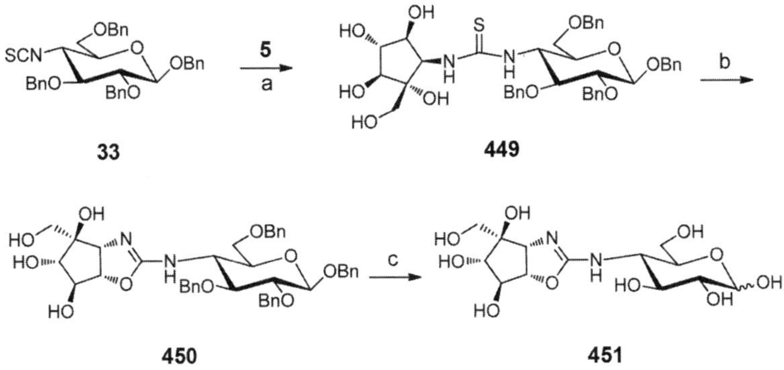

SCHEME 66. (a) H_2O, THF, 24 °C, 3 days, 100%; (b) 2-chloro-3-ethylbenzoxazolium tetrafluoroborate, MeCN, 0 °C under N_2, 40 min; then Et_3N, 0 °C, 50 min, 58%; (c) MeOH, 20% $Pd(OH)_2$/C, 24 °C, H_2, 60 °C, 1 h, 39%.

prepared conventionally, [23] starting from methyl 2,3,4-tri-*O*-benzyl-6-deoxy-6-fluoro-α-D-glucopyranoside (**456**).[121] The α (**457**) and β anomers (**458**) were obtained in yields of 27% and 23%. Coupling of **5**[103] and **457** proceeded smoothly to give the thiourea **459** (93%), which on cyclization, followed by deprotection, afforded the isourea analogue **460** (86%), the structure of which was confirmed as its hepta-*N,O*-acetyl derivative.

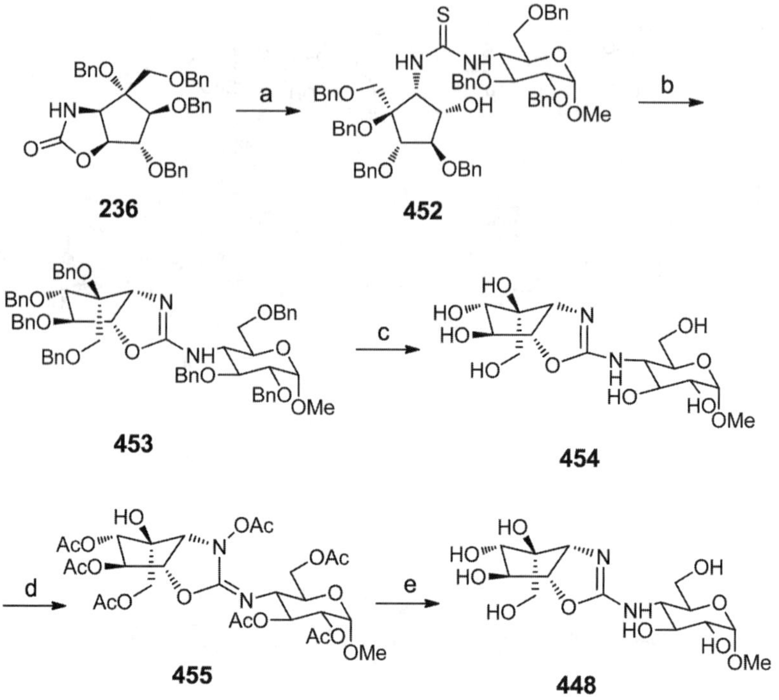

SCHEME 67. (a) 1. 2 M aq. KOH, EtOH, reflux, 12 h; 2. Sugar isothiocyanate, Et$_3$N, MeOH, 90% two steps; (b) yellow HgO, THF, sieves, 95%; (c) 1. H$_2$, 20% Pd(OH)$_2$, EtOH; (d) Ac$_2$O, Py, 75% two steps; (e) NaOMe, MeOH, 95%.

3. Synthesis of 1-Thiatrehazolin

The trehazolin analogue, 1-thiatrehazolin (**463**), has been synthesized from carbohydrate precursors by employing an intramolecular nucleophilic displacement-reaction for construction of the thiazoline ring (Scheme 69).[121] 1-Thiatrehazolin was found to be a very potent, slow, tight-binding trehalase inhibitor.[121] Treatment of the thiourea derivative **387**[79] with triflic anhydride and pyridine effected clean cyclization to afford the 2-aminothiazoline **461** in 93% yield, with concomitant inversion of stereochemistry at the center bearing the secondary hydroxyl group via the intramolecular S$_N$2 displacement of a transient triflate by the vicinal thiocarbonyl group. Acetal hydrolysis of **461** using p-TsOH afforded **462**, which was subjected to debenzylation using sodium/NH$_3$ to produce 1-thiatrehazolin (**463**) in 83% net yield from **387**.

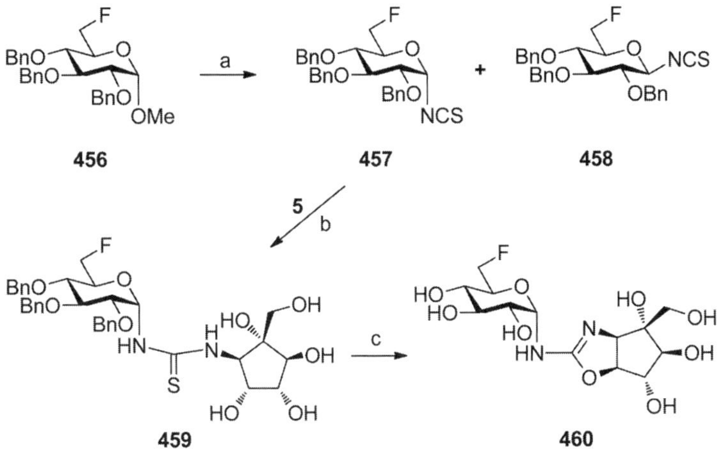

SCHEME 68. (a) 1. 2 M aq. HCl, AcOH, 24 h, 80 °C, 47%; 2. Acetylation, 97%; 3. 1,4-dioxane saturated with HCl, 1 h, 40 °C; 4. KSCN, n-Bu$_4$NBr, MS 4A, CH$_3$CN, rt, then reflux, 5 h, **457** (27%), **458** (23%); (b) 75% aq. THF, 3 h, rt, 93%; (c) 1. yellow HgO, diethyl ether, 44 h, rt, 98%; 2. Na/NH$_3$, THF, −78 °C, 15 min, 86%.

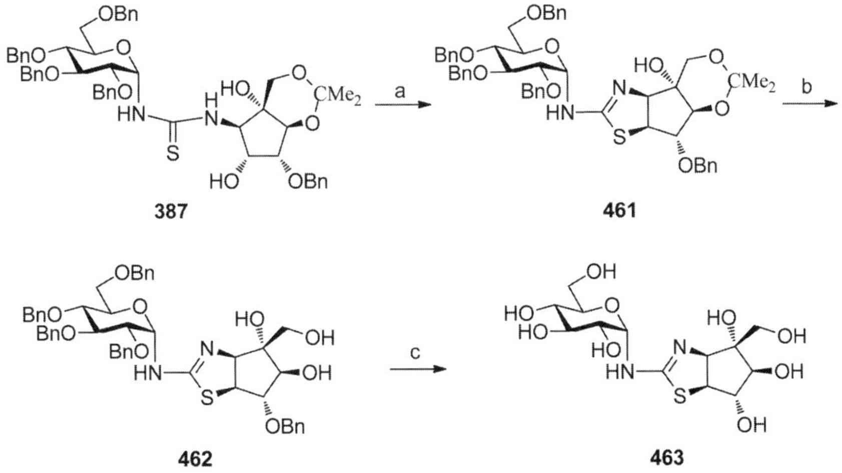

SCHEME 69. (a) Tf$_2$O, Py, CH$_2$Cl$_2$, −40 °C to rt, 93%; (b) cat. p-TsOH, CH$_2$Cl$_2$/MeOH, rt, 89%; (c) Na, NH$_3$/THF, −35 °C, quant.

4. Deoxynojirimycin Analogues

The 1-deoxynojirimycin-trehalamine fused compounds **469**, **470**, **479**, and **480**, which may be considered as as pseudodisaccharides, have been reported (Schemes 70 and 71).[116,117] The suitably protected trehazolin aminocyclitol moiety **370** was prepared as shown in section VI.13. Treatment of **370** with carbon disulfide, Et$_3$N and 2-chloro-1-methylpyridinium iodide gave the isothiocyanate **464**, which was condensed with tetra-*O*-benzyl-1-deoxynojirimycin (**465**)[122] in the presence of triethylamine to furnish the thiourea derivative **466**. Treatment of **466** with Bu$_4$NF gave the pentol **467**, which was treated with 2-chloro-3-ethylbenzoxazolium tetra-fluoroborate and Et$_3$N to give the 1-deoxynojirimyci-*N*-trehalamine fused oxazoline compound **468**, which upon hydrogenation gave a 4:1 equilibrium mixture of **469** and **470**. The IC$_{50}$ value for the biological activity of the mixture of **469** and **470** toward rat intestinal maltase is 0.68 μg/mL.

Synthesis of a nojirimycin derivative (**472**) was conducted by condensation of **465** with *N*-(*tert*-butoxycarbonyl)glycine, using DCC as a condensing reagent to give

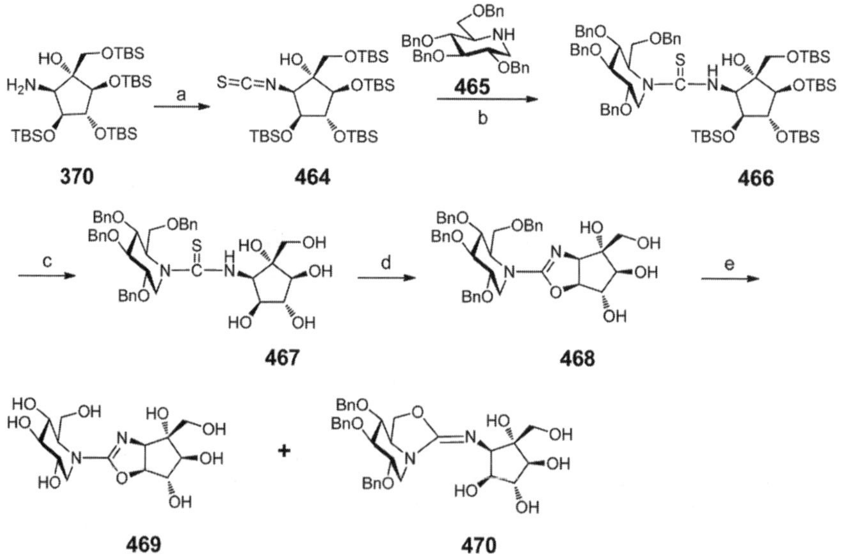

SCHEME 70. (a) CS$_2$, Et$_3$N, 2-chloro-1-methylpyridinium iodide, CH$_2$Cl$_2$, 24 °C, 2.5 h, 87%; (b) Et$_3$N, THF, 60–65 °C, 3 h, 68% (recovery of **464**, 22% and **465**, 19%); (c) *n*-Bu$_4$NF, THF, 24 °C 3 h, 93%; (d) 2-chloro-3-ethylbenzoxazolium tetrafluoroborate, Et$_3$N, MeCN, 0 °C, 10 min, 95%; (e) H$_2$, Pd(OH)$_2$/C, MeOH, 60 °C, 40 min, a 4:1 equilibrium mixture of **469** and **470**, 32%.

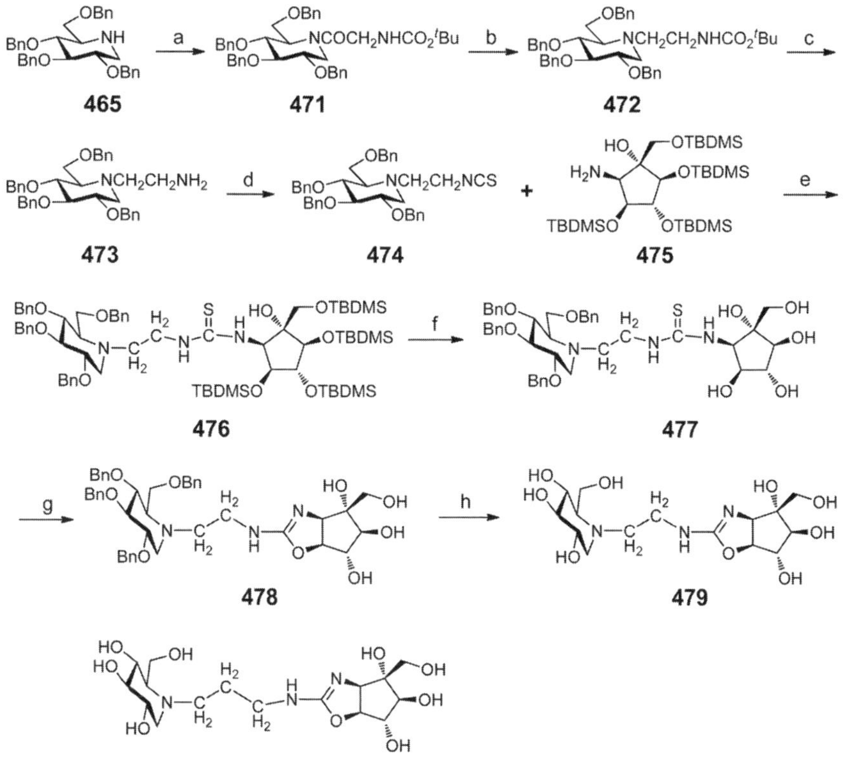

TREHALASE INHIBITORS: SYNTHESIS OF TREHALOZIN AND ANALOGUES 107

SCHEME 71. (a) t-BuO$_2$CNHCH$_2$CO$_2$H, DCC, CH$_2$Cl$_2$, 24 °C, 16 h; **b.** BH$_3$–THF complex, 24 °C, 16 h, two steps 62%; (c) CF$_3$CO$_2$H, CH$_2$Cl$_2$, 24 °C, 30 min; (d) CS$_2$, Et$_3$N, 2-chloro-1-methylpyridinium iodide, CH$_2$Cl$_2$, 24 °C, 2.5 h, two steps 41%; (e) catalytic Et$_3$N, THF, 20–25 °C, 2 days, 87%; (f) 10% HCl–MeOH, MeOH, 24 °C, 16 h, quantitative; (g) 2-chloro-3-ethylbenzoxazolium tetrafluoroborate, 0–5 °C, 1 h, then Et$_3$N, MeCN, 0 °C, 30 min, 67%; (h) H$_2$, Pd(OH)$_2$/C, MeOH, 65 °C, 8 h, 72%.

amide **471**. Reduction of **471** with BH$_3$–THF complex gave the tertiary amine **472** (Scheme 71).[116,117] Removal of of the t-BOC group of **472** with TFA gave the primary amine **473**. Isothiocyanate formation from **473** using CS$_2$, Et$_3$N, and 2-chloro-1-methylpyridium iodide yielded **474**. Reaction of the isothiocyanate **474** with amine **475**, using Et$_3$N as a catalyst gave thiourea **476**, which was also obtainable by the condensation of isothiocyanate **475** and amine **473**. Desilylation of the tetrasilylated compound **476** yielded **477**. Treatment of **476** with 2-chloro-3-ethylbenzoxazolium tetrafluoroborate and Et$_3$N gave aminooxazoline **478**, which was deprotected by hydrogenation to give **479**. Compound **480** was also synthesized by similar treatment of isothiocyanate **475** with N-(3-aminopropyl)-1-deoxy-2,3,4,6-tetra-O-benzyl-nojirimycin, which was also obtained from **465** and N-($tert$-butoxycarbonyl)-β-alanine. The IC$_{50}$ values for the biological activity of **479** and **480** toward rat intestinal maltase were 4.2 and 1.5 μg/mL, respectively.

In conclusion, the synthesis of trehazolin and its analogues generally requires a disubstituted thiourea bearing a glucosyl residue on one of the two nitrogen atoms, while the other has a cyclopentyl residue. The latter must possess at least one hydroxyl group on the carbon adjacent to the one linked to the nitrogen atom in order to effect cyclization to the oxazoline ring and form the trehazolin analogue. The formation of such thiourea derivatives can be achieved by reaction of a glycosyl amine (or isothiocyanate) with a 2-thiocyanato-(or 2-amino)-cyclopentol. This article has highlighted the various approaches that have been reported for the total synthesis of trehazolin and its analogues. Efforts toward developing a "green" chemistry approach for the synthesis of such important molecules constitute an ongoing challenge.

Acknowledgments

The continued support of the Alexander von Humboldt Stiftung in Germany and valuable discussions with Prof. Dr. Valentine Whittmann at Konstanz University, Germany, are highly appreciated.

References

1. E. S. H. El Ashry, N. Rashed, and A. H. S. Shobier, Glycosidase inhibitors and their chemotherapeutic value, part 1, *Pharmazie*, 55 (2000) 251–262.
2. E. S. H. El Ashry, N. Rashed, and A. H. S. Shobier, Glycosidase inhibitors and their chemotherapeutic value, part 2, *Pharmazie*, 55 (2000) 331–348.
3. E. S. H. El Ashry, N. Rashed, and A. H. S. Shobier, Glycosidase inhibitors and their chemotherapeutic value, part 3, *Pharmazie*, 55 (2000) 403–415.
4. E. S. H. El Ashry and A. El Nemr, Synthesis of Naturally Occurring Nitrogen Heterocycles from Carbohydrates, (2005) Blackwell, , Oxford, UK, http://www.blackwellpublishing.com/more_reviews.asp?ref=9781405129343&site=1.
5. E. S. H. El Ashry (Ed.), Heterocycles from Carbohydrate Precursors, in the Series. Topics in heterocycles (Gupta, Ed), Springer, Germany, Vol. 7, 2007.
6. A. Vasella, G. J. Davies, and M. Böhm, Glycosidase mechanisms, *Curr. Opin. Chem. Biol.*, 6 (2002) 619–629.
7. R. A. Dwek, Glycobiology: Towards understanding the function of sugars, *Chem. Rev.*, 96 (1996) 683–720.

8. B. Winchester and G. W. J. Fleet, Amino-sugar glycosidase inhibitors: Versatile tools for glycobiologists, *Glycobiology*, 2 (1992) 199–210.
9. G. P. Koushal and A. D. Elbein, Glycosidase inhibitors in study of glycoconjugates, *Methods Enzymol.*, 230 (1994) 316–329.
10. A. D. Elbein, Inhibitors of the biosynthesis and processing of N-linked oligosaccharide chains, *Annu. Rev. Biochem.*, 56 (1987) 497–534.
11. A. D. Elbein, Y. T. Pan, I. Pastuszak, and D. Carroll, New insights on trehalose: A multifunctional molecule, *Glycobiology*, 13 (2003) 17R–27R.
12. O. Ando, M. Nakajima, M. Kifune, H. Fang, and K. Tanzawa, Trehazolin, a slow, tight-binding inhibitor of silkworm trehalase, *Biochem. Biophys. Acta*, 1244 (1995) 295–302.
13. J. S. Clegg and D. R. Evans, The physiology of blood trehalose and its function during flight in the blowfly, *J. Exp. Biol.*, 38 (1961) 771–792.
14. O. Ando, H. Satake, K. Itoi, A. Sato, M. Nakajima, S. Takahashi, H. Haruyama, Y. Ohkuma, T. Kinoshita, and R. Enokita, Trehazolin, a new trehalase inhibitor, *J. Antibiot.*, 44 (1991) 1165–1168.
15. R. Aono, M. Kobayashi, H. Nakajima, and H. Kobayashi, A close correlation between improvement of organic solvent tolerance levels and alteration of resistance toward low levels of multiple antibiotics in *Escherichia coli*, *Biosci. Biotechnol. Biochem.*, 59 (1995) 213–218.
16. S. V. Kyosseva, Z. N. Kyossev, and A. D. Elbein, Inhibitors of pig kidney trehalase, *Arch. Biochem. Biophys.*, 316 (1995) 821–826.
17. J. Defaye, H. Driguez, B. Henrissat, and E. Bar-Guilloux, in J. J. Marshall, (Ed.), *Mechanism of Saccharide Polymerization and Depolymerization,* Academic Press, New York, 1980, pp. 331–353.
18. J. Defaye, H. Driguez, B. Henrissat, and E. Bar-Guilloux, Stereochemistry of the hydrolysis of α,α-trehalose by trehalase, *Nouv. J. Chim.*, 4 (1980) 59–68.
19. N. Asano, A. Kato, and K. Matsui, Two subsites on the active center of pig kidney trehalase, *Eur. J. Biochem.*, 240 (1996) 692–698.
20. J. Defaye, H. Driguez, B. Henrissat, and E. Bar-Guilloux, Stereochemistry of the hydrolysis of α,α-trehalose by trehalase, determined by using a labelled substrate, *Carbohydr. Res.*, 124 (1983) 265–273.
21. K. H. Clifford, Stereochemistry of the hydrolysis of trehalose by the enzyme trehalase prepared from the flesh fly sarcophaga barbata, *Eur. J. Biochem.*, 106 (1980) 337–340.
22. B. L. Rhinehart, K. M. Robinson, C. H. R. King, and P. S. Liu, Castanospermine-glucosides as selective disaccharidase inhibitors, *Biochem. Pharmacol.*, 39 (1990) 1537–1543.
23. C. Uchida, H. Kitahashi, S. Watanabe, and S. Ogawa, Synthesis of trehazolin analogues containing modified sugar moieties, *J. Chem. Soc. Perkin Trans.*, 1 (1995) 1707–1717.
24. B. S. Sacktor and E. Wormser-Shavit, Regulation of metabolism in working muscle in vivo. I. Concentrations of some glycolytic, tricarboxylic acid cycle, and amino acid intermediates in insect flight muscle during flight, *J. Biol. Chem.*, 241 (1966) 624–631.
25. L. I. Hecker and A. S. Sussman, Activity and heat stability of trehalase from the mycelium and ascospores of neurospora, *J. Bacteriol.*, 115 (1973) 582–591.
26. L. I. Hecker and A. S. Sussman, Localization of trehalase in the ascospores of neurospora: Relation to ascospore dormancy and germination, *J. Bacteriol.*, 115 (1973) 592–599.
27. H. Inoue and C. Shimoda, Induction of trehalase activity on a nitrogen-free medium: A sporulation-specific event in the fission yeast, *Schizosaccharomyces pombe*, *Mol. Gen. Genet.*, 183 (1981) 32–36.
28. J. M. Thevelein, J. A. den Hollander, and R. G. Shulman, Changes in the activity and properties of trehalase during early germination of yeast ascospores: Correlation with trehalose breakdown as studied by in vivo ^{13}C NMR, *Proc. Natl. Acad. Sci. USA*, 79 (1982) 3503–3507.
29. J. M. Thevelein and K. A. Jones, Reversibility characteristics of glucose-induced trehalase activation associated with the breaking of dormancy in yeast ascospores, *Eur. J. Biochem.*, 136 (1983) 583–587.

30. B. Sacktor, Trehalase and the transport of glucose in the mammalian kidney and intestine, *Proc. Natl. Acad. Sci. USA*, 60 (1968) 1007–1014.
31. A. Elbein, Inhibitors of the biosynthesis and processing of N-linked oligosaccharide chains, *Annu. Rev. Biochem.*, 56 (1987) 497–534.
32. O. Ando, M. Nakajima, K. Hamano, K. Itoi, S. Takahashi, Y. Takamatsu, A. Sato, R. Enokita, T. Okazaki, H. Haruyama, and T. Kinoshita, Isolation on trehalamine, the aglycon of trehazolin, from microbial broths and characterization of trehazolin related compounds, *J. Antibiot.*, 46 (1993) 1116–1125.
33. S. Murao, T. Sakai, H. Gibo, T. Nakayama, and T. Shin, A novel trehalase inhibitor, trehalostatin, produced by *Amycolatopsis trehalostatica* HG-25, *Agric. Biol. Chem.*, 55 (1991) 895–897.
34. T. Nakayama, T. Amachi, S. Murao, T. Sakai, T. Shin, P. T. M. Kenny, T. Iwashita, M. Zagorski, H. Komura, and K. Nomoto, Structure of trehalostatin: A potent and specific inhibitor of trehalase, *J. Chem. Soc. Chem. Commun.* (1991) 919–921.
35. S. Ogawa, C. Uchida, and Y. Yuming, Synthesis of aminocyclitol moieties of trehalase inhibitors, trehalostatin and trehazolin, *J. Chem. Soc. Chem. Commun.* (1992) 886–888.
36. Y. Kobayashi, H. Miyazaki, and M. Shiozaki, Synthesis and absolute configurations of trehazolin and its aglycon, *J. Am. Chem. Soc.*, 114 (1992) 10065–10066.
37. Y. Kobayashi, H. Miyazaki, and M. Shiozaki, Synthesis and absolute configuration of trehazolin aminocyclitol moiety, *Tetrahedron Lett.*, 34 (1993) 1505–1506.
38. Y. Kobayashi, H. Miyazaki, and M. Shiozaki, Synthesis of trehazolin, trehalamine, and the aminocyclitol moiety of trehazolin: Determination of absolute configuration of trehazolin, *J. Org. Chem.*, 59 (1994) 813–822.
39. R. P. Gibson, T. M. Gloster, S. Roberts, R. A. J. Warren, I. S. de Gracia, A. Garcia, J. L. Chiara, and G. J. Davies, Molecular basis for trehalase inhibition revealed by the structure of trehalase in complex with potent inhibitors, *Angew. Chem. Int. Ed.*, 46 (2007) 4115–4119.
40. C. Uchida, H. Kimura, and S. Ogawa, Potent glycosidase inhibitors, *N*-phenyl cyclic isourea derivatives of 5-amino- and 5-amino-1-C-(hydroxymethyl)-cyclopentane-1,2,3,4-tetraols, *Bioorg. Med. Chem. Lett.*, 4 (1994) 2643–2648.
41. Y. Kobayashi, Chemistry and biology of trehazolins, *Carbohydr. Res.*, 315 (1999) 3–15.
42. A. Boiron, P. Zillig, D. Faber, and B. Giese, Synthesis of trehazolin from D-glucose, *J. Org. Chem.*, 63 (1998) 5877–5882.
43. T. Ito, N. Roongsawang, N. Shirasaka, W. Lu, P. M. Flatt, N. Kasanah, C. Miranda, and T. Mahmud, Deciphering pactamycin biosynthesis and engineered production of new pactamycin analogues, *ChemBioChem*, 10(13), (2009) 2253–2265.
44. K. A. L. De Smet, A. Weston, I. N. Brown, D. B. Young, and B. D. Robertson, Three pathways for trehalose biosynthesis in mycobacteria, *Microbiology*, 146 (2000) 199–208.
45. N. Avonce, A. Mendoza-Vargas, E. Morett, and G. Iturriaga, Insights on the evolution of trehalose biosynthesis, *BMC Evol. Biol.*, 6 (2006) 109–123. http://www.biomedcentral.com/1471-2148/6/109.
46. H. N. Murphy, G. R. Stewart, V. V. Mischenko, A. S. Apt, R. Harris, M. S. B. McAlister, P. C. Driscoll, D. B. Young, and B. D. Robertson, The OtsAB pathway is essential for trehalose biosynthesis in *Mycobacterium tuberculosis*, *J. Biol. Chem.*, 280 (2005) 14524–14529.
47. T. Dairi, Studies on biosynthetic genes and enzymes of isoprenoids produced by actinomycetes, *J. Antibiot.*, 58(4), (2005) 227–243.
48. J. D. Carroll, I. Pastuszak, V. K. Edavana, Y. T. Pan, and A. D. Elbein, A novel trehalase from *Mycobacterium smegmatis* purification, properties, requirements, *FEBS J.*, 274 (2007) 1701–1714.
49. B. E. Ledford and E. M. Carreira, Total synthesis of (+)-trehazolin: Optically active spirocycloheptadienes as useful precursors for the synthesis of aminocyclopentitols, *J. Am. Chem. Soc.*, 117 (1995) 11811–11812.

50. T. Mikami, H. Asano, and O. Mitsunobu, Acetal-bond cleavage of 4,6-O-alkylidenehexopyranosides by diisobutylaluminium hydride and by lithium triethylborohydride-TiCl$_4$, *Chem. Lett.*, 16 (1987) 2033–2036.
51. Y. Kobayashi, M. Shiozaki, and O. Ando, Syntheses of trehazolin derivatives and evaluation as glycosidase inhibitors, *J. Org. Chem.*, 60 (1995) 2570–2580.
52. M. J. Camarasa, P. Fernández-Resa, M. T. García-Lopez, F. G. de las Heras, P. P. Méndez- Castrillón, and A. S. Felix, A new procedure for the synthesis of glycosyl isothiocyanates, *Synthesis* (1984) 509–510.
53. S. Knapp, A. B. J. Naughton, and T. G. M. Dhar, Intramolecular amino delivery reactions for the synthesis of valienamine and analogues, *Tetrahedron Lett.*, 33 (1992) 1025–1028.
54. M. Ek, P. J. Garegg, H. Hultberg, and S. Oscarson, Reductive ring openings of carbohydrate benzylidene acetals using borane-trimethylamine and aluminium chloride. Regioselectivity and solvent dependance, *J. Carbohydr. Chem.*, 2 (1983) 305–311.
55. N. Chida, M. Ohtsuka, K. Nakazawa, and S. Ogawa, Total synthesis of antibiotic hygromycin A, *J. Org. Chem.*, 56 (1991) 2976–2983.
56. A. Berecibar, C. Granjean, and A. Siriwardena, Synthesis and biological activity of natural aminocyclopentitol glycosidase inhibitors: Mannostatins, trehazolin, allosamidins, and their analogues, *Chem. Rev.*, 99 (1999) 779–844.
57. L. Qiao and J. C. Vederas, Synthesis of a C-phosphonate disaccharide as a potential inhibitor of peptidoglycan polymerization by transglycosylase, *J. Org. Chem.*, 58 (1993) 3480–3482.
58. R. E. Ireland and L. Liu, An improved procedure for the preparation of the Dess-Martin periodinane, *J. Org. Chem.*, 58 (1993) 2899.
59. R. O. C. Norman, R. Purchase, and C. B. Thomas, Oxidation of some derivatives of hydroxylamine, *J. Chem. Soc. Perkin Trans.*, 1 (1972) 1701–1704.
60. R. H. Weiss, E. Furfine, E. Hausleden, and D. W. J. Dixon, Oxidation of N,N'-dialkyl-1,2-bis (hydroxylamines), *J. Org. Chem.*, 49 (1984) 4969–4972.
61. I. S. de Gracia, H. Dietrich, S. Bobo, and J. L. Chiara, A highly efficient pinacol coupling approach to trehazolamine starting from D-glucose, *J. Org. Chem.*, 63 (1998) 5883–5889.
62. E. Decoster, J.-M. Lacombe, J.-L. Strebler, B. Ferrari, and A. A. Pavia, Une autre approche á la Synthèse de dérivés benzyles des monosaccharides reducteurs. Etude des equilibres pyrannose-furannose par R.M.N. du carbone-13, *J. Carbohydr. Chem.*, 2(3), (1983) 329–341.
63. P. Girard, J. L. Namy, and H. B. Kagan, Divalent lanthanide derivatives in organic synthesis. 1. Mild preparation of samarium iodide and ytterbium iodide and their use as reducing or coupling agents, *J. Am. Chem. Soc.*, 102 (1980) 2693–2698.
64. J. L. Chiara and Á. García, Control of diastereoselectivity in C=O/C=N reductive cyclization using an intramolecularly tethered hydrazone, *Synlett*, 17 (2005) 2607–2610.
65. J. L. Chiara, J. Marco-Contelles, N. Khira, P. Gallego, C. Destabel, and M. Bernabe, Intramolecular reductive coupling of carbonyl-tethered oxime ethers promoted by samarium diiodide: A powerful method for the stereoselective synthesis of aminocyclopentitols, *J. Org. Chem.*, 60 (1995) 6010–6011.
66. J. Marco-Contelles, P. Gallego, M. Rodriguez-Fernandez, N. Khiar, C. Destabel, M. Bernanbe, A. Marinez-Grau, and J. L. Chiara, Synthesis of aminocyclitols by intramolecular reductive coupling of carbohydrate derived δ- and ε-functionalized oxime ethers promoted by tributyltin hydride and samarium diiodide, *J. Org. Chem.*, 62 (1997) 7397–7412.
67. H. Yuasa, J. Tamura, and H. Hashimoto, Synthesis of per-O-alkylated 5-thio-D-glucono-1,5-lactones and transannular participation of the ring sulphur atom of 5-thio-D-glucose derivatives on solvolysis under acidic conditions, *J. Chem. Soc. Perkin Trans.*, 1 (1990) 2763–2769.
68. G. D. McAllister and R. J. K. Taylor, The synthesis of polyoxygenated, enantiopure cyclopentene derivatives using the Ramberg-Backlund rearrangement, *Tetrahedron Lett.*, 42 (2001) 1197–1200.

69. B. Bernet and A. Vasella, Carbocyclische verbindungen aus monosacchariden. I. Umsetzungen in der glucosereihe, *Helv. Chim. Acta*, 62 (1979) 1990–2016.
70. G. A. Lee, Simplified synthesis of unsaturated nitrogen-heterocycles using nitrile betaines, *Synthesis* (1982) 508–509.
71. Y. Kobayashi, H. Miyazaki, M. Shiozaki, and H. Haruyama, Synthesis of 5-*epi*-trehazolin (trehalostatin), *J. Antibiot.*, 47 (1994) 932–938.
72. H. Miyazaki, Y. Kobayashi, M. Shiozaki, O. Ando, M. Nakajima, H. Hanzawa, and H. Haruyama, Synthesis of a 2-aminohexahydrobenzoxazole analogue related to trehazolin, *J. Org. Chem.*, 60 (1995) 6103–6109.
73. K. Fuji, S. Nakano, and E. Fujita, An improved method for methoxymethylation of alcohols under mild acidic conditions, *Synthesis* (1975) 276–277.
74. K. Sato, N. Kubo, R. Takada, A. Aqeel, H. Hashimoto, and J. Yoshimura, Convenient synthesis of hex-5-enopyranosides, *Chem. Lett.*, 17 (1988) 1703–1704.
75. R. J. Ferrier, Unsaturated carbohydrates. Part 21. A carbocyclic ring closure of a hex-5-enopyranoside derivative, *J. Chem. Soc. Perkin. Trans.*, 1 (1979) 1455–1458.
76. R. J. Ferrier and S. R. Haines, A route to functionalised cyclopentanes from 6-deoxyhex-5-enopyranoside derivatives, *Carbohydr. Res.*, 130 (1984) 135–146.
77. R. Blattner, R. J. Ferrier, and S. R. Haines, Unsaturated carbohydrates. Part 28. Observations on the conversion of 6-deoxyhex-5-enopyranosyl compounds into 2-deoxyinose derivatives, *J. Chem. Soc. Perkin Trans.*, 1 (1985) 2413–2416.
78. N. Chida, M. Ohtsuka, K. Ogura, and S. Ogawa, Synthesis of optically active substituted cyclohexenones from carbohydrates by catalytic Ferrier rearrangement, *Bull. Chem. Soc. Jpn.*, 64 (1991) 2118–2121.
79. I. S. de Gracia, S. Bobo, M. D. Martin-Ortega, and J. L. Chiara, A concise and highly efficient synthesis of trehazolin and trehalamine starting from D-mannose, *Org. Lett.*, 1 (1999) 1705–1708.
80. T. Kametani, K. Kawamura, and T. Honda, New entry to the C-glycosidation by means of carbenoid displacement reaction. Its application to the synthesis of showdomycin, *J. Am. Chem. Soc.*, 109 (1987) 3010–3017.
81. S. Bobo, I. S. de Gracia, and J. L. Chiara, A concise synthesis of a trehazolamine epimer with moderate α-mannosidase inhibitory activity starting from D-mannose, *Synlett* (1999) 1551–1554.
82. J. Gelas and D. Horton, Kinetic acetonation of D-mannose: Preparation of 4,6-mono- and 2,3:4,6-di-*O*-isopropylidene-D-mannopyranose, *Carbohydr. Res.*, 67 (1978) 371–387.
83. T. K. M. Shing, D. A. Elsley, and J. G. Gillhouley, A rapid entry to carbocycles from carbohydrates via intramolecular nitrone cycloaddition, *J. Chem. Soc. Chem. Commun.* (1989) 1280–1282.
84. J. Marco-Contelles, C. Destabel, P. Gallego, J. L. Chiara, and M. Bernabe, A new synthetic approach to the carbocyclic core of cyclopentane-type glycosidase inhibitors: Asymmetric synthesis of aminocyclopentitols via free radical cycloisomerization of enantiomerically pure alkyne-tethered oxime ethers derived from carbohydrates, *J. Org. Chem.*, 61 (1996) 1354–1362.
85. B. Lee and T. Nolan, Sugars with potential antiviral activity—I: A new method for the preparation of glycorufanoyl chlorides and the synthesis of a mannosyl nucleoside, *Tetrahedron*, 23 (1967) 2789–2794.
86. J. G. Buchanan, A. D. Dunn, and A. R. Edgar, Nucleoside studies. Part II. Pentofuranosylethynes from 2,3-*O*-isopropylidene-D-ribose, *J. Chem. Soc. Perkin Trans.*, 1 (1975) 1191–1200.
87. J. L. Chiara, C. Destable, P. Gallego, and J. Marco-Contelles, Cleavage of N−O bonds promoted by samarium diiodide: Reduction of free or N-acylated O-alkylhydroxylamines, *J. Org. Chem.*, 61 (1996) 359–360.
88. M. Seepersaud and Y. Al-Abed, Studies directed toward the synthesis of carba-D-arabinofuranose, *Tetrahedron Lett.*, 41 (2000) 7801–7803.

89. M. Seepersaud and Y. Al-Abed, An enantioselective approach to trehazolin: A concise and efficient synthesis of the aminocyclopentitol core, *Tetrahedron Lett.*, 42 (2001) 1471–1473.
90. M. Shiozaki, M. Arai, Y. Kobayashi, A. Kasuya, S. Miyamoto, Y. Furukawa, T. Takayama, and H. Haruyama, Stereocontrolled syntheses of 6-epi-trehazolin and 6-epi-trehalamine from D-ribonolactone, *J. Org. Chem.*, 59 (1994) 4450–4460.
91. J. J. Baldwin, A. W. Raab, K. Mensler, B. H. Arison, and D. E. McClure, Synthesis of (R)- and (S)-epichlorohydrin, *J. Org. Chem.*, 43 (1978) 4876–4878.
92. J. M. Klunder, T. Onami, and K. B. Sharpless, Arenesulfonate derivatives of homochiral glycidol: versatile chiral building blocks for organic synthesis, *J. Org. Chem.*, 54 (1989) 1295–1304.
93. L. E. Overman, Thermal and mercuric ion catalyzed [3,3]-sigmatropic rearrangement of allylic trichloroacetimidates. 1,3-Transposition of alcohol and amine functions, *J. Am. Chem. Soc.*, 96 (1974) 597–599.
94. R. P. Elliot, A. Hui, A. J. Fairbanks, R. J. Nash, B. G. Winchester, G. Way, C. Smith, R. B. Lamont, R. Storer, and G. W. J. Fleet, Highly substituted cis-β-cyclopentane amino acids: An approach to the synthesis of trehazolin analogues, *Tetrahedron Lett.*, 34 (1993) 7949–7952.
95. R. P. Elliott, G. W. J. Fleet, L. Pearce, C. Smith, and D. J. Watkin, A reductive aldol strategy for the synthesis of very highly substituted cyclopentanes from sugar lactones, *Tetrahedron Lett.*, 32 (1991) 6227–6230.
96. S. J. Angyal, M. E. Tate, and S. D. Gero, Cyclitols. Part IX. Cyclohexylidene derivatives of *myo*-inositol, *J. Chem. Soc.* (1961) 4116–4122.
97. C. Uchida, T. Yamagishi, and S. Ogawa, Total synthesis of the trehalase inhibitors trehalostatin and trehazolin, and of their diastereoisomers. Final structural confirmation of the inhibitor, *J. Chem. Soc. Perkin Trans.*, 1 (1994) 589–602.
98. C. Uchida, T. Yamagishi, and S. Ogawa, Total synthesis and trehazlase-inhibitory activity of trehalostatin and its diastereoisomer, *Chem. Lett.* (1993) 971–974.
99. T. Suami, K. Tadano, S. Nishiyama, and F. W. Lichtenthaler, Aminocyclitols. 30. Unambiguous synthesis of seven aminocyclopentanetetrols, *J. Org. Chem.*, 38 (1973) 3691–3696.
100. S. J. Angyal and S. D. Gero, Convenient preparation of a crystalline derivative of meso-tartraldehyde, *Aust. J. Chem.*, 18 (1965) 1973–1976.
101. R. Ahluwalia, S. J. Angyal, and B. M. Luttrell, Cyclitols. XXXI. Synthesis of amino- and nitro-cyclopentanetetrols, *Aust. J. Chem.*, 23 (1970) 1819–1829.
102. S. Ogawa and C. Uchida, Synthesis of aminocyclitol moieties of trehalase inhibitors, trehalostatin and trehazolin. Confirmation of the correcte structure of the inhibitor, *J. Chem. Soc. Perkin Trans.*, 1 (1992) 1939–1942.
103. C. Uchida, H. Kitahashi, T. Yamagishi, Y. Iwaisaki, and S. Ogawa, Synthesis of trehazolin analogues containing modified aminocyclitol moieties, *J. Chem. Soc. Perkin Trans.*, 1 (1994) 2775–2785.
104. V. E. Marquez, M. Lim, M. S. Khan, and B. Kaskar, (4R,5R)-(-)-3-(Benzyloxymethyl)-4,5-O-isopropylidene-2-cyclopentenone-4,5-diol. An optically active α-, β-unsaturated cyclopentenone for the synthesis of neplanocin A and other cyclopentene carbocyclic nucleosides, *Nucleic Acid Chem.*, 4 (1991) 27–35.
105. S. Knapp, A. Purandare, K. Rupitz, and S. G. Withers, A (1→4)-"trehazoloid" glucosidase inhibitor with aglycon selectivity, *J. Am. Chem. Soc.*, 116 (1994) 7461–7462.
106. M. Shiozaki, Y. Kobayashi, M. Arai, and H. Haruyama, Synthesis of 6-*epi*-trehazolin from D-ribonolactone: Evidence for the non-existence of a 5,6-ring fused structural isomer of 6-*epi*-trehazolin, *Tetrahedron Lett.*, 35 (1994) 887–890.
107. M. Shiozaki, M. Arai, Y. Kobayashi, A. Kasuya, S. Miyamoto, Y. Furukawa, T. Takayama, and H. Haruyama, Stereocontrolled syntheses of 6-epi-trehazolin and 6-epi-trehalamine from D-ribonolactone, *J. Org. Chem.*, 59 (1994) 4450–4460.

108. E. J. Corey, U. Koelliker, and J. Neuffer, Methoxymethylation of thallous cyclopentadienide. Simplified preparation of a key intermediate for the synthesis of prostaglandins, *J. Am. Chem. Soc.*, 93 (1971) 1489–1490.
109. J. Li, F. Lang, and B. Ganem, Enantioselective approaches to aminocyclopentitols: A total synthesis of (+)-6-*epi*-trehazolin and a formal total synthesis of (+)-trehazolin, *J. Org. Chem.*, 63 (1998) 3403–3410.
110. M. T. Crimmins and E. A. Tabet, Formal total synthesis of (+)-trehazolin. Application of an asymmetric aldol-olefin metathesis approach to the synthesis of functionalized cyclopentenes, *J. Org. Chem.*, 66 (2001) 4012–4018.
111. B. M. Trost, J. D. Chisholm, S. T. Wrobleski, and M. Jung, Ruthenium-catalyzed alkene-alkyne coupling: Synthesis of the proposed structure of amphidinolide A, *J. Am. Chem. Soc.*, 124 (2002) 1242–12421.
112. M. Akiyama, T. Awamura, K. Kimura, Y. Hosomi, A. Kobayashi, K. Tsuji, A. Kuboki, and S. Ohira, Stereocontrolled synthesis of the aminocyclitol moiety of (+)-trehazolin via C-H insertion reaction of alkylidenecarbene, *Tetrahedron Lett.*, 45 (2004) 7133–7136.
113. G. Mehta and N. Mohal, A norbornyl route to cyclopentitols via novel regiospecific fragmentation of a 2,7-disubstituted norbornane, *Tetrahedron Lett.*, 40 (1999) 5791–5794.
114. G. Mehta and N. Mohal, Norbornyl route to cyclopentitols: Synthesis of trehazolamine analogues and the purported structure of salpantiol, *Tetrahedron Lett.*, 40 (1999) 5795–5798.
115. X. Feng, E. N. Duesler, and P. S. Mariano, Pyridinium salt photochemistry in a concise route for synthesis of the trehazolin aminocyclitol, trehazolamine, *J. Org. Chem.*, 70 (2005) 5618–5623.
116. M. Shiozaki, O. Ubukata, H. Haruyama, and R. Yoshiike, Synthesis of 1-deoxynojirimycin-trehalamine fused compound and its related compounds, *Tetrahedron Lett.*, 39 (1998) 1925–1928.
117. M. Shiozaki, R. Yoshiike, O. Ando, O. Ubukata, and H. Haruyama, Syntheses of 1-deoxynojirimycin-trehalamine-fused and -linked compounds and their biological activities, *Tetrahedron*, 54 (1998) 15167–15182.
118. S. Ogawa and C. Uchida, Total synthesis of trehalase inhibitor, trehazolin, *Chem. Lett.* (1993) 173–176.
119. Y. Kobayashi and M. Shiozaki, Synthesis of trihazolin β-anomer, *J. Antibiot.*, 47 (1994) 243–246.
120. C. Uchida, T. Yamagishi, H. Kitahashi, Y. Iwaisaki, and S. Ogawa, Further chemical modification of trehalase inhibitor trehazolin: Structure and inhibitory-activity relationship of the inhibitor, *Bioorg. Med. Chem.*, 3 (1995) 1605–1624.
121. P. Ková, H. J. C. Yeh, and G. L. Jung, The reaction of 2,3,4-trii-*O*-benzyl-D-glucose with diethylaminosulfur trifluoride (Dast), *J. Carbohydr. Chem.*, 6 (1987) 423–439.
122. H. S. Overkleeft, J. van Wiltenburg, and U. K. Pandit, A facile transformation of sugar lactones to azasugars, *Tetrahedron*, 50 (1994) 4215–4224.

POLYSACCHARIDES OF THE RED ALGAE

Anatolii I. Usov

N. D. Zelinsky Institute of Organic Chemistry, Leninskii Prospect 47, Moscow, Russian Federation

I. Introduction	116
II. Storage Carbohydrates: Floridean Starch	117
III. Neutral Structural Polysaccharides: Cellulose, Mannans, and Xylans	119
IV. Sulfated Galactans	122
1. General Information and Nomenclature	122
2. Isolation of Galactans	127
3. Chemical Methods of Structural Analysis	129
4. Physicochemical Methods of Structural Analysis	145
5. Enzymatic and Immunological Methods	155
6. Polysaccharides of the Agar Group	162
7. Polysaccharides of the Carrageenan Group	165
8. DL-Hybrids	168
9. Biological Activity of Sulfated Galactans	169
V. Sulfated Mannans	171
VI. Glycuronans	173
VII. Heteropolysaccharides of Unicellular and Freshwater Red Algae. Miscellaneous Polysaccharides	174
VIII. Algal Systematics and Polysaccharide Composition	177
References	179

Abbreviations

ADPG, adenosine 5′-(α-D-glucopyranosyl diphosphate); BTSA, N,O-bis(trimethylsilyl)acetamide; BTSTFA, N,O-bis(trimethylsilyl)trifluoroacetamide; CID, collision-induced dissociation; CTMS, chlorotrimethylsilan; DEAE, diethylaminoethyl; DP, degree of polymerization; EI, electron impact; ESI, electrospray ionization; FAB, fast-atom bombardment; FTIR, Fourier-transform infrared; GLC, gas–liquid chromatography; HIV, human immunodeficiency virus; HPAEC, high-performance anion-exchange chromatography; HPLC, high-performance liquid chromatography; IR, infrared; IUBMB, International Union of Biochemistry and Molecular Biology; IUPAC, International Union of Pure and Applied Chemistry; MALDI, matrix-assisted laser desorption/ionization; Me_2SO, dimethyl sulfoxide; MS, mass spectrometry; MTSTFA, N-methyl-N-(trimethylsilyl)trifluoroacetamide; NMR, nuclear magnetic resonance; PAD, pulsed-amperometric detection; Pmg, pyrenemethylguanidine; PSD, post-source decay; TOF, time of flight; TSIM, N-(trimethylsilyl)imidazole; UDPG, uridine 5'-(α-D-glucopyranosyl diphosphate); UV, ultraviolet

I. INTRODUCTION

Red algae are phylogenetically the oldest division of lower plants and contain about 5000 species, mainly marine multicellular organisms (macrophytes). They differ from all the other plants by the polysaccharide composition of their cell walls and the intercellular matrix. Most of them contain unique sulfated galactans, and two groups of these polysaccharides, known as agars and carrageenans, are prepared industrially from the algae on a large scale and find wide practical application as powerful gelling and stabilizing agents.[1,2] Chemical investigation of red algal polysaccharides was begun in the middle of 20th century. Agarose, a non-sulfated agar component, was isolated[3] in 1937, and its structural analysis was completed[4] in 1956. A very peculiar monosaccharide, 3,6-anhydrogalactose, was discovered at that time as a unique component of red algal galactans, and the specific properties of this monosaccharide component were shown to determine importantly the chemical behavior of the corresponding biopolymers. Chemical data on the red algal galactans obtained during this initial stage of investigation were reviewed by Mori in Volume 8 of this series.[5] A procedure for fractionation of carrageenan[6] was suggested in 1953, followed by further development of chemical, physicochemical, and enzymatic methods of structural analysis of galactans, which permitted determination of the structures of several carrageenans and formulation of a very important hypothesis of "masked repeating structure," that is, inherent in all of the galactans[7] and plays the determining role in their gel-forming properties.[8] During this period, the brilliant monograph of E. Percival and R. H. McDowell was published,[9] and this book retains its significance even now. A new phase of investigations began with the introduction of ^{13}C nuclear magnetic resonance (NMR) spectroscopy for the structural analysis of galactans.[10,11] This approach proved exceptionally fruitful, and so it is now the obligatory component of the set of modern methods used for elucidation of galactan structures, despite the considerable improvement of many chemical procedures. These methodological achievements made it possible to determine the structures of sulfated galactans that are much more complex, to obtain new evidence on the possible structural diversity in these biopolymers and to find interesting correlations between the polysaccharide structure and the taxonomical position of algae. There are two excellent surveys on the chemical nature of red algal polysaccharides[12,13] and a review on their biological role in algal cell walls,[14] although these papers were written 20–30 years ago. Several modern reviews are restricted to the chemistry of red algal galactans in accordance with their principal practical importance.[15–18] Gelling galactans are very popular models for the study on polysaccharide conformations and molecular interactions leading to formation of gels, but these physicochemical investigations are not discussed in the present article. The corresponding information may be found in many reviews (see, e.g., refs. 8,19–25), and in a vast number of experimental papers published mainly in *Carbohydrate Polymers*, *Food Hydrocolloids*, *Journal of Applied Phycology*, *Food Chemistry*, and various related journals. This article is devoted to the description of modern approaches for elucidating the primary structures of red algal polysaccharides, and of contemporary views on the structural diversity of these biopolymers.

II. Storage Carbohydrates: Floridean Starch

Red algae biosynthesize two kinds of unique storage carbohydrates. They contain several low-molecular glycosides, which function not only as primary photosynthetic reserve products but also as osmoregulators (see ref. 26). Floridoside (2-O-α-D-galactopyranosylglycerol, **1**) is the most widely distributed substance of this series.[27] In addition, two diastereoisomeric isofloridosides, 1-O-α-D-galactopyranosyl-D-glycerol (D-isofloridoside, **2**) and 1-O-α-D-galactopyranosyl-L-glycerol (L-isofloridoside,**3**), were found in representatives of the order Bangiales,[28] whereas species of the order Ceramiales may contain 2-O-α-D-mannopyranosyl-D-glyceric acid (digeneaside, **4**).[27] Recent publications describe some improved chromatographic procedures for isolation from crude extracts and quantitative determination of floridoside,[29,30] isofloridosides,[28] and digeneaside[31] as well as complete signal assignments in their ^{1}H and ^{13}C NMR spectra.[28,30–32] The structure and chirality of natural floridoside[33,34] and digeneaside[35] were confirmed by single-crystal X-ray diffraction analysis.

The high-molecular-mass storage product of the red algae is known as floridean starch,[36] although "floridean glycogen" was recommended as the more correct term.[12] According to both names, the polysaccharide is built up of α-D-glucopyranose residues forming (1→4)-linked chains with branches at position 6. In the earlier work on the structure of a preparation isolated from *Dilsea edulis*, it was shown by several methods that the polysaccharide contained some (1→3)-linkages.[37,38] The most important evidence supporting this conclusion was formation of some nigerose (3-O-α-D-glucopyranosyl-D-glucose) after the action of amylolytic enzymes.[39] Apparently, this conclusion was not disproved by subsequent experiments, but at the same time, it was not confirmed by investigation of floridean starches isolated from other sources.[40,41] Therefore, floridean starch is usually regarded as the polysaccharide, structurally identical to amylopectins and glycogens, but having an intermediate degree of branching and average linear chain length of about 14–18 glucose residues.[41] It differs from starches of green algae and plants by the absence of amylose and by its localization in the cells: like animal glycogen, floridean starch is biosynthesized and placed in the cytosol, whereas real starch is formed and stored in plastids. Enzymatic formation of both glycogen and floridean starch uses the same precursor, UDPG, rather than ADPG, which is the starting material for starch biosynthesis.[42] It is suggested that further investigation of the biochemical and molecular features of floridean starch synthesis may offer new insight into the metabolic strategies of photosynthetic eukaryotes.[42,43]

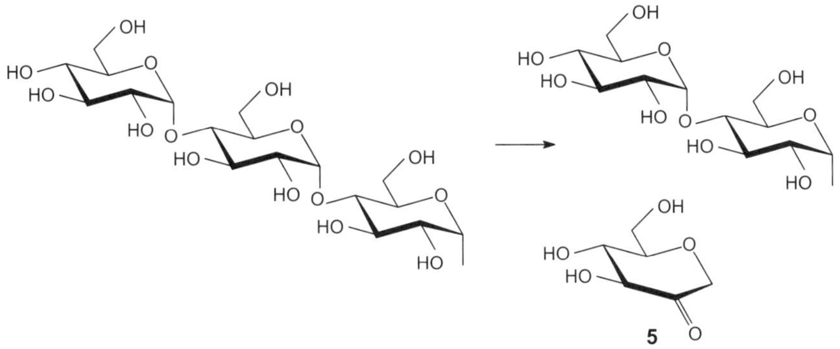

SCHEME 1. Action of α-(1→4)-glucan lyase on floridean starch (the monosaccharide derivatives, which are split from the nonreducing ends of the polysaccharide chains probably in the form of 2-hydroxyglucal, are tautomerized to the more stable 1,5-anhydro-D-fructose, **5**).

Storage glucans in the most primitive unicellular red algae, which are phylogenetically far from the multicellular seaweeds, may have different structural features. For example, some amylose was detected in "semi-amylopectins" stored by species of *Porphyridium*, *Rhodosorus*, and *Rhodella*, whereas species of *Galdieria* and *Cyanidium* contain glycogens.[44,45] When growing under heterotrophic conditions, *Galdieria maxima* can synthesize and store a big amount of an extremely highly branched α-glucan of the glycogen type.[46]

Floridean starch is hydrolyzed by amylolytic enzymes,[47] and amylolysis may be used for quantitative determination of the floridean-starch content in the algae,[46] which varies considerably depending on the algal species and growth conditions.[36] Recently, a new enzyme capable of cleaving the polysaccharide chain by an elimination reaction, giving rise to 1,5-anhydro-D-fructose (**5**, Scheme 1) was found in the red algae.[48] The same enzymatic activity was also found in fungi[49] and animals,[50] but was not detected in plants. The enzymatic conversion of starch into 1,5-anhydro-D-fructose may be performed on a large scale, thus rendering this unusual monosaccharide available for many interesting chemical and biochemical transformations (see ref. 51).

III. NEUTRAL STRUCTURAL POLYSACCHARIDES: CELLULOSE, MANNANS, AND XYLANS

Cellulose, a linear (1→4)-linked β-D-glucan,[52] is present in the cell walls of the majority of red algae in moderate amounts (usually less than 10%). Determination of cellulose in earlier works was based on the relative insolubility of the polysaccharide in aqueous solutions,[53] but residues left after exhaustive extraction of algal biomass often gave mannose and xylose together with glucose upon acid hydrolysis. Nevertheless, the chemical structure of red algal cellulose was confirmed by methylation analysis of a preparation from *Porphyra tenera*[54] and by the ^{1}H NMR spectrum of the permethylated polysaccharide from *Rhodymenia pertusa*.[55] Solutions of three cellulose preparations (isolated from *Gracilaria longa*, *Gracilaria dura*, and *Hypnea musciformis*) in dimethylacetamide–LiCl gave ^{13}C NMR spectra coinciding with the spectrum of authentic cellulose of higher plants in the same solvent.[56] Solid-state ^{13}C NMR spectroscopy and wide-range X-ray scattering were used to characterize the crystalline forms and microfibril dimensions of cellulose in algal cell walls, giving evidence on some difference in molecular packing of cellulose molecules in Bangiophyceae and Florideophyceae.[57,58] Biosynthesis of the red algal cellulose was studied mainly in several species of the order Bangiales.[59,60] It was shown that cellulose microfibrils, like those of land plants, are synthesized by clusters of cellulose synthase enzymes ("terminal complexes") in the plasma membrane, but the resulting microfibrils differ from those of land plants in their dimensions and morphology (see ref. 61).

A linear mannan structurally similar to cellulose was firstly isolated from *Porphyra umbilicalis*[62] and was then found in *P. tenera*[54] and *Bangia atropurpurea*.[63] Later on, it was stated that two generations of bangialean algae, namely, conchocelis and generic phases, differ considerably in the structural polysaccharide composition of their cell walls. The conchocelis phases of *B. atropurpurea*[64] and *P. tenera*[54] contained cellulose and minor amounts of mannan. In contrast, the generic phases of the same species contained $(1\rightarrow 4)$-linked β-D-mannan as the main structural component,[54,65] whereas cellulose was not detected. Similar results were obtained in the study of the cell-wall composition of *P. leucosticta* and *P. umbilicalis*.[66,67]

Small amounts of xylose are usually formed upon acid hydrolysis of red algal cell walls, but there are two orders, Palmariales and Nemaliales, where xylans may be the main polysaccharide components. Thus, the Atlantic species *Palmaria* (formerly *Rhodymenia*) *palmata* contains a water-soluble xylan (rhodymenan). Its structure was carefully investigated by chemical methods and enzymatic hydrolysis.[68,69] It was shown that the polysaccharide contains linear chains of 4-linked β-D-xylopyranose residues interspersed by single 3-linked β-D-xylopyranose residues. The proportion between $(1\rightarrow 3)$- and $(1\rightarrow 4)$-linkages was about 1:4, but less-soluble fractions extracted by alkali contained more 4-linked β-D-xylopyranose residues, up to pure $(1\rightarrow 4)$-β-D-xylan.[70] The same structure was established by chemical methods for a xylan isolated from the closely related Pacific species *Palmaria* (formerly *Rhodymenia*) *stenogona* and confirmed by the ^{13}C NMR spectrum of the polysaccharide.[71] The latter method was then used to detect the presence of similar xylans in other representatives of the order Palmariales, such as *Halosaccion glandiforme*[72] and *Leptosarca simplex*,[73] also known as *Palmaria decipiens*.[74] The ^{13}C NMR spectra of xylan solutions were again described in detail,[75] whereas similar spectra in the solid state were used to investigate the conformation and interactions of xylan with other polysaccharide components in the algal cell walls.[76]

Since *P. palmata* is an edible alga constituting a potential protein source in the human diet and is readily available in large amounts, its chemical composition was carefully investigated.[77] Several attempts to improve the digestibility of the algal proteins were made, including elimination of xylan by physical processes and fermentation.[78–80] In this connection, a reinvestigation of the xylan chemical structure was conducted, and the results were interpreted as some evidence of the rather regular distribution of $(1 \to 3)$-linkages along the chains of mixed-linkage molecules and of the presence of some covalent linkages between xylans and charged (sulfated and/or phosphorylated) xylogalactoproteins.[81,82]

Mixed-linkage $(1 \to 3,1 \to 4)$-β-D-xylans similar to those present in Palmariales were found also in Nemaliales. The examples were the polysaccharides from *Audouinella floridula* (as *Rhodochorton floridulum*),[70] *Nothogenia fastigiata* (as *Chaetangium fastigiatum*),[83,84] *Nothogenia erinacea* (as *Chaetangium erinaceum*),[85] *Nemalion vermiculare*,[86,87] *Galaxaura rugosa* (as *Galaxaura squalida*),[88] and *Liagora valida*.[89] Their structures were elucidated by chemical methods, including the preparation of a large series of oligosaccharide fragments,[85] and by ^{13}C NMR spectra.[72,87–89] Matrix-assisted ultraviolet laser desorption/ionization time-of-flight mass spectrometry (UV-MALDI-TOF MS) in the positive- and negative-ion modes was applied to three xylan fractions isolated from *N. fastigiata* to establish optimal conditions for obtaining the spectra and to calculate the molecular masses of the fractions.[90] The structure of *N. erinacea* xylan was carefully reinvestigated[91] by enzymatic hydrolysis. Two enzymes differing in substrate specificity were used, namely, xylanases from *Cryptococcus adeliae* and from *Thermomyces lanuginosus*, representatives of glycoside hydrolase families 10 and 11, respectively. The main enzymolysis products, mixed-linkage trisaccharides, and tetrasaccharides, respectively, as well as numerous higher oligosaccharides produced as the result of partial degradation, were isolated by high-performance anion-exchange chromatography (HPAEC)-PAD and identified by NMR spectroscopy and MS. The observed oligosaccharide pattern indicated an irregular distribution of single $(1 \to 3)$-linkages along the $(1 \to 4)$-linked backbone. The same results were obtained by the authors with rhodymenan, the xylan from *P. palmata* (see above), and hence, the opinion on its regular structure[81,82] was not confirmed.

A linear (1 → 4)-β-D-xylan was obtained from *Scinaia hatei* (the order Nemaliales) by alkali extraction after exhaustive treatment of the biomass with hot water.[92] Chemical sulfation of the polysaccharide gave rise to derivatives possessing strong antiviral activity. Water-insoluble xylans of different structure containing linear chains of 3-linked β-D-xylopyranose residues were detected in several species of the genus *Porphyra*.[54,70] They formed microfibrils in the cell walls of the generic phase of these algae, which do not contain cellulose.

IV. SULFATED GALACTANS

1. General Information and Nomenclature

As mentioned in Section I, the main polysaccharide components of the red algae are sulfated galactans, which are produced industrially on a large scale and have great practical use. As a result, the overwhelming majority of papers dealing with the polysaccharide composition of the red algae are devoted to structural analysis or investigation of the physicochemical properties of sulfated galactans. Now it is known[12,13,15] that sulfated galactans usually have a linear backbone built up of alternating 3-linked β-D-galactopyranose and 4-linked α-galactopyranose residues. The latter have the L-configuration in the agar group of polysaccharides, but the D-configuration in carrageenans. In addition, 4-linked residues may be present, in part or completely, as 3,6-anhydro derivatives. Hence, there are four types of disaccharide repeating units (**6–9**) in the molecules of ancestor galactans. Hydroxyl groups in these repeating units may be methylated, sulfated (Scheme 2), and sometimes, they are substituted by single monosaccharide residues (D-xylose, 4-O-methyl-L-galactose, D-glucuronic acid), whereas 3-linked β-D-galactopyranose residues may carry 4,6-O-(1-carboxyethylidene) group (six-membered cyclic acetal of pyruvic acid). The formation of 3,6-anhydrogalactose residues *in vivo* proceeds as simultaneous elimination of a sulfate group and a proton of the 3-OH group from 4-linked α-galactose 6-sulfate residues under the action of specific enzymes.[93] Similar transformation *in vitro* may be readily effected by alkaline treatment of a polysaccharide.[94] Other substituents are probably introduced during the biosynthetic process into the previously prepared polymeric ancestor. As a result, the primary backbone regularity may be masked by an uneven distribution of substituents, and the final biosynthetic product should be regarded as a "molecular hybrid" containing several different repeating units, as for example, porphyran **10** (Scheme 2).

Agaran, **6**

Agarose, **7**

Carrageenan, **8**

Carrageenose, **9**

Moreover, different molecules of the same biopolymer may vary in the proportion and distribution of these units, and the whole substance should be considered as a continuum of molecules transitional between two or more "extreme" structures.[95] Hence, for complete elucidation of the primary structure, it is necessary to identify the repeating units present in a galactan molecule and to determine their distribution or sequence. It is clear that the procedure of structural analysis becomes much more difficult for less regular polysaccharides.[16]

Additional difficulties may be found in the study on so-called "deviating" structures, such as polysaccharides where the rule of strict alternation of D- and L-galactose derivatives or α-(1→3) and β-(1→4) linkages along the backbone is infringed. Several examples of such molecules are described next. The terms "carragar" or "DL-hybrid" were suggested for agar–carrageenan hybrid structures,[96] but in most cases, the description of these polysaccharides contains no unambiguous evidence that the material is a real hybrid and not a mixture, different components of which possess the regularity typical of classical agars and carrageenans, respectively.

The nomenclature of red algal galactans used in most of the papers is rather arbitrary. The term "agar," or "agar–agar," is known from ancient times. "Agarose" and "agaropectin" were introduced by analogy with amylose and amylopectin, when

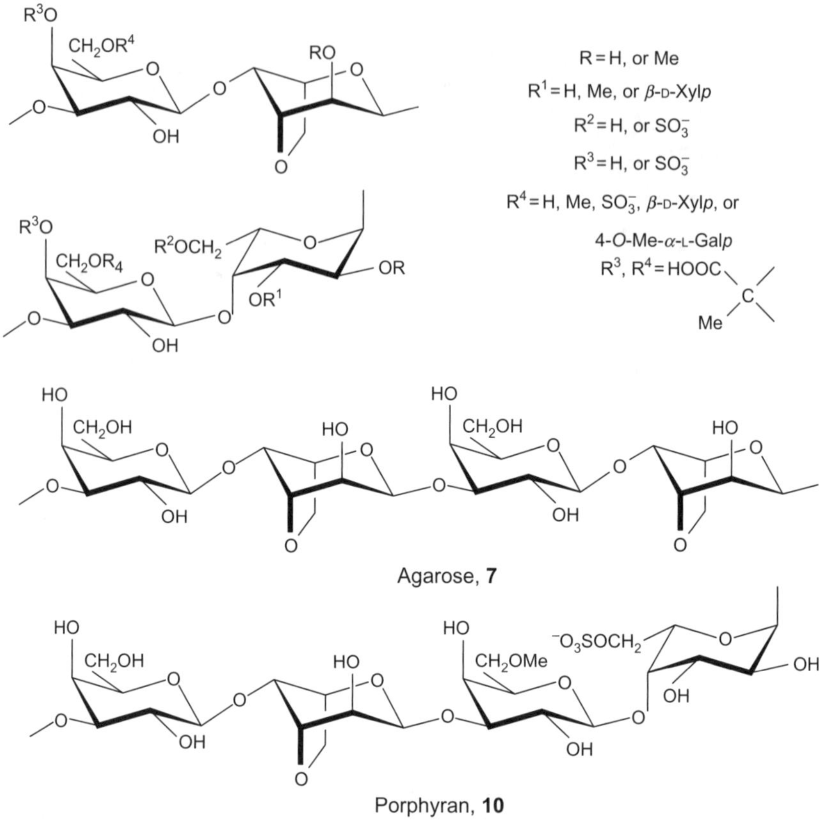

SCHEME 2. Structural diversity in the polysaccharides of the agar group.

agar, like starch, was separated into two components.[97] The term "carrageenan" (formerly "carrageenin") was derived from the name of the Ireland district known as the source of the raw material ("Irish moss," *Chondrus crispus*) for production of this polysaccharide.[98] Many polysaccharides were named, before elucidation of their chemical nature, according to their biological source (porphyran[99] **10** from algae of the genus *Porphyra*, similarly odonthalan[100] from *Odonthalia*, furcellaran[101] from *Furcellaria*, corallinan[102] from *Corallina*, and so on). Fractions obtained after the first separation of carrageenan into two components were named κ and λ,[6] and then all of the new carrageenans were designated by Greek letters without any system (Scheme 3). The galactan nomenclature, based on their chemical structures,[103] was

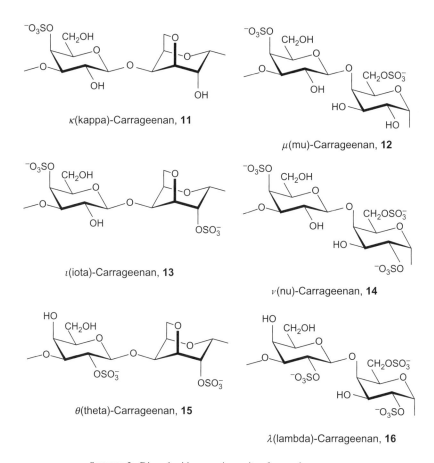

SCHEME 3. Disaccharide repeating units of several carrageenans.

suggested in 1994. It retains the trivial names of four parent structures (**6–9**). Other repeating units could be described as the result of substitution (O-methylation, sulfation, and so on) of these structures. A shorthand notation system was also proposed, which is especially useful for designation of hybrid structures containing several different repeating units (Tables I and II). Nevertheless, discussing many earlier works on the red algal galactans, we are obliged in this chapter to use the old names of several polysaccharides. Similarly, the old names of several red algae are indicated here according to the original papers cited, in spite of the fact that these algae have been renamed as the result of recent corrections of algal taxonomy.

TABLE I
Examples of the Shorthand Terminology System for Galactan Building Units and Sequences with Their IUPAC Names[103]

Letter Code	Full IUPAC Name
D	4-Linked α-D-galactopyranosyl (typical for carrageenans)
DA	4-Linked 3,6-anhydro-α-D-galactopyranosyl (typical for carrageenans)
G	3-Linked β-D-galactopyranosyl
L	4-Linked α-L-galactopyranosyl (typical for agars)
LA	4-Linked 3,6-anhydro-α-L-galactopyranosyl (typical for agars)
M	O-Methyl
P	4,6-O-(1-carboxyethylidene)
S	Ester sulfate
L2M,6S	4-Linked 2-O-methyl-α-L-galactopyranosyl 6-sulfate
G6[L4M]	3-Linked 6-O-(4-O-methyl-α-L-galactopyranosyl)-β-D-galactopyranosyl
-G4S-DA-G-	→3)-β-D-galactopyranosyl-4-sulfate-(1→4)-3,6-anhydro-α-D-galactopyranosyl-(1→3)-β-D-galactopyranosyl-(1→

TABLE II
Disaccharide Repeating Units of the Carrageenan Group of Polysaccharides[103]

Greek Name	Greek Letter	Reference (Named by)	Proposed Name	Letter Code
			Carrageenan	G-D
alpha	α	Zablakis and Santos[104]	Carrageenose 2-sulfate	G-DA2S
beta	β	Greer and Yaphe[105]	Carrageenose	G-DA
gamma	γ	Greer and Yaphe[105]	Carrageenan 6-sulfate	G-D6S
delta	δ	Zablakis and Santos[104]	Carrageenan 2,6-disulfate	G-D2S,6S
theta	θ	Anonymous[106]	Carrageenose 2,2′-disulfate **15**	G2S-DA2S
iota	ι	Anderson et al.[107]	Carrageenose 2,4′-disulfate **13**	G4S-DA2S
kappa	κ	O'Neill[108]	Carrageenose 4′-sulfate **11**	G4S-DA
lambda	λ	Dolan and Rees[109]	Carrageenan 2,6,2′-trisulfate **16**	G2S-D2S,6S
mu	μ	Anderson et al.[110]	Carrageenan 6,4′-disulfate **12**	G4S-D6S
nu	ν	Stancioff and Stanley[111]	Carrageenan 2,6,4′-trisulfate **14**	G4S-D2S,6S
xi	ξ	Penman and Rees[112]	Carrageenan 2,2′-disulfate	G2S-D2S
omicron		Craigie[13]	Carrageenan 2,4′-disulfate	G4S-D2S
pi	π	DiNinno et al.[113]	4′,6′-O-(1-carboxyethylidene)-carrageenan 2,2′-disulfate	GP,2S-D2S
psi	ψ	Craigie[13]	Carrageenan 6,6′-disulfate	G6S-D6S
omega	ω	Mollion et al.[114]	Carrageenose 6′-sulfate	G6S-DA

The structural analysis of red algal galactans is similar in many respects to that of other groups of natural polysaccharides, but has some important peculiarities. First of all, the possible presence of both D- and L-galactose derivatives in the same molecule requires determination of the absolute configurations of all the galactose derivatives not only in the native polysaccharide but also in the products of its chemical modifications. The procedures for chemical analyses should take into account the acid lability of 3,6-anhydrogalactose and the rapid hydrolysis of its glycosidic bond. The presence of sulfate groups requires the use of effective desulfation procedures while retaining the carbohydrate moiety of polymers intact. Therefore, the analytical procedures described here emphasize these specific features of the red algal galactans.

2. Isolation of Galactans

Sulfated galactans are readily soluble in water, and therefore, the extraction of algae with water followed by precipitation of polysaccharides with ethanol or 2-propanol is used as the standard procedure for isolation of galactans. Solubilization of highly sulfated fractions proceeds even at room temperature. Most carrageenans are usually extracted at 80–90 °C, but heating up to 120 °C under pressure has been recommended for complete extraction of agar.[115–117] Highly methylated agar fractions may be isolated by preliminary treatment of algal biomass with ethanol–water mixtures.[118]

Treatment of the biomass with dilute acid at low temperature removes the polyvalent cations and facilitates subsequent extraction of galactans with a buffer solution,[119] but care should be taken to prevent the possible degradation of acid-sensitive polysaccharides. Industrial procedures for isolation of agar and carrageenan usually involve a preliminary alkaline treatment of the raw material to remove some extraneous matter and simultaneously to modify the polysaccharides. During this modification, the 4-linked α-galactopyranose 6-sulfate residues are transformed into 3,6-anhydro-α-galactopyranose residues, and the gel-forming ability of galactans is usually increased.[120] It is clear that alkaline treatment should be excluded from the procedures directed to the isolation of unmodified native galactans.

Gel-forming extracts may be purified using the freezing–thawing procedure.[115,117,121] Agar preparations thus obtained may contain floridean starch, which can be removed by amylolysis.[122] Agar is usually a mixture of neutral (agarose) and sulfated fractions. The neutral agarose was first isolated from such a mixture by using a very tedious procedure that included acetylation, fractional solubilization in chloroform (acetylated agarose is soluble in chloroform), and deacetylation.[3,97,115] Later on, such cationic detergents, as cetyltrimethylammonium bromide (Cetavlon, Cetrimide) or cetylpyridinium chloride,[123,124] as well as ion exchangers[125,126] were widely used to separate neutral and sulfated polysaccharides. The content of neutral agarose and the distribution of sulfated molecules according to their charge density can be clearly demonstrated by using anion-exchange chromatography on DEAE-Sephadex.[95,117,127–130]

As in the case of agarose, the establishment of a fractionation procedure for carrageenans[6] was the prerequisite for successful elucidation of their structures. The formation of gels in this class of polysaccharides is cation dependent, and the first essentially regular galactan of the carrageenan group, called κ-carrageenan (11), was isolated from the aggregate of polysaccharides as a gel formed in the presence of potassium cations. Later, the fractionation in potassium chloride solutions became the most common procedure for preparative isolation of carrageenans.[116,117,131–133] It was also suggested as an analytical tool for characterization of complex mixtures of carrageenans.[134–136] Ion-exchange chromatography may also be used for carrageenan fractionation,[137,138] but it should be noted that a high sulfate content results in very firm adsorption of carrageenans on anion exchangers, and separation may be accompanied by considerable irreversible loss of the material, even if concentrated salt solutions are used as eluants.[139] Analytical separation of high-molecular-mass carrageenans according to their charge density may be also effected by using capillary electrophoresis.[140]

Size-exclusion chromatography has been widely used for the determination of molecular weights and molecular-mass distribution in red algal galactans. The corresponding procedures have been described for agarose and its analogues,[141] and for several carrageenans.[142–144] The method is popular in studies on conformational transitions and intermolecular association in galactans,[145,146] but is especially valuable for investigation of partial degradation of polysaccharides by acid or enzymatic hydrolysis (see ref. 18 and Section IV.5 here).

There are several analytical procedures devoted to the determination of sulfated galactans in such complex mixtures as raw materials or food preparations,[147] where carbohydrates of different chemical structures may predominate. The methods specific for red algal polysaccharides are based on color reactions of 3,6-anhydrogalactose in strongly acid media, for example, with resorcinol[148] or with thiobarbituric acid,[149] as well as on the interaction of sulfated polysaccharides with such cationic dyes, as Ruthenium Red[150] or Methylene Blue.[151] It is clear that colorimetric methods are sensitive to the 3,6-anhydrogalactose content and degree of sulfation, and therefore, different galactans should give different responses in these procedures. For example, acid degradation of 3,6-anhydrogalactose depends markedly on the presence of sulfate at C-2, and hence, κ- and ι-carrageenans require different calibrations.

3. Chemical Methods of Structural Analysis

a. Analysis of Composition.—To determine the monosaccharide composition of galactans, it is necessary to perform total acid hydrolysis, for example, by treatment with 1 M sulfuric acid (6 h at 100 °C) or with 2 M trifluoroacetic acid (1–2 h at 121 °C). Under these conditions, all of the glycosidic linkages are split and sulfate and pyruvic acid residues are liberated. Unfortunately, 3,6-anhydrogalactose derivatives are usually degraded completely. The most popular procedure for quantitative determination of monosaccharides in hydrolyzates is gas–liquid chromatography[152] (GLC) of the corresponding alditol acetates[153] or acetylated aldononitriles[154] using *myo*-inositol as the internal standard. GLC is often combined with MS for more-reliable identification of different components of the mixture.[155] GLC can be used also for determination of the absolute configuration of galactose residues by transforming them into glycosides with optically active secondary alcohols (2-butanol[156] or 2-octanol[157]). Similar analyses of 2-octylglycosides[158] or reductive amination products formed in the reaction with optically active 1-amino-2-propanol or α-methylbenzylamine[159] were suggested for determination of the absolute configuration of mono-*O*-methylgalactoses. D-Galactose may be selectively determined by enzymatic procedures, as by using D-galactose dehydrogenase.[160] L-Fucose dehydrogenase was suggested for similar determination of L-galactose.[161]

Even in the earlier period of investigation, it was noted that commercial agar contains, besides D-galactose, small amounts of L-galactose.[162,163] Later on, galactans were found that contain much more L-galactose, up to the D:L proportion of close to unity.[164–166] 6-*O*-Methyl-D-galactose was discovered for the first time in a galactan from *Porphyra capensis*,[167] and then it was detected in many other representatives of the agar group polysaccharides.[168–170] Other mono-*O*-methylgalactoses present in galactans may have either the D- or L-configuration. For instance, 2-*O*-methyl-D-galactose was shown to be the main methylated monosaccharide in polysaccharides from *Aeodes orbitosa*[171] and *Pachymenia carnosa*,[172] whereas small amounts of the L enantiomer were detected in many agar-like polysaccharides. 4-*O*-Methyl-L-galactose was also found in many agars.[169,173,174] It is evident that this monosaccharide cannot occupy the usual place of 4-linked L-galactose derivatives in the backbone and should be a side-chain substituent.[175,176] 4-*O*-Methyl-D-galactose was found in a galactan from *Grateloupia elliptica*.[177] All of the possible four positional isomers of mono-*O*-methylgalactose were detected in a xylogalactan from *Corallina officinalis*,[178] three of them belonging to the L and one (6-*O*-methyl-D-galactose) to the D series.

Small amounts of D-xylose were found in commercial agar[169] and subsequently also in many other galactans, especially in polysaccharides isolated from algae of the order Ceramiales.[179–181] The highest content of D-xylose was registered in sulfated xylogalactans occurring in calcareous algae of the family Corallinaceae.[164–166,178] Traces of an unidentified mono-O-methylpentose were detected in commercial agar;[169] later on, 4-O-methylxylose was unambiguously identified as a galactan component.[182] D-Glucuronic acid was also described as a minor component of agar[183] and of several other galactans.[139,184]

In contrast to agars, carrageenans were considered for a long time to be galactans devoid of L-galactose and methylated galactoses, but having much higher degree of sulfation. In fact, L-galactose should be absent from carrageenans according to their definition. Nevertheless, several minor polysaccharide fractions containing L-galactose were isolated from typical carrageenophytes.[185–190] Moreover, carrageenans containing O-methylgalactoses[191,192] as well as carrageenans practically devoid of sulfate[193] were found in some new sources.

Pyruvic acid was first found in agar[194] and then in several carrageenans.[113,195,196] It may be determined quantitatively in the form of its 2,4-dinitrophenylhydrazone,[197] but an enzymatic procedure with the available lactate dehydrogenase seems to be more convenient.[198]

Quantitative determination of 3,6-anhydrogalactose in the presence of other sugars is based on its higher sensitivity to acids,[149,199–201] the most popular procedure[148] being the color reaction with resorcinol in hydrochloric acid. Unfortunately, these analytical procedures provide no possibility for determining the absolute configuration of the monosaccharide and to distinguish between 3,6-anhydrogalactose and its 2-O-methyl derivative. Therefore, a search for methods capable of splitting the glycosidic bonds of 3,6-anhydrogalactose without degradation of the liberated monosaccharide was conducted during the whole period of investigation of red algal galactans. One of the first such procedures was acid mercaptolysis,[115] namely, treatment of polysaccharides with ethanethiol and concentrated hydrochloric acid in the cold, giving rise to crystalline diethyl dithioacetals of the monosaccharides. Using mercaptolysis, 3,6-anhydro-L-galactose was identified as an agar constituent,[202] whereas the D enantiomer of the same monosaccharide was found in κ-carrageenan[203] and furcellaran.[101] Acid methanolysis[115] was used as another similar approach. Under the conditions of methanolysis, 3,6-anhydrogalactose is transformed mainly into its dimethyl acetal, whereas galactose gives rise to a mixture of methyl galactosides. In spite of generating more complex reaction mixtures, the experimental conditions of methanolysis are much more convenient than those of mercaptolysis, and the former procedure was widely used for isolation and identification of 3,6-anhydrogalactose and 2-O-methyl-3,6-anhydrogalactose dimethyl acetals from a great number of red algal galactans.[204–206] Later on the methanolysis conditions were optimized in order to obtain quantitative results in the determination of monosaccharide components of gelling carrageenans and agarose by HPLC.[207] At the same time, an improved procedure for anhydrous mercaptolysis of agar was also suggested for almost quantitative preparation of 3,6-anhydrogalactose diethyl dithioacetal from red algal galactans.[208]

Glycosidic bonds of 3,6-anhydrogalactose residues are hydrolyzed by acids 100–1000 times faster than those of galactopyranose residues. It was proposed to use this property for stabilization of liberated 3,6-anhydrogalactose residues by employing a two-step hydrolysis with the intermediate reduction[209] or oxidation[210] to obtain finally acid-stable 3,6-anhydrogalactitol or 3,6-anhydrogalactonic acid, respectively. Both approaches were tested for several galactans[100,209,210] but did not find wide application. Oxidative hydrolysis of galactans followed by esterification of the resulting 3,6-anhydrogalactonic acid with optically active 2-butanol and subsequent GLC analysis was suggested as a procedure for determination of the absolute configuration of 3,6-anhydrogalactose in algal polysaccharides.[211]

A keen interest in reductive hydrolysis was displayed again when a reducing agent was found that was stable enough in acid media. It was borane-4-methylmorpholine complex, which gave the possibility of simultaneously effecting hydrolysis of polysaccharides and reduction of the liberated 3,6-anhydrogalactose.[212] Using this reducing agent, two research group independently suggested analytical procedures for complete quantitative determination of the monosaccharide composition of red algal galactans, including separate estimation of 3,6-anhydrogalactose and 3,6-anhydro-2-O-methyl-galactose. One procedure uses the rapid reduction, by borane-4-methylmorpholine, of galactan fragments, having 3,6-anhydrogalactose at the reducing end, which are formed at the first step of the reaction. Subsequent hydrolysis results in degradation of the reductant so that the other monosaccharides are liberated in nonreduced form (Scheme 4). Their transformation into acetylated aldononitriles and simultaneous acetylation of 3,6-anhydrogalactitol (**17**) and 3,6-anhydro-2-O-methyl-galactitol (**18**) gives rise to a mixture that may be analyzed by GLC.[213,214] Another procedure requires transformation of all of the sugars into acetylated alditols, and hence, reduction is performed in two stages.[215] Later on, this procedure was modified in order to use HPAEC without derivatization for estimation of the resulting alditols.[216] The reductive hydrolysis procedure is now generally favored, especially as it gave the possibility to obtain data on the polysaccharide composition of algae by analysis of small samples of biomass prior to the isolation of polysaccharides.[217]

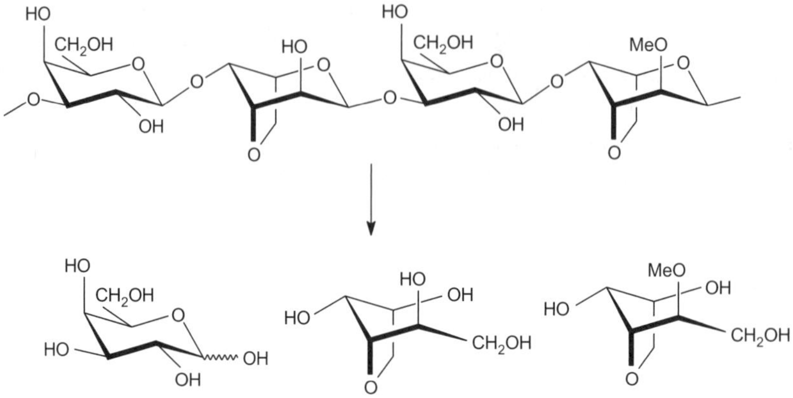

SCHEME 4. Total reductive hydrolysis of a galactan of the agar group containing 3,6-anhydro-L-galactose and its 2-methyl ether (borane-4-methylmorpholine complex in 2 M CF_3CO_2H, 8 h at 100 °C, according to refs.[213,214]

Determination of the sulfate content requires the preliminary hydrolysis of galactans by the action of 1 M HCl at 100 °C for 6 h (hydrolysis in 2 M trifluoroacetic acid can also be used for simultaneous determination of monosaccharides and sulfate). Quantitative estimation of sulfate in hydrolyzates can be performed by many methods known from inorganic analytical chemistry,[218,219] but the most popular procedure in the case of galactans is turbidimetry of stabilized suspensions of $BaSO_4$. It was suggested in the early 1960s[220] and is widely used at present in the slightly modified form.[221]

b. Partial Acid Hydrolysis and Related Procedures.—The higher sensitivity of glycosidic bonds of 3,6-anhydrogalactose to the action of acids provides a useful possibility for effecting the partial hydrolysis of galactans and to use the resulting oligosaccharides for structural analysis. It should be noted, however, that sulfate groups at different positions may have a great influence on the comparative stability of glycosidic linkages. In the autohydrolysis of carrageenans, it was shown that linkages between 3,6-anhydro-α-galactose 2-sulfate and β-galactose 4-sulfate or between α-galactose 2,6-disulfate and β-galactose 4-sulfate are split first, and the next step is a loss of sulfate groups at C-2, whereas 3,6-anhydro-α-D-galactosidic linkages are cleaved with a lower reaction rate.[222–226] Therefore, mild hydrolysis may give rise to rather complex mixtures of fragments where sulfate groups are mainly retained. The hydrolysis products having 3,6-anhydrogalactose residues at the reducing end are not very convenient for subsequent isolation and identification. The procedure used in the earliest investigations[227] was subsequently improved by application of modern chromatographic isolation methods.[228] As a result, the partial acid hydrolysis was successfully applied for preparation of agarobiose, agarotetraose, carrabiose 4'-sulfate, and carratetraose 4',4'''-disulfate for mass-spectrometric analysis.[229] Odd-numbered oligosaccharides containing galactose residues at both reducing and nonreducing ends can be also prepared by partial acid hydrolysis of agarose.[229,230]

To obtain more stable products, it was suggested to reduce the oligosaccharides before separation,[209] but this two-step procedure did not find wide application, neither did partial oxidative hydrolysis[231] and transformation of oligosaccharides into phenylosotriazoles[232] or reductive amination products with 2-aminopyridine.[233] Partial mercaptolysis, affording crystalline diethyl dithioacetals of neutral disaccharides (agarobiose or carrabiose) as the result of simultaneous splitting of polysaccharide chains and desulfation, was successfully used during the earliest stage of structural analysis of agar,[6,115] κ-carrageenan,[108] and furcellaran.[101] Partial methanolysis, which is also accompanied by effective desulfation, was used even more frequently.[115] It was applied to typical agars[234] and carrageenans[235] as well as to some more complex polysaccharides.[236,237] In particular, it was found possible to use partial methanolysis for elucidation of the site of pyruvic acid residues in galactan backbones.[238,239] The conditions of partial methanolysis giving rise to maximum yields of agarobiose and carrabiose dimethyl acetals were carefully determined and used to elucidate the degree of regularity of polysaccharide molecules, namely, the length of blocks built up of contiguous residues of these disaccharides.[240,241] The different chromatographic mobilities of agarobiose and carrabiose dimethyl acetals made it possible to determine the absolute configuration of 3,6-anhydrogalactose residues in polysaccharides by using chromatograms of their partial methanolysis products.[207,242]

Partial reductive hydrolysis turned out to be an even more convenient approach for the partial fragmentation of galactans.[214,243] Mild hydrolytic conditions were selected, wherein practically only the 3,6-anhydrogalactosidic linkages are split, and sulfate groups are mostly retained. Simultaneous reduction with borane-4-methylmorpholine gives rise to stable reaction products that are suitable for effective separation and identification (Scheme 5). Agarobiitol (**19**) and carrabiitol (**20**) can be readily distinguished by using GLC of their acetates, thus providing chromatographic information about the absolute configuration of the parent 3,6-anhydrogalactose.[214] This evidence may be obtained after hydrolysis of algal biomass prior to the isolation of polysaccharides, and in this particular case, a preliminary alkaline treatment of biomass may be useful for obtaining higher yields of agarobiitol and carrabiitol.[217]

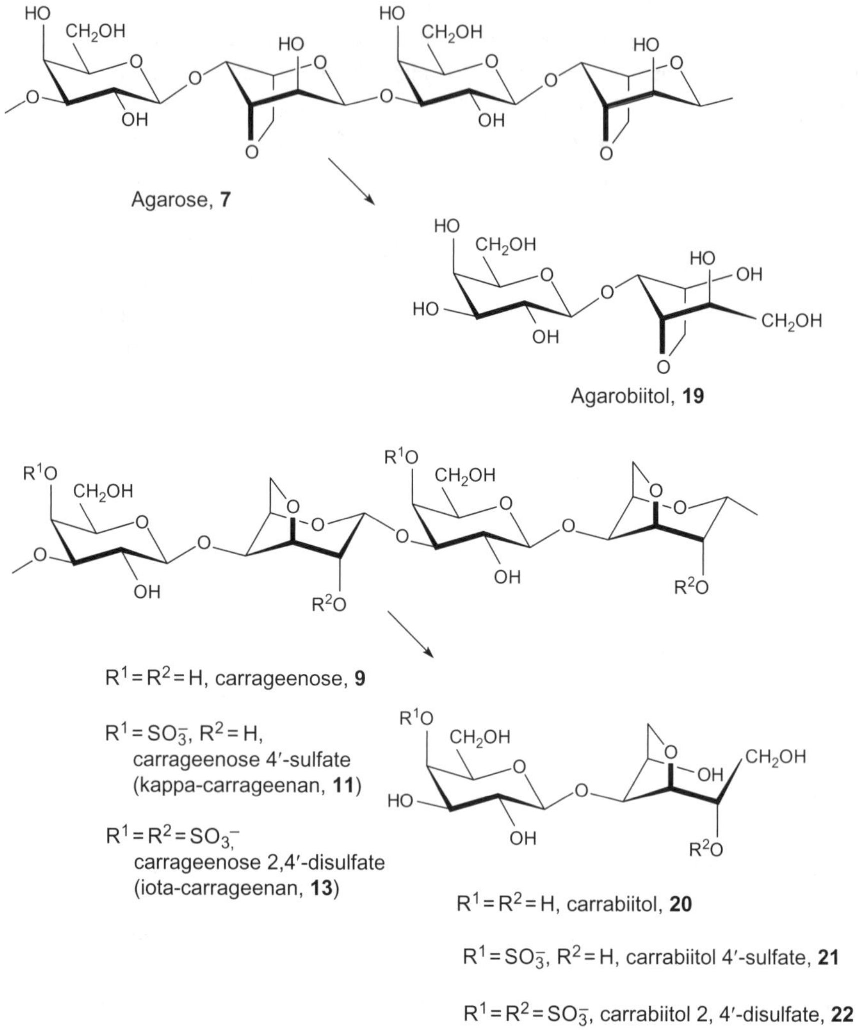

SCHEME 5. Partial reductive hydrolysis of agarose and 3,6-anhydrogalactose-containing carrageenans (borane-4-methylmorpholine complex in 0.5 M CF$_3$CO$_2$H, 8 h at 65 °C, according to refs. 214 and 213).

If sulfated oligosaccharides, such as the carrabiitol derivatives **21** and **22**, are formed as the result of partial reductive hydrolysis directed to determination of the absolute configuration of 3,6-anhydrogalactose, they may be desulfated and acetylated simultaneously.[244] Naturally methylated agars give rise to methylated disaccharide alditols.

Thus, agarobiitol, its 2-O-methyl-, 6'-O-methyl-, and 2,6'-di-O-methyl derivatives were obtained after reductive hydrolysis of a polysaccharide isolated from *Gelidiella acerosa*. These fragments were resolved as acetates by GLC, and MS was used to confirm their structures.[245] Mild reductive hydrolysis conditions made it possible to retain pyruvic acid residues and sulfate groups in the resulting fragments, and to use the charged oligosaccharides for characterization by capillary electrophoresis, mass spectrometry, and NMR spectroscopy.[246,247] It should be noted that the stability of several linkages depends on the position of some sulfate groups: thus, θ-carrageenan [G2S-DA2S]$_n$ requires more prolonged hydrolysis time, evidently due to the presence of a 2-sulfate group in the 3,6-anhydro-α-D-galactose residue, whereas 2-sulfate in the neighboring β-D-galactose residue is lost under the partial hydrolysis conditions.[247] Higher oligosaccharides, which arise from those parts of the polysaccharide molecule deviating from regularity, are especially important for the structural analysis. A good example is the isolation of two reduced sulfated tetrasaccharides (**25** and **26**) and one reduced branched pentasaccharide with a side β-D-xylopyranose residue (**27**), together with agarobiitol (**19**), its 2-methyl ether **23** and 2'-sulfate **24**, from the partial reductive hydrolysis products of a complex galactan from *Laurencia nipponica*[181] (Scheme 6).

Partial depolymerization also may be used for polysaccharides devoid of 3,6-anhydrogalactose. Highly sulfated galactans are usually treated under acetolysis conditions with a mixture of sulfuric acid, acetic acid, and acetic anhydride, wherein the desulfation rate prevails considerably over the rate of glycosidic bond splitting. Acetylated oligosaccharides devoid of sulfate, which are the products of partial acetolysis, are deacetylated and then separated by chromatographic methods. Such an approach was used in the structural analysis of λ-carrageenan,[248] but unfortunately in this early work a wrong structure with two contiguous (1→3)-linkages between α-D-galactopyranose residues was ascribed to the isolated trisaccharide. Several years later, during a revision of these data it was shown that (1→4)-linkages are split by acetolysis slightly more readily than (1→3)-linkages. It was therefore possible to isolate two disaccharides and two trisaccharides, corresponding to the alternation of two these linkages along the backbone of λ-carrageenan, but only one tetrasaccharide containing two more-stable and one more labile bonds was obtained.[249] Later on, a similar approach was used for the investigation of other galactans.[250–252] In particular, it helped to demonstrate the presence of deviation from the usual alternating sequence of D- and L-galactose derivatives in several polysaccharides. For example,[253] after partial acetolysis of a galactan from *Grateloupia divaricata*, that contained D- and L-galactose in the ratio of about 4:1 (Scheme 7), disaccharides **28** and **29** and trisaccharides **30** and **31** were obtained,

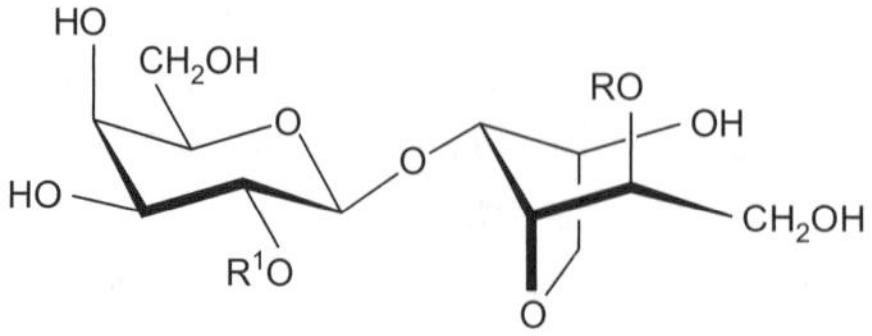

R = R¹ = H, agarobiitol, **19**

R = Me, R¹ = H, 2-O-methyl-agarobiitol, **23**

R = H, R¹ = SO$_3^-$, agarobiitol 2'-sulfate, **24**

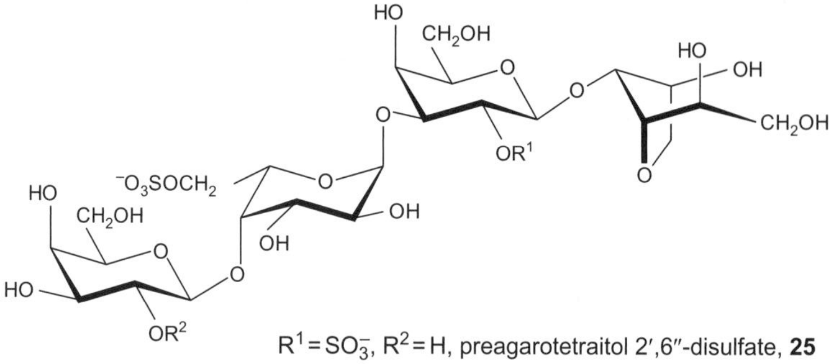

R¹ = SO$_3^-$, R² = H, preagarotetraitol 2',6''-disulfate, **25**

R¹ = H, R² = SO$_3^-$, preagarotetraitol 2''',6''-disulfate, **26**

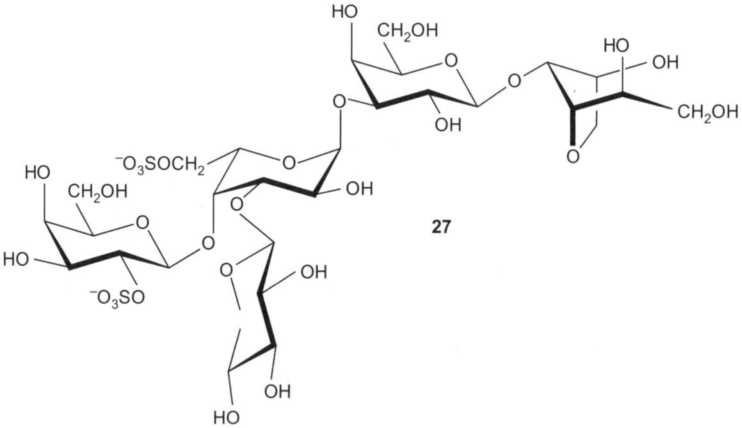

27

SCHEME 6. Reduced oligosaccharides obtained after partial reductive hydrolysis of a sulfated galactan from *Laurencia nipponica*[181] (the term "preagarotetraitol" is used for a biogenetic precursor of agarotetraitol containing L6S instead of LA).

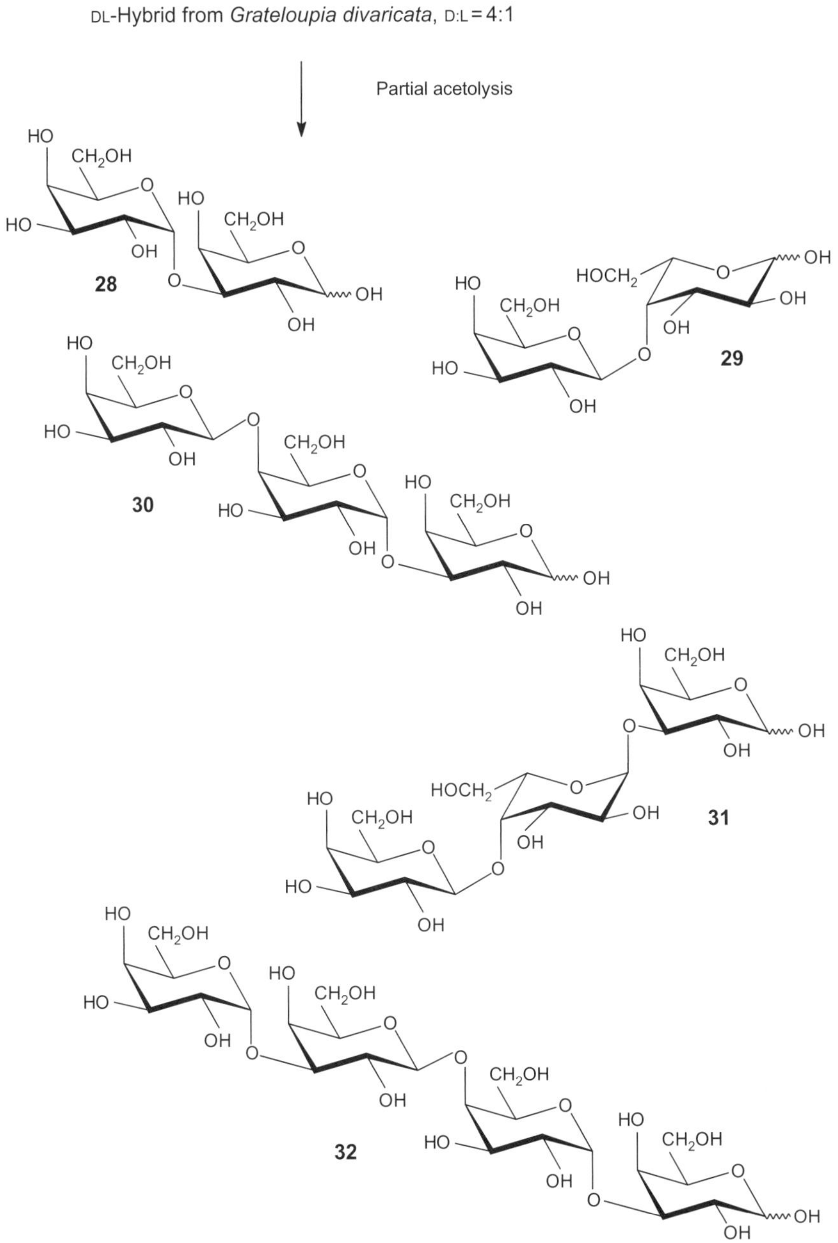

SCHEME 7. Oligosaccharides obtained after partial acetolysis of a complex sulfated galactan from *Grateloupia divaricata*.[253]

which were structurally related to either agaran or carrageenan. Unfortunately, the only tetrasaccharide (**32**) isolated was built up of D-galactose residues, and hence, the presence of an agar–carrageenan hybrid structure, strictly speaking, remains to be proved in this case.

Hydrolysis with aqueous acids is also suitable for the partial fragmentation of galactans devoid of 3,6-anhydrogalactose.[172,184,254,255] The stability of glycosidic bonds under these conditions may differ from their stability under acetolysis,[249] and so the application of both procedures was recommended for obtaining the complementary sets of fragments. The stability of sulfate groups in aqueous acids is comparable to that of glycosidic bonds, and therefore, partial hydrolysis may lead to sulfated monosaccharides, but the yields are usually small.[256] Isolation and identification of sulfated monosaccharides have been sometimes used to localize sulfate groups in polysaccharides.[257–261] Primary, secondary axial, and secondary equatorial sulfate groupings may be distinguished according to their stability to aqueous acids (the proportion of half-hydrolysis periods is approximately 2:1:0.1, respectively),[262] but determination of the position of sulfate groups based on their hydrolysis rates did not find wide application.

Partial depolymerization of galactans by the action of specific enzymes is described later in Section IV.5.

c. Desulfation.—In many cases, elucidation of galactan structures requires desulfation (see ref. 263). Sulfate groups may be split by acid hydrolysis, but in most cases, the process is accompanied by considerable degradation of the carbohydrate chains.[264] Treatment with a solution of HCl in absolute methanol, resulting in the formation of methyl sulfate (MeOSO$_3$H) by acid-catalyzed transesterification was suggested for sulfated glycosaminoglycans, which are not very amenable to acid depolymerization.[265] Sometimes, the procedure was used for galactans having low 3,6-anhydrogalactose content,[109,266] but marked degradation of the polysaccharides was observed because of simultaneous methanolysis of glycosidic bonds. This method is useless for more acid-labile polysaccharides, which should be treated under the so-called solvolytic desulfation conditions, namely, the heating of a solution of a pyridinium salt of galactan in dimethyl sulfoxide[267,268] (Scheme 8A), often in the presence of small amounts of water or methanol.[269,270] Solvolytic desulfation is probably based on the reversibility of the well-known sulfation of alcohols with such complexes as SO$_3$·Me$_2$SO or SO$_3$·C$_5$H$_5$N, leading to acid sulfates. Recently, a variant of the solvolytic procedure was suggested, wherein a sodium salt of a polysaccharide is heated in dimethyl sulfoxide in the presence of organic acid and a substance capable of binding tightly the liberated sulfuric anhydride (e.g., As$_2$O$_3$).[271] Solvolytic desulfation is greatly accelerated in a microwave oven.[272] It should be kept

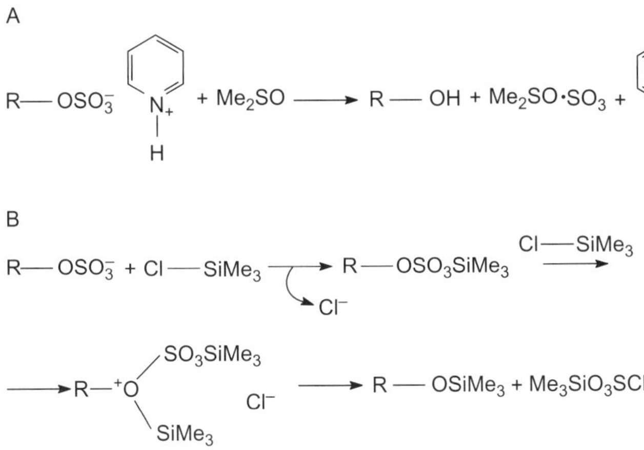

A

R—OSO$_3^-$![pyridinium] + Me$_2$SO ⟶ R—OH + Me$_2$SO·SO$_3$ + ![pyridine]

B

R—OSO$_3^-$ + Cl—SiMe$_3$ ⟶ R—OSO$_3$SiMe$_3$ $\xrightarrow{\text{Cl—SiMe}_3}$
                              ↖ Cl$^-$

⟶ R—$\overset{+}{\text{O}}\underset{\text{SiMe}_3}{\overset{\text{SO}_3\text{SiMe}_3}{\diagup}}$ Cl$^-$ ⟶ R—OSiMe$_3$ + Me$_3$SiO$_3$SCl

SCHEME 8. Proposed mechanisms of desulfation by solvolysis of pyridinium salts of sulfates in dimethyl sulfoxide[267] (A) and by treatment of sulfates with chlorotrimethylsilane[282] (B).

in mind that the mechanism of solvolysis is not fully understood, and examples of deep depolymerization under solvolytic desulfation conditions have been described, even for several acid-stable polysaccharides.[273] Nevertheless, the method is widely used in the structural analysis of red algal galactans, although partial depolymerization should result in inevitable partial loss of the structural information on the acid-labile part of the molecule.[274]

Desulfation may also be carried out by the action of silylating reagents in pyridine.[275] Comparison of desulfation of the model compounds methyl α-D-galactopyranoside 3- and 6-sulfate under solvolytic and silylating conditions made it possible to classify the silylating reagents into three groups. Some of them, such as N-(trimethylsilyl)imidazole (TSIM), were inactive and blocked solvolytic desulfation. Others, for instance, N,O-bis(trimethylsilyl)acetamide (BTSA) and N,O-bis(trimethylsilyl)trifluoroacetamide (BTSTFA), were selective toward desulfation of primary sulfates, whereas members of the third group, for example, chlorotrimethylsilane (CTMS), were nonselective desulfating agents.[276] Partial regioselective 6-desulfation with BTSA was successfully applied to sulfated galactans from *Porphyra yezoensis*,[277] *Gloiopeltis complanata*,[278] and *Joculator maximus*,[165] whereas the best results with heparin were obtained by using N-methyl-N-(trimethylsilyl)trifluoroacetamide (MTSTFA).[279] Treatment with the nonspecific CTMS was used for complete desulfation of a polysaccharide from *Georgiella confluens*, where other procedures were ineffective[280] as well as for desulfation of galactans from *Pterocladiella capillacea*.[281] Conditions of desulfation by CTMS were carefully investigated in the transformation of κ-carrageenan [D4S-DA]$_n$ into "carrageenose" ([D-DA]$_n$, also known as β-carrageenan.[282] The best reaction conditions, giving the highest decrease of sulfate content with lowest degradation and maximum product recovery, were established (heating of the pyridinium salt at 100 °C during 3 hours, and using CTMS–sulfate molar ratio of 50:1), and all signals in the NMR spectra of carrageenose were assigned. Unfortunately, the desulfation procedure was accompanied by marked degradation of the polysaccharide. A reaction mechanism was proposed leading to the trimethylsilyl ether of chlorosulfonic acid (Scheme 8B), but its presence in the reaction mixtures was not confirmed.

Specific enzymatic hydrolysis of sulfate would appear very attractive, but no sulfatases capable of splitting sulfate from polysaccharides by a hydrolytic mechanism have yet been found.

An alternative route for preparation of desulfated polysaccharides is alkaline treatment. Sulfate groups are usually regarded as stable to alkaline hydrolysis, but may be split, by analogy with the corresponding sulfonic esters,[283] in the course of an elimination reaction, which requires suitable stereochemical conditions for intramolecular nucleophilic substitution[256] (Scheme 9). According to the stereochemistry of galactose, sulfate groups are stable at any position of a 3-linked galactose residue, whereas 4-linked galactose 6-sulfate (**33**) is readily transformed into 4-linked 3,6-anhydrogalactose (**34**) by moderate heating of a galactan in 1 M NaOH, usually in the presence of some NaBH$_4$ to prevent oxidation and "peeling" of the molecules from the reducing end.[94] As already mentioned, this process is similar to the enzymatic formation of 3,6-anhydrogalactose residues[93,284–286] (the term "sulfohydrolase," often used in the literature for the corresponding enzyme, is incorrect, since the release of sulfate proceeds by elimination and not by hydrolysis) and is widely used in industry to improve the gel-forming properties of natural galactans during their extraction from the algae.[287] An interesting eco-friendly alternative to alkali treatment may be a light-deprived regime for cultured seaweeds, resulting in an increase of 3,6-anhydrogalactose content in both agarophytes[288] and carrageenophytes[289] with concomitant decrease of the amount of contaminating floridean starch.[290]

Alkaline elimination of sulfate was carefully investigated using carrageenans,[291–294] porphyrans,[295] and more-complex DL-hybrids[296,297] as models. It was stated that the rate of cyclization of 4-linked α-galactose 6-sulfate residue is mainly determined by the position of sulfate in the preceding 3-linked β-D-galactose residue (A-unit), whereas the presence of sulfate or a methyl group at position 2 of the reacting residue has minor or no effect. Sulfate at C-4 increases and sulfate at C-2 decreases the cyclization rate, as compared to an A-unit devoid of sulfate.

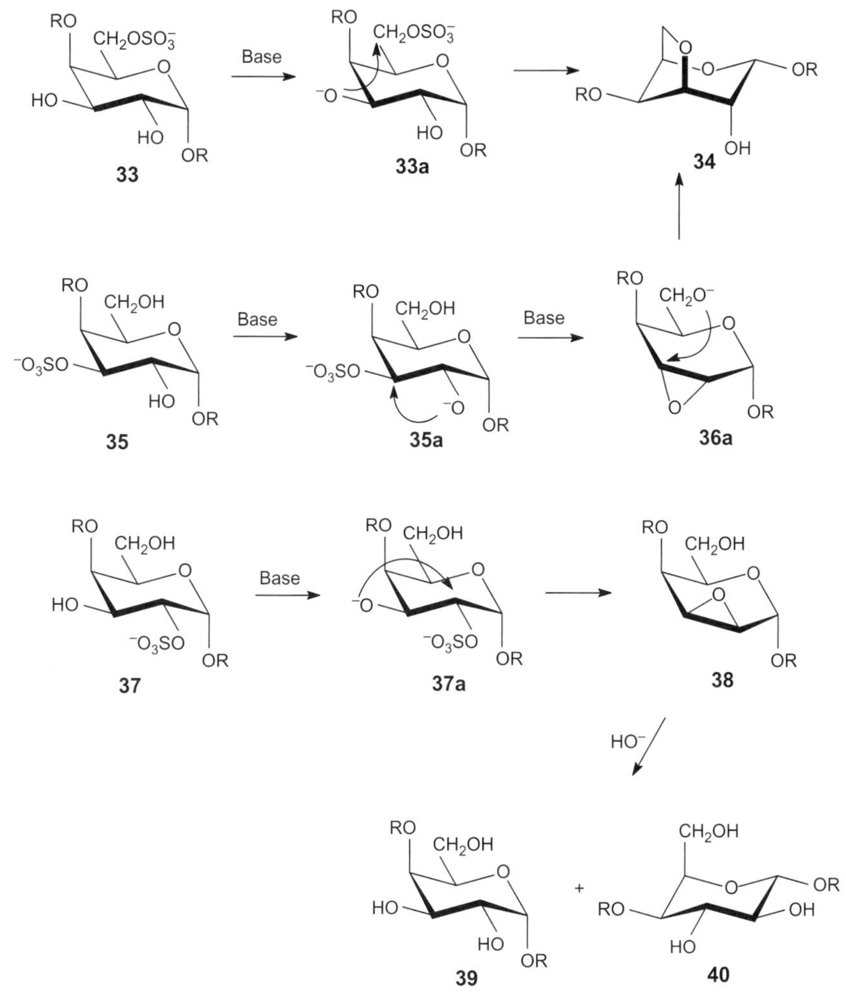

SCHEME 9. Possible routes of intramolecular desulfation of isomeric (1→4)-linked D-galactose sulfates under basic conditions.

In accordance to this observation, λ-carrageenan [G2S-D2S,6S]$_n$ showed the lowest cyclization rate, and partially cyclized μ/ν-carrageenans (precursors of κ/ι-carrageenans, both containing G4S)—the highest ones, whereas porphyrans [G-L6S]$_n$ had intermediate rate-constants. Interestingly, single μ-carrabiose (G4S-D6S) units do not differ from short blocks of such units in the conversion rate.[298] Alkaline modification is greatly accelerated in a microwave oven.[299]

Alkaline modification of galactans has been successfully used many times to differentiate between alkali-labile and alkali-stable sulfate groups. A special procedure was suggested that made it possible to determine the absolute configuration of cyclizable α-galactose 6-sulfate units by a combination of alkaline treatment of a polysaccharide with subsequent partial acid hydrolysis, reductive amination with chiral (S)-α-methylbenzylamine of the resulting oligosaccharides having 3,6-anhydrogalactose residues at the reducing end, and total hydrolysis.[300] The diastereoisomeric derivatives of 3,6-anhydrogalactose or its 2-methyl ether thus formed are then separated and quantified as acetates, using GLC. Interestingly, this procedure disclosed a small amount of the unexpected α-D-galactose 6-sulfate in a sample of porphyran (**10**).[300]

Theoretically, sulfate groups at positions 3 (**35**) and 2 (**37**) of 4-linked galactose may also be split in the presence of bases by intramolecular reactions of the corresponding anions **35a** and **37a** (probably under more drastic conditions) to form 2,3-anhydro sugars of different stereochemistry due to the obligatory configurational inversion at carbon atoms bearing sulfate groups[256] (Scheme 9). Moreover, subsequent intramolecular reaction of anion **36a**, formed from a 2,3-anhydro sugar having the *gulo* configuration, should give the 3,6-anhydrogalactose derivative **34**, whereas intermolecular opening of 2,3-anhydro sugar **38** having the opposite *talo* configuration by hydroxide may give both galactose (**39**) and idose (**40**). Alkaline elimination of secondary sulfate groups was observed for 4-linked L-iduronic acid 2-sulfate residues in heparin and heparan sulfate[301] and for 4-linked xylose 2-sulfate residues in a polysaccharide isolated from the green seaweed.[302] Red algal galactans having sulfate at positions 2 and 3 of 4-linked galactose residues were found either in agar or in carrageenan groups, but the possibility of their desulfation under alkaline conditions remains to be investigated.

d. Methylation Analysis.—This is a classical approach for elucidating the positions of substituents in carbohydrate chains. It was applied to agar at the earliest stage of investigation of its structure.[303,304] By using this method, it was demonstrated, and then repeatedly confirmed,[4,97] that D-galactose residues in agarose are linked through position 3, whereas 3,6-anhydro-L-galactose residues are linked through position 4. Similarly in the earliest works, it was shown that methylation of non-fractionated carrageenan followed by hydrolysis afforded mainly 2,6-di-*O*-methyl-D-galactose.[305]

Methylation of polysaccharides is carried out in the presence of strong bases. Sulfate groups, depending on their position, may be split to give anhydro sugars or remain intact (see above, Section IV.3.c). Splitting is possible, but only if the spatial position of sulfate is favorable for intramolecular attack by alkoxide according to the bimolecular nucleophilic substitution mechanism. Examples of alkali-labile sulfate are 4-linked galactose 6-sulfate residues having position 3 unsubstituted as well as 4-linked galactose 3-sulfate residues having free positions 2 and 6 (both residues are transformed into 3,6-anhydrogalactose after alkaline elimination of sulfate)[256] (Scheme 9). The presence of such structural components should be detected before methylation using alkaline modification of a galactan.[94,291,293,306–308] Sulfate groups in other positions in galactans are usually regarded as stable under the methylation conditions. This feature results in ambiguity in the interpretation of the methylation data, since the positions of inter-residue linkages cannot be distinguished from the positions of sulfate groups in the set of partially methylated monosaccharides formed after hydrolysis of a methylated galactan. To solve this problem, the results of methylation of the native galactan and of its desulfated derivative should be compared.[109,309,310] An alternative approach uses data on the position of sulfate obtained by physicochemical methods of analysis[112] (see later).

Several natural red algal galactans, such as porphyran (**10**), have high content of O-methyl groups. Since it is not possible to split these groups without destruction of the polysaccharide molecules, comparison of polymeric products of total methylation of agarose and porphyran was once used to prove their structural similarity (porphyran was preliminary alkali modified).[99] To localize the natural O-methyl groups, trideuteriomethylation or ethylation may be used instead of methylation.[311]

In spite of its considerable age, methylation analysis retains completely its importance and is used practically in every new work devoted to structural investigation of galactans. The experimental technique of the method has been improved many times[312] and requires several milligrams of polysaccharide. The methylation reaction is carried out in dimethyl sulfoxide by treatment with methyl iodide in the presence of methylsulfinyl carbanion $CH_3SOCH_2^-$ (the Hakomori procedure[313]) or powdered sodium hydroxide,[314,315] the good solubility of the starting material in dimethyl sulfoxide being a prerequisite of effective substitution. To improve the solubility of highly sulfated polysaccharides in Me_2SO, it was suggested to transform their sodium salts into triethylammonium salts.[215] It is advisable to split methylated polysaccharides containing 3,6-anhydrogalactose residues under the conditions of reductive hydrolysis.[215,244] The composition of the resulting mixtures of partially methylated monosaccharide derivatives is usually analyzed in the form of acetylated alditols by GLC and GLC–MS.[155] In discussing the quantitative interpretation of the data obtained, it should be kept in mind that a part of the sulfate content may sometimes be split off by the alternative mechanism of alkaline hydrolysis followed by methylation of hydroxyl groups liberated.[166,307,308] It was suggested that some unidentified reducing impurities may catalyze desulfation in the presence of alkali,[316] but the conditions of this undesirable side reaction and sensitivity to alkaline hydrolysis of sulfate groups occupying different positions in galactans need to be additionally investigated further.

e. Periodate Oxidation.—The ideal agarose (**7**) and carrageenose (**9**) have no diol groupings capable of reacting with periodate (excepting end residues). Therefore, Smith degradation (periodate oxidation followed by reduction and controlled mild acid hydrolysis) of these polysaccharides was used only to eliminate residues deviating from the ideal formulas (in particular, to destroy some 4-linked D-galactose or its 6-sulfate present in κ- and ι-carrageenans) and to obtain the polymeric fragments of more-regular structure for investigation of their physicochemical properties.[317]

In contrast, in the molecules of agaran (**6**) and carrageenan (**8**), every second monosaccharide residue should be oxidized, and the method may be used as an illustration of the regular structure of the carbohydrate chain. In fact, the expected 2-O-β-D-galactopyranosyl-D-threitol was isolated after Smith degradation of a desulfated carrageenan from *Tichocarpus crinitus* having low 3,6-anhydrogalactose content.[318] A similar result was obtained after oxidation of a desulfated polysaccharide from *G. divaricata*,[319] but more-careful analysis of the Smith degradation products showed that two closely related compounds are formed, probably, D- and L-threitol galactosides (Usov and Barbakadze, unpublished), since both D- and L-galactose in the ratio of 4:1 are present in the starting polysaccharide. The yields of galactosyl-threitol from both polysaccharides were markedly lower than those expected, probably because of incomplete oxidation as the result of intramolecular acetal formation in partially oxidized molecules. Such a process is well known for many polysaccharides,[320] but was not investigated in detail in the case of red algal galactans.

Sulfate groups are usually considered as completely stable under conditions of periodate oxidation (splitting of sulfate, which was detected during the oxidation of free monosaccharide sulfates, may be explained as the result of overoxidation of intermediate hydroxymalondialdehyde sulfate[321,322]). Hence, sulfate groups at positions 2 or 3 in 4-linked galactopyranose residues should protect the backbone from the action of periodate. This fact was successfully used for elimination of terminal D-xylopyranose residues from the molecules of sulfated xylogalactans isolated from calcareous algae of the family Corallinaceae.[166,323] At the same time, it is possible that the stability of sulfate in partially oxidized polysaccharide derivatives may be lower than that in the starting compounds, and the rather unusual results of periodate oxidation of several complex galactans[324] may probably be explained by partial splitting of sulfate in the course of the reaction.

4. Physicochemical Methods of Structural Analysis

Several physicochemical methods of investigation, such as measurement of optical rotation (including optical rotation dispersion and circular dichroism) rheological methods, and X-ray analysis of fibers and films, have played a very important role for determination of conformations of galactan molecules and characterization of their supramolecular structures.[8,19] As mentioned in Section I, only physicochemical methods used for determination of primary structures of polysaccharides are considered in this chapter.

a. Vibrational (Infrared and Raman) Spectroscopy.—Infrared (IR) spectra provide valuable information about the structure of carbohydrates.[325-327] In the field of red algal galactans, they were used mainly to distinguish the polysaccharides of the agar group from carrageenans and to determine the position of sulfate groups. The method is very attractive, since only a few milligrams of the substance are necessary to record the spectrum. The solubility of a sample is of little importance, and hence, it is possible to analyze either the isolated polysaccharide or the starting algal biomass. The well-known difficulties in unambiguous assignment of absorption bands to the definite types of molecular vibration should be mentioned as a shortcoming of the method.

In the early 1960s, it was stated that sulfate groups in polysaccharides may be detected by an intense absorption band at about 1240–1250 cm^{-1}. In addition, IR spectra of sulfated carbohydrates display less-intense absorption bands in the region of 800–850 cm^{-1}, the exact position of which depends on the nature of the sulfate group environment: primary, secondary equatorial, and secondary axial sulfates absorb at 820, 830, and 850 cm^{-1}, respectively.[328,329] Later, it was shown[330] that the formally axial sulfate at position 2 of a 3,6-anhydrogalactose residue has a specific absorption band at 805 cm^{-1}. IR spectra therefore provide the possibility of localizing sulfate groups in a galactan molecule. They were often used for this purpose, especially for characterization of κ-carrageenans (containing an axial sulfate at C-4 of D-galactose residues only), ι-carrageenans (containing an additional sulfate at position 2 of 3,6-anhydrogalactose), and their hybrids and biogenetic precursors (containing primary sulfate groups).[112,242]

Attribution of a galactan to the agar or carrageenan group using IR spectra seems to be more complicated problem. It was first suggested to use the proportion of intensities of absorption bands at 895–900 and 930–940 cm^{-1} for this purpose.[242] The latter band was connected with the presence of 3,6-anhydrogalactose residues in a polysaccharide molecule. The nature of the former band was not determined, but it was more intense in authentic agars, as compared with carrageenans. Later on, it was shown that the absorption band at 930–940 cm^{-1} may be caused not only by 3,6-anhydrogalactose but also by galactose 4-sulfate residues.[331] The latter authors suggested use of the proportion of intensities of sulfate absorption bands and the absorption band of C–H vibration at 1920 cm^{-1} as a quantitative measure of total sulfate and of content of sulfate groups occupying different positions in a galactan molecule.

The recent development of Fourier-transform infrared (FTIR) spectroscopy has led to great advances in application of the method to the structural analysis of polysaccharides. Thus, the second derivative mode in the FTIR spectra was shown to provide more information by increasing the resolution of bands as compared to the usual absorption curve. It enabled the unambiguous identification of low-intensity absorption bands and to determine more correctly their spectral positions. By this technique, the problem of attribution of a galactan to the agar or carrageenan group according to its IR spectrum was reinvestigated. It was shown[332] that agars have characteristic absorption bands at 790 and 713 cm^{-1}, whereas carrageenans demonstrate specific absorption at 1160, 1140, 1100, 1070, 1040, 1008, 610, and 580 cm^{-1}. FTIR spectra were used to localize carrageenans in algal tissues[333–335] and to identify carrageenans rapidly in dry powdered biomass.[336] In the latter case, diffuse reflectance spectroscopy was suggested instead of the usual transmittance spectroscopy. FTIR spectra were also used for identification of carrageenans in food compositions[337] and for quantitative determination of κ-, ι-, and λ-carrageenans in blends by special calibration methods,[338–340] sometimes in combination with ^{13}C NMR spectroscopy.[341]

IR spectral information may be effectively supplemented by analysis of Raman spectra of polysaccharides,[342,343] which register the same molecular vibrations, but with a different selection mode. Usually, the Raman spectra are more sensitive to skeletal vibrations as compared to side-group (for example sulfate) vibrations. Detailed assignment of Raman spectra of carrageenans has been made,[344,345] and the method was applied to characterization of phycocolloids obtained from several red algae of the Portuguese coast.[346,347] In contrast to IR, Raman spectroscopy can be implemented directly on the production line by using optical fibers, thus enabling fast monitoring of the carrageenan content of industrial batches. A corresponding quantitative method for measuring the content of five different carrageenans (ι, κ, λ, μ, and ν) in powder samples was developed, which was based mainly on specific treatment of Raman spectral data.[348]

b. Mass Spectrometry.—Mass spectra are widely used in carbohydrate chemistry to obtain information on the molecular mass and the structure of different derivatives of mono- and oligosaccharides.[349–351] Their application for the determination of number and position of methyl groups in partially methylated alditol or aldononitrile acetates has been mentioned.[155]

In a paper devoted to a study of the behavior of 3,6-anhydrogalactose derivatives under electron-impact (EI) mass-spectrometric conditions, the spectra of many compounds commonly obtained as the result of splitting of red algal galactans (methyl glycosides, dimethyl acetals, diethyl dithioacetals, alditols, including the corresponding derivatives of carrabiose, namely, 3,6-anhydro-4-O-β-D-galactopyranosyl-D-galactose) were analyzed. To obtain volatile derivatives, these compounds were transformed into their trimethylsilyl ethers or peracetates. It was shown that the presence of a 3,6-anhydro ring results in the appearance of characteristic peaks in mass spectra of these compounds, and this fact can be used for their unambiguous mass-spectrometric identification.[352]

In the field of natural red algal galactans, MS was used for the first time in the structural analysis of odonthalan.[353] Galactose and 6-O-methylgalactose, the hydrolysis products of the polysaccharide, were separated by column chromatography and identified by MS in the form of their di-O-isopropylidene derivatives. Cyclic derivatives of this type had been shown previously to be especially suitable for mass-spectrometric characterization of the structure and (in certain instances) stereochemistry of monosaccharides.[349,354] A combination of GLC and MS was also used for identification of methyl 4,6-O-(1-methoxycarbonylethylidene)-D-galactosides (as their trifluoroacetates) in the methanolysis products of π-carrageenan from *Petrocelis middendorfii*.[355]

Subsequently, EI-MS was used repeatedly for characterization of disaccharides obtained after splitting of 3,6-anhydrogalactosidic bonds in galactan molecules. When these disaccharides contained methylated sugars, O-methyl groups can be localized according to the mass-spectral data. For example, dimethyl acetals of agarobiose and 6′-O-methylagarobiose were identified in the form of acetates in the partial methanolysis products of odonthalan.[100] The presence of agarobiose and 2-O-methylagarobiose was similarly detected in the partial methanolysis products of a galactan isolated from *Rhodomela larix*.[356] Mass spectra were also used for elucidation of structures of partially methylated and reduced disaccharides obtained after partial reductive hydrolysis of methylated galactans.[244] The method was applied to the investigation of oligogalactosides having degree of polymerization 2–4, which were obtained as acetolysis products of a galactan from *G. divaricata*[253] (Scheme 7). These oligosaccharides were transformed into acetylated N-p-tolylglycosylamines, which gave intense peaks of molecular ions as well as peaks of fragments resulting from the consecutive splitting of monosaccharide residues from both the reducing and nonreducing ends of molecules.

Less-volatile higher or sulfated oligosaccharides may be analyzed effectively by using fast-atom bombardment MS (FAB-MS).[351] In the case of sulfated oligosaccharides, mass spectra give the possibility for simultaneously determining the number of monosaccharide residues and sulfate groups. FAB-MS was effectively used in the structural analysis of the partial reductive hydrolysis products derived from galactans isolated from *L. nipponica*[181] (Scheme 6) and *L. coronopus*.[357]

Modern soft-ionization methods, such as electrospray ionization (ESI-MS)[358] and matrix-assisted laser desorption/ionization time-of-flight MS (MALDI-TOF MS),[359–361] are especially attractive for determination of molecular mass and structure of underivatized sulfated oligosaccharides,[362] but their use may be complicated due to the tendency of such compounds to lose sulfate in MALDI-TOF MS and to form clusters and multiply charged ions in ESI-MS. The latter method was applied to characterization of neocarrabiose-type oligosaccharide fragments (from di- to dodeca-saccharides) enzymatically produced from κ-carrageenan and isolated by size-exclusion chromatography.[363] The negative-ion mode was used for registration of the spectra, and the structure of the major ions was confirmed by collision-induced dissociation (CID) mass spectra. ESI-MS was also used to characterize sulfated and pyruvated disaccharide alditols obtained by reductive hydrolysis of a complex galactan.[246] The enzymatic digests of κ-carrageenans (up to tetratriacontasaccharide) were separated by anion-exchange chromatography and characterized off-line with ESI-MS in the positive-ion mode.[364] Later on, it was possible to use on-line liquid chromatography coupled with ESI-MS in the negative-ion mode to analyze complex mixtures of sulfated oligosaccharides,[365,366] whereas high-sensitivity negative-ion electrospray tandem MS was suggested for sequence determination of sulfated carrageenan fragments.[367]

MALDI-TOF mass spectra of neutral oligosaccharides, such as agarose fragments,[368] were used without problem to determine the molecular size, whereas the behavior of sulfated oligosaccharides is highly dependent on the nature of the matrix, conditions of ionization, and mode of registration of mass spectra, where it is possible to observe positive or negative ions in linear, reflectron, or post-source decay (PSD) mode. The mass spectra of model sulfated carrageenan fragments under MALDI-TOF conditions were carefully analyzed.[369] Application of different matrices for MALDI-TOF MS has been compared, and several new compounds, such as *nor*-harmane[370] and pyrenemethylguanidine (pmg),[371] were recommended as supplements to classical MALDI-TOF matrices. The latter compound gave especially interesting results, since it binds tightly to every sulfate group as a specific cation. In the positive mode, all sulfate groups are bound to this cation, whereas the negative-ion analysis indicates that one of the sulfate groups always stays free. The mass spectra of every sulfated oligosaccharide followed the general formula of n peaks in the negative mode and $(n+1)$ peaks in the positive mode, where n is the degree of sulfation. The mass difference between neighboring peaks in both ionization modes corresponds to a SO_3-pmg cleavage with an H substituting the SO_3–pmg pair (net loss of 353 mass units). The highest m/z values follow the general formula $m/z = [M_{OS} + (n+1) \times M_{pmg} + H]^+$ in the positive mode and $m/z = [M_{OS} + (n-1) \times M_{pmg} - H]^-$ in the negative mode, where M_{OS} and M_{pmg} are the exact masses of the fully protonated oligosaccharide and the guanidine derivative, respectively. Usually, the peaks with the lowest m/z values correspond to the fully desulfated ion in the positive mode and to a monosulfated ion in the negative mode.

Recently, both oligosaccharides and oligosaccharide alditols, fragments of agaroses and carrageenans, were additionally investigated by MALDI-TOF MS in order to evaluate the prompt (accompanying the ionization process) and PSD fragmentation processes using different matrices and registration modes.[229] Sulfated alditols showed in the negative-ion mode the molecular ion as $[M - Na]^-$ together with some species formed by prompt fragmentation (mainly desulfation), while the sulfated oligosaccharides showed mainly glycosidic prompt fragmentation with some desulfation. Non-sulfated aldoses and alditols, which could only be analyzed in the positive-ion mode ($[M+Na]^+$), did not show any prompt fragmentation, but aldoses yielded cross-ring fragmentation in the PSD mode. The best results were obtained by using 2,5-dihydroxybenzoic acid and/or *nor*-harmane as matrices for all the compound studied.

It may be concluded that, in spite of the complicated mass-spectrometric behavior of sulfated oligosaccharides, the method has been successfully used many times for characterization of fragmentation products of red algal galactans, especially of enzymolysis fragments (additional examples will be given below).

c. Nuclear Magnetic Resonance Spectroscopy.—The introduction of NMR spectroscopy in the practice of structural analysis of carbohydrate derivatives had the greatest influence on the development of this field of investigation,[372–376] red algal galactans being an excellent example.[377,378]

(i) ^1H NMR Spectroscopy. In the early 1970s, it was stated in the numerous model experiments that the presence of 3,6-anhydrogalactose[379] and several substituents, such as O-methyl[380] or sulfate groups[381] and cyclic acetals of pyruvic acid,[382] in a galactan molecule results in appearance of characteristic signals in ^1H NMR spectra. Based on these data, an analytical procedure was suggested,[383] which offered the possibility of evaluating the contents of 3,6-anhydrogalactose, 6-O-methylgalactose, galactose 6-sulfate, and pyruvic acid by analysis of the ^1H NMR spectra of acid depolymerization products of some agar-like polysaccharides. Unfortunately, it was not possible for a long time to obtain satisfactory resolution of resonances in the spectra of high-molecular-mass natural galactans. To avoid the undesirable side reactions (desulfation and degradation of 3,6-anhydrogalactose) during the acid depolymerization, it was suggested to perform enzymatic hydrolysis of agars with

α- or β-agarase[384,385] before of the registration ^{1}H NMR spectra. Well-resolved ^{1}H NMR spectra of several high-molecular-mass polysaccharides, namely, agarose, sulfated agarose from *Gloiopeltis cervicornis*, and κ- and ι-carrageenans, were obtained and interpreted only in 1977 (a 300-MHz spectrometer was used and spectra were recorded at 80 °C).[386] In the same year, ^{13}C NMR spectra of galactans were shown to be much more convenient and informative. As a result, subsequent works used ^{1}H NMR spectra mainly for characterization of oligosaccharides obtained after enzymatic or acid hydrolysis of galactans, whereas the polymeric substances were analyzed by ^{13}C NMR spectroscopy. Nevertheless, ^{1}H NMR spectroscopy has many advantages because of its higher sensitivity, higher rate of registration, and higher accuracy of quantitative measurements. It was successfully used alongside ^{13}C NMR spectra for the characterization of natural polysaccharides. For example, based on this method, agar from *G. dura* was shown to differ from commercial agar in its content of pyruvate and methylated monosaccharides (6-*O*-methylgalactose and 3,6-anhydro-2-*O*-methylgalactose).[387] Complete interpretation of resonances in the ^{1}H NMR spectrum of a polysaccharide obtained after periodate oxidation and desulfation of a xylogalactan from *Corallina pilulifera* was accomplished by two-dimensional spectroscopy and used to elucidate the agaran-like structure of its backbone. At the same time, analysis of the spectrum made it possible to attribute the L-configuration to 2-*O*- and 3-*O*-methylgalactose residues and the D-configuration to 6-*O*-methylgalactose residues.[166] The content of floridean starch and the proportion of different carrageenan structures were evaluated in galactan fractions isolated from *Furcellaria lumbricalis*[122,388] and *Gigartina scottsbergii*.[389] Analysis of the anomeric region of ^{1}H NMR spectra was suggested as a simple method for the quantitative determination of κ-, ι-, and λ-carrageenans in industrial blends.[390]

Introduction of NMR spectrometers of higher sensitivity, and two-dimensional ^{1}H–^{1}H and ^{1}H–^{13}C correlation procedures, made it possible to assign unambiguously all of the resonances in the ^{1}H NMR spectra of neo-κ-carrabiose (DA-G4S),[391] neo-ι-carrabiose (DA2S-G4S),[392,393] neo-λ-carrabiose (D2S,6S-G2S),[394] and their higher analogues. Signals in the spectra of odd-numbered κ-carrageenan fragments containing G4S at both the reducing and nonreducing ends were also assigned.[395] The proton NMR spectra played an important role in studies on the conformational behavior of κ-carrageenan[396] and physicochemical properties of κ/ι-hybrids[397,398] and were effectively used for characterization of hybrid carrageenans by the action of specific enzymes[399,400] (additional examples see later).

(ii) ^{13}C *NMR Spectroscopy.* The ^{13}C NMR spectra of several well-known red algal galactans (agarose, κ- and ι-carrageenans) were obtained and analyzed for the

first time simultaneously and independently by two research groups.[10,11] The signal assignments in these spectra were made by comparison with the spectra of numerous model compounds, such as O-methylated and sulfated monosaccharide derivatives, including methyl 3,6-anhydro-α-D-galactopyranoside,[401,402] and it was shown that ^{13}C NMR spectra of galactans have indisputable advantages relative to ^1H NMR spectra due to greater range of chemical shifts for carbon resonances and the very high sensitivity of chemical shift values to even small changes in the chemical environment of carbon atoms. In particular, substitution at the oxygen atom of a monosaccharide residue (glycosylation, methylation, or sulfation) results in the downfield shift of the resonance of the nearest carbon atom by up to 6–11 ppm (the so-called α-effect of substitution), whereas the signals of two neighboring carbons are shifted upfield by up to 2–4 ppm (β-effects). Formation of a 3,6-anhydro ring, which may be regarded as the intramolecular substitution simultaneously at O-3 and O-6, causes conformational inversion of the pyranose ring and changes considerably the chemical shifts of all carbon atoms, as compared with the parent galactopyranose residue. As a result, the spectra of regular galactans, such as agarose, κ- or ι-carrageenan, show twelve well-resolved signals corresponding to twelve carbon atoms of the disaccharide repeating unit (Fig. 1).

An additional important peculiarity of ^{13}C NMR spectra of red algal galactans involves dependence of the chemical shift of the C-1 resonance of a 4-linked galactose residue on its absolute configuration. It was detected for the first time for

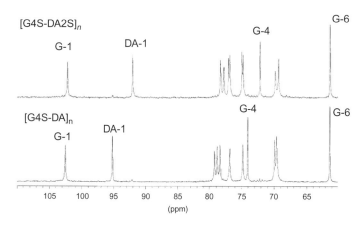

FIG. 1. ^{13}C NMR spectra of κ- [G4S-DA]$_n$ and ι-carrageenans [G4S-DA2S]$_n$. Note that sulfation at C-2 of 3,6-anhydrogalactose residue does not practically alter the positions of G-1 and G-6 signals, but results in considerable shifts of resonances of neighboring DA-1 and remote G-4 carbons (adapted from ref. 378).

agarose and κ- and ι-carrageenans[10,11] and was showed later for polysaccharides devoid of 3,6-anhydrogalactose.[403] This property makes it possible to attribute unknown galactans to the agar or carrageenan groups at the earliest stage of investigation, according to the position of signals in the anomeric resonance region of the ^{13}C NMR spectrum. Interestingly, the sensitivity of ^{13}C NMR spectra to the absolute configuration of monosaccharide residues in diastereoisomeric disaccharides, which was first detected for red algal galactans, turned out to be a common feature.[404] It has been assumed as a basis for determination of the absolute configuration of components of oligosaccharide repeating units in polysaccharides by using ^{13}C NMR spectra (provided that the absolute configuration of at least one monosaccharide is known).[405]

Detailed consideration of ^{13}C NMR spectra of natural galactans belonging either to the agars[406] or to the carrageenans[407] showed the method to have great possibilities for rapid and nondestructive analysis of algal polysaccharides. The simplest case is the direct identification of a galactan isolated from a new source by coincidence of its ^{13}C NMR spectrum with the spectrum of a known polymer[408] (complete interpretation of ^{13}C NMR spectra for the majority of then known galactans was published[409] in 1980). The presence of non-carbohydrate substituents, such as O-methyl groups or pyruvic acid residues, can be readily detected from the signals of the corresponding carbon atoms. Signals of O-methyl groups usually occupy the region 57–60 ppm, that is, they are situated upfield as compared with the signals of carbon atoms of monosaccharide residues.[410] In the case of pyruvic acid acetals, the chemical shift of the C-methyl group is extremely important (24.4–26.1 ppm for its equatorial and 15.0–18.7 ppm for its axial position[411,412]), providing the possibility of determining the configuration of the asymmetric ketal center. Positions of O-methyl and sulfate groups can be elucidated by the known substitution effects or by comparison of the spectra of sulfated and desulfated galactans. The ^{13}C NMR spectra of hybrid molecules usually appear as a superposition of spectra of the corresponding repeating units. Such spectra enable identification of each repeating unit, as for κ–ι- or κ–β-carrageenans, or even to find unknown units, as in the detection of carrabiose 6′-sulfate residues in a galactan from *Phyllophora nervosa*[407] (Fig. 2). ^{13}C NMR spectra also provide the possibility to monitor the chemical modifications of polysaccharides. In particular, they were used to illustrate the transformation of porphyran into agarose[63,179,413] or of μ- and ν-carrageenans into κ- and ι-carrageenans, respectively,[190] by the action of alkali.

Later on, several studies were devoted to verification and correction of the primary signal assignments in the ^{13}C NMR spectra of galactans and their fragments. Oligosaccharides obtained after enzymatic hydrolysis of galactans were commonly used as model compounds. Different papers contain spectral information concerning

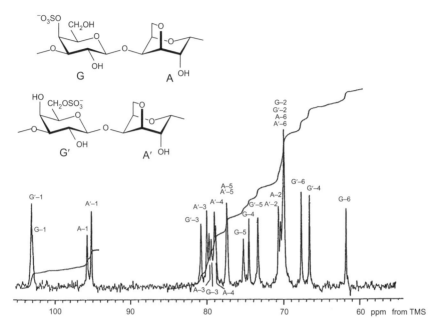

FIG. 2. The ^{13}C NMR spectrum of a κ–ω-hybrid carrageenan G4S-DA+G6S-DA from *Phyllophora nervosa* (adapted from ref. 408).

oligomers of κ-carrageenan,[414] ι-carrageenan,[415] and agarose[416] as well as sulfated and desulfated polysaccharides of the agar group[417] and methylated, sulfated, and pyruvated derivatives of agarose and its oligomers.[418] The spectrum of desulfated λ-carrageenan was published[409] in 1980, but the satisfactorily resolved spectra of native λ-carrageenan were obtained and interpreted much later.[419,420]

At the same time, ^{13}C NMR spectra were used elucidating the structures of more- and more-complex galactans isolated from new species of the red seaweeds. All the papers of this kind are not mentioned, but several important early works should be cited, which gave evidence on the polysaccharide composition of numerous red seaweeds from the Sea of Japan,[179] on the structural diversity of agar-like polysaccharides isolated from different species of the genus *Gracilaria*,[421,422] and on the structure of carrageenans isolated from some new sources.[423,424]

The direct interpretation of ^{13}C NMR spectra for complex hybrid structures may be very difficult. Sometimes, alkaline treatment may be helpful, as it eliminates the alkali-labile sulfate and increases the 3,6-anhydrogalactose content and, hence, the degree of polysaccharide regularity.[424] If the sulfate groups are stable to alkali, it is necessary to split the polysaccharide partially and investigate the resulting fragments. For instance, analysis of ^{13}C NMR spectra of oligosaccharides obtained after partial reductive hydrolysis of a sulfated xylogalactan from *L. nipponica* (Scheme 6) made it possible to explain the spectrum of the parent polysaccharide, which contained more than 10 signals in the anomeric region.[181] In the case of two other xylogalactans isolated from *C. pilulifera* (Fig. 3) and *Bossiella cretacea*, well-resolved ^{13}C NMR spectra were obtained only for several products of their chemical modifications by desulfation and Smith degradation.[166,323] A combination of ^{13}C NMR spectroscopy and enzymatic hydrolysis also proved very fruitful, and this approach will be discussed next.

It is known that bacterial polysaccharides built up of repeating oligosaccharide units can be analyzed by computer programs, which generate all possible ^{13}C NMR spectra for the given monosaccharide composition and allow elucidation of the structure of repeating unit by comparison of simulated spectra with the real spectrum of polysaccharide.[425–428] These programs cannot be used directly for red algal galactans having masked regular structures. Although calculated spectra for most of

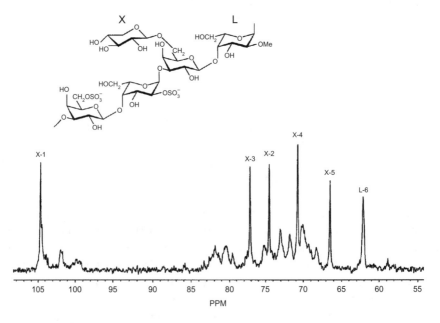

FIG. 3. The ^{13}C NMR spectrum of a sulfated xylogalactan from the calcareous red alga *Corallina pilulifera* (X—signals of the corresponding carbon atoms of nonreducing unsubstituted β-D-xylopyranosyl branches, L-6—signal of C-6 of 4-linked α-L-galactopyranose residues of the backbone, adapted from ref. 166).

the possible disaccharide units of carrageenans were published in the literature,[429] the calculation may not be regarded as completely successful, since several predicted spectra were shown to differ from the experimental ones.[419,430]

In 2000, Miller and Blunt published three papers[431–433] under the common title "New ^{13}C NMR methods for determining the structure of algal polysaccharides." The authors suggested analyzing the galactan structures by comparing the spectra of native, desulfated and methylated polysaccharides and to apply some mathematical treatment of the results using matrix manipulations and nomenclature of the set theory. This approach was then used to obtain structural information on a large series of complex polysaccharides.[434–449] In spite of the impressive results of this work, it should be borne in mind that the authors' approach had several disadvantages. The polysaccharides are *a priori* regarded as regular polymers built up of disaccharide repeating units, and hence, there is no possibility to control this structural feature. Spectral characteristics of these diads used for the mathematical treatment of the spectra were taken from different papers and given by different authors without any account of the possible influence of the neighboring diads of other chemical structure on the position of signals (now such influence is confirmed experimentally[400]). Therefore, the authors' conclusions on the galactan structures, especially on the unusual ones, evidently need confirmation by independent chemical methods. Detailed criticism of this approach was published in a special letter[450] together with the author's reply.[451]

Much more effective is the modern spectral technique based on the analysis of two-dimensional ^{1}H–^{1}H and ^{1}H–^{13}C correlation spectra. It was used to reinvestigate the signal assignment in the oligosaccharide fragments of different galactans[246,393,395] and in the structural analysis of many complex polysaccharides[166,196] (additional examples may be found later).

5. Enzymatic and Immunological Methods

a. Agarolytic Enzymes.—Agar is used as the gel-forming component of bacterial cultivation media due to its stability to the enzymatic systems of most microorganisms. Nevertheless, some of them, mainly bacteria of marine origin, can liquefy the agar media and, hence, are able to produce agarolytic enzymes. It is evident that these bacteria play an extremely important role in marine ecosystems, since they are capable of utilizing the organic compounds of biomass of dead red algae.[452] Agarase preparations were obtained from *Pseudomonas kyotoensis*,[453] *Pseudomonas atlantica*,[454–456] *Cytophaga* sp.,[457–459] and several other bacteria.[460–465] Sometimes, agarase activity was detected in other organisms, such as actinomycetes,[466] protozoa,[467,468] and herbivorous gastropods,[469,470] although recently it was demonstrated that galactanase activities in abalones[471] and herbivorous fishes[472] originate mainly from the resident gut bacteria. Standard procedures of preparative enzymology are suitable for isolation of agarase. It should be noted that gels of cross-linked agarose, the conventional carriers for gel-permeation chromatography of biopolymers, may be used as affinity sorbents in this particular case.[473–475]

The bacterial agarolytic complex may contain several different enzymes. One class of them is especially important. Its representatives, known as β-agarases, are endo-enzymes capable of hydrolyzing β-D-galactosidic linkages in agarose and related polysaccharides to produce oligosaccharides of the neoagarobiose series [3,6-anhydro-α-L-galactopyranosyl-$(1 \rightarrow 3)$-D-galactose and its di-, tri-, and tetra-mer]. Hence, in contrast to the action of acids, glycosidic linkages of 3,6-anhydro-L-galactose are retained under the conditions of enzymatic hydrolysis with β-agarase and can be characterized by identification of the enzymolysis products. The enzymatic formation of oligosaccharides, which are complementary to the corresponding oligomers obtained as a result of chemical fragmentations, was regarded as the final argument in elucidation of the structure of agarose as a regular linear polymer built up of repeating disaccharide units.[453,476]

Later on, two groups of investigators studied the enzymatic hydrolysis of porphyran. Valuable data were obtained on both the specificity of agarase action and the structure of polysaccharide. When porphyran was treated with an extracellular β-agarase from *Cytophaga* sp., neoagarotetraose was obtained as the major reaction product.[477–479] It was stated that 6-O-methyl groups in D-galactose residues diminish the reaction rate 4–5 fold, whereas sulfate at position 6 of L-galactose residues protects the nearest linkages from the enzymolysis and gives rise to higher sulfated oligosaccharides having the minimum degree of polymerization (DP) of 8–10. An enzyme isolated from *P. atlantica* (β-agarase I) is somewhat more specific, since it can split only β-galactosidic linkages of unsubstituted agarobiose fragments.[456,480] Treatment of porphyran with this enzyme afforded as the main products two neutral oligosaccharides (6^3-O-methylneoagarotetraose and $6^3,6^5$-di-O-methylneoagarohexaose), two monosulfated tetrasaccharides [6-sulfate-α-L-galactopyranosyl-$(1 \rightarrow 3)$-β-D-galactopyranosyl-$(1 \rightarrow 4)$-3,6-anhydro-α-L-galactopyranosyl-$(1 \rightarrow 3)$-D-galactopyranose and its 6^3-O-methyl derivative], and a high-molecular-mass sulfated and methylated fraction having an average DP of 40. Structures of porphyran fragments were elucidated by using ^{13}C NMR spectroscopy. These results illustrate the mode of distribution of different structural components along the linear backbone of porphyran.[481]

An enzyme with new specificity acting on porphyran was found recently in the marine bacterium *Zobellia galactanivorans* and named β-porphyranase A.[482] In contrast to previously used β-agarases, which produce predominantly non-sulfated oligosaccharides of the neoagarobiose series and hybrid oligosaccharides, the new enzyme produced sulfated oligosaccharides of the α-L-Gal$p$6S-(1→3)-β-D-Gal$p$ series, the disaccharide L6S-G being the major final product. These enzymolysis products were characterized by chromatographic behavior and by NMR spectra, and a combination of enzymatic degradation by both β-agarase B[483] and β-porphyranase A[482] from *Z. galactanivorans* was used to revise the primary structure of porphyran from *P. umbilicalis*.

The commercially available β-agarase from *P. atlantica* is a valuable tool for structural analysis of agar-like polysaccharides, especially in combination with the study of reaction products by NMR spectroscopy. Data on the structure of galactans from *R. larix* and *Polysiphonia morrowii* may be mentioned as examples. In the first case, apart from neoagarobiose and neoagarotetraose, mono-*O*-methyl derivatives of these oligosaccharides having methyl group at position 2 of the 3,6-anhydro-L-galactose residue were obtained after enzymatic hydrolysis.[484] In the second case, using β-agarase, blockwise distribution of substituents was demonstrated in partially sulfated agarose, and agarose 6-sulfate (**43**) was isolated as a high-molecular enzyme-resistant fraction, together with oligosaccharides **41** and **42**, the expected degradation products of neutral blocks having the agarose structure[485] (Scheme 10).

The action of β-agarase from *P. atlantica* on an agar fraction enriched with pyruvic acid residues resulted in isolation of two new oligosaccharides, $4^3,6^3$-*O*-(1-carboxyethylidene)neoagarotetraose and $4^5,6^5$-*O*-(1-carboxyethylidene)neoagarohexaose. This fact demonstrated that the presence of a pyruvylidene group, even in the disaccharide unit neighboring to the unit attacking by enzyme, does not prevent the polysaccharide from undergoing the enzymatic hydrolysis.[486]

Other components of the agarase complex were isolated from *P. atlantica*, namely, hydrolases of *p*-nitrophenyl α-D-galactopyranoside,[487] neoagarobiose,[488] and neoagarotetraose.[489] Since these enzymes use only low-molecular fragments of polysaccharides as substrates, they had no application in the structural analysis of galactans. A bacterial enzyme (α-agarase) capable of hydrolyzing the glycosidic linkages of 3,6-anhydro-α-L-galactose residues, that is, to split agarose molecule in a manner similar to the action of common acids, was isolated from *Alteromonas agarlyticus*.[384,475] Oligosaccharides of the agarobiose series derived by the action of this enzyme on agarose were characterized using ^{13}C NMR spectroscopy.[490]

Recently, many new bacterial strains capable to degrade agars have been isolated, the corresponding genes were cloned and sequenced, and several agarases were overexpressed and purified to homogeneity. The enzymes were used to determine their

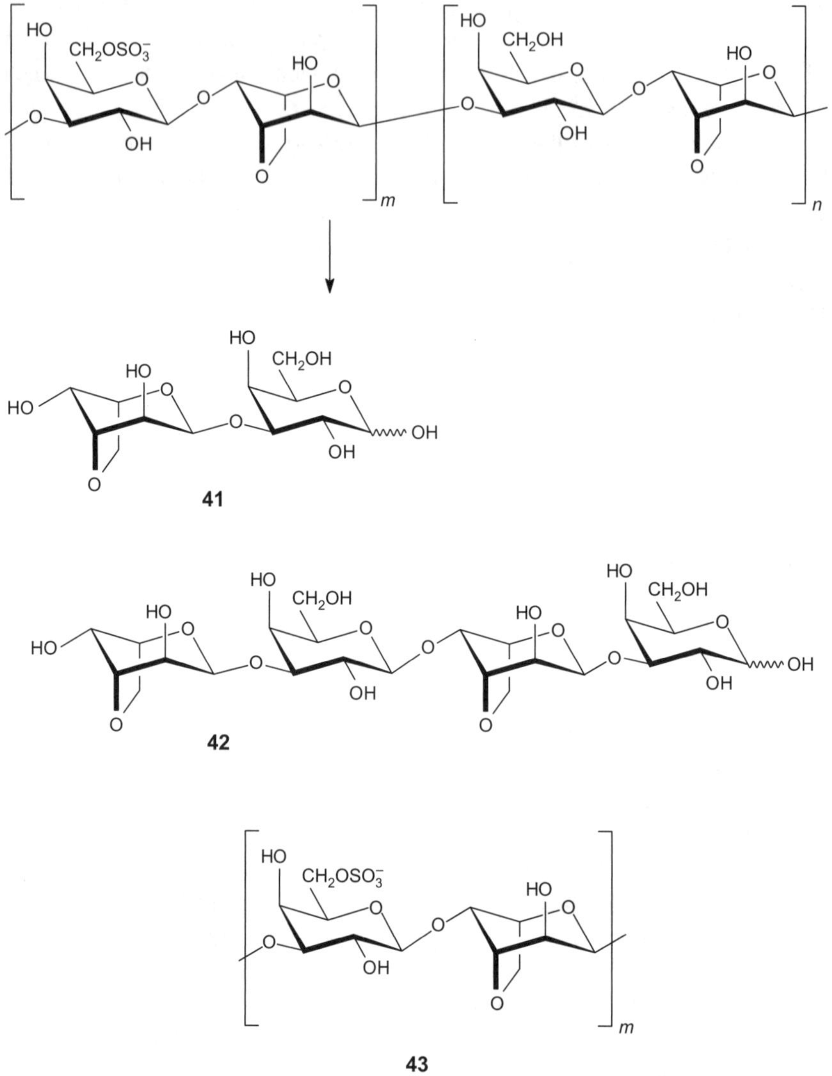

SCHEME 10. Neoagarobiose (**41**), its dimer **42**, and agarose 6′-sulfate (**43**) obtained after enzymatic hydrolysis of a polysaccharide from *Polysiphonia morrowii* by bacterial β-agarase.[485]

molecular structure and to investigate their mechanism of action (see the review, ref. 452). In addition to structural analysis of agar-like polysaccharides, the available recombinant agarases were suggested as reagents for the production of algal protoplasts[491,492] and for preparative isolation of neoagaro-oligosaccharides that may find applications in the food industry and in medicine.[493-495] Previous methods of isolation of neoagaro-oligosaccharides[461] were considerably improved by optimization of experimental conditions and application of modern chromatographic procedures, which made it possible to generate from agarose and isolate in preparative amounts individual oligomers of various sizes.[221]

b. Enzymes Acting on Carrageenans.—Enzymes capable of splitting the carbohydrate chains of carrageenans, like agarases, have been found mainly in marine bacteria[496-498] (see the review, ref. 452), but the presence of carrageenase activity was also detected in the digestive tract of gastropods[499] and sea urchins.[500]

Carrageenases can be obtained from the bacterium *Pseudomonas carrageenovora*, which produces extracellular enzymes specific to different carrageenans, depending on the structure of polysaccharide used in cultivation medium as the only carbon source.[501,502] The ability of algal extracts to be split by a κ-carrageenase preparation from this bacterium was used in an early paper[503] as an important chemical factor for classification of algae. Later on, the action of this enzyme on κ-carrageenan from *C. crispus* was studied in detail.[501,504] As in the case of action of β-agarase on agar, only β-galactosidic linkages were split and the polysaccharide was converted with a yield of about 80% into a mixture of homologous oligosaccharides of the neocarrabiose series, the smallest of which proved to be O-α-3,6-anhydro-D-galactopyranosyl-$(1 \rightarrow 3)$-D-galactose 4-sulfate (neocarrabiose 4-sulfate). In accordance with the hybrid nature of the starting polysaccharide, an enzyme-resistant fraction (20%) was obtained, which contained ι- and μ-carrageenan units. The latter fraction, after alkaline treatment, acquired the ability to be additionally hydrolyzed by the action of κ-carrageenase.

More-detailed investigation of enzymes of *Pseudomonas carrageenovora* led to isolation of κ-carrageenase in the electrophoretically homogeneous state[505] and to detection of two other enzymes, namely, glycosulfatase[506,507] and neocarratetraose 4-sulfate β-hydrolase.[508] Both enzymes use oligosaccharide fragments of κ-carrageenan as substrates: glycosulfatase can split sulfate from neocarrabiose 4-sulfate as well as one of the two sulfate groups (that which is nearest to the nonreducing end) from neocarratetraose disulfate. The resulting neocarratetraose monosulfate is a substrate for β-hydrolase. The mutual action of all three enzymes was used in a procedure suitable for preparation of several grams of neocarrabiose from κ-carrageenan.[509]

An ι-carrageenase capable of hydrolyzing β-galactosidic linkages in ι-carrageenan molecules was also isolated from a marine bacterium.[510] As already mentioned, the oligosaccharide products of enzymatic hydrolysis of κ- and ι-carrageenans were carefully characterized by NMR spectroscopy.[391,393,414,415] A combination of specific enzymolysis and NMR spectroscopy was successfully applied to the investigation of polysaccharides from *Kappaphycus alvarezii* (as *Eucheuma cottonii*),[511,512] *Eucheuma nudum*,[513,514] *Eucheuma gelatinae*,[515] *H. musciformis*,[516] *F. lumbricalis*,[399] and several other algae. A large-scale procedure for obtaining the κ-carrageenase from *P. atlantica* was described,[517] and effective chromatographic methods for the separation of carrageenan oligosaccharides released by ι- and κ-carrageenase were developed.[364,392] In the structural analysis of λ-carrageenan from *Gigartina skottsbergii*, λ-carrabiose oligosaccharides were prepared by the action of λ-carrageenase from *Pseudoalteromonas carrageenovora*. In addition to typical oligosaccharides [D2S,6S-G2S]$_n$, a heptasulfated tetrasaccharide D2S,6S-G2S,4S-D2S,6S-G2S was obtained, demonstrating the mode of oversulfation of the polysaccharide.[394]

As in the case of agarase, κ-carrageenase from *Pseudoalteromonas carrageenovora* and ι-carrageenase from *Alteromonas fortis* were cloned, and the pure recombinant enzymes were used for the investigation of protein structures and mechanism of enzymatic reactions[518,519] (see the review, ref. 452). Based on the results of enzymatic degradation of κ-carrageenan in aqueous solution in the presence of iodide, a single-helix ordered conformation of polysaccharide molecules was proposed, in contrast to the previously considered double-helix model.[520] The composition and distribution of carrabiose moieties in hybrid κ-/ι-carrageenans extracted from *G. skottsbergii*, *Chondracanthus chamissoi*, and *C. crispus* were reinvestigated using both carrageenases, and the presence of three modes of diad distribution (blocks of κ-carrabiose, κ-carrabiose-rich fraction containing randomly distributed ι-carrabiose units and ι-carrabiose-rich fraction) was demonstrated, the proportions of each fraction being related to the botanical origin and to the place of growth of the seaweed.[400] A similar approach was applied to the structural analysis of carrageenans extracted with water or alkali from *Mazzaella laminarioides*, *Sarcothalia crispata*, and *Sarcothalia radula*, where distinct species-specific differences in composition and distribution of carrabiose moieties in different polysaccharides were demonstrated. Interestingly, μ- and ν-carrabiose units, the biogenetic precursors of κ- and ι-carrabiose, were localized in κ-carrabiose- and ι-carrabiose-rich segments, respectively.[521] Several higher hybrid oligosaccharides of the ι-/ν-type were isolated and carefully characterized by NMR spectra after the action of ι-carrageenase from *A. fortis* on a hybrid ι-/ν-carrageenan water-extracted from *Eucheuma denticulatum*.[522] This paper describes the complete assignment of proton and carbon resonances in the NMR spectra of single ν-carrabiose fragment inserted into ι-carrageenan chains.

c. Enzymes Releasing Sulfate.—The formation of 3,6-anhydrogalactose residues from galactose 6-sulfate during the biosynthesis of galactans is accompanied by the release of sulfate. The corresponding enzymes were found firstly in *P. umbilicalis* containing porphyran,[284,523] and then in several algae containing carrageenans, namely, in *Gigartina stellata*,[93] *C. crispus*,[285,524] *Calliblepharis jubata*.[286] A similar process of agar biosynthesis was studied also in *Gracilaria chilensis*.[525] The term "sulfohydrolase" used for the corresponding enzymes in most of the cited publications seems to be incorrect, since the reaction is not hydrolysis, but elimination of sulfate. The term "galactose-6-sulfurylase" recommended by the IUBMB Enzyme Nomenclature (EC 2.5.1.5.) seems to be even more unsuccessful, since the term "sulfuryl" is used solely by inorganic chemists and is not relevant to organic ester sulfates. Genuine sulfate hydrolases, or sulfatases, use as substrates only low-molecular carbohydrate sulfates: the glycosulfatase from *Pseudomonas carrageenovora* already mentioned [506,507] may be cited as an example. Sulfatases capable of splitting sulfate groups from polymeric galactans with retention of their carbohydrate chains might be of indispensable value for the structural analysis of polysaccharides. Such sulfatases acing on sulfated glycosaminoglycans are widely distributed in *Nature*, play an important physiological role, and are effectively used for elucidation of glycosaminoglycan structures.[526] Unfortunately, similar enzymes acting on polymeric red algal galactan sulfates are practically unknown.

d. Immunological Methods.—Carrageenans have antigenic activity and cause the formation of antibodies in laboratory animals.[527,528] It was stated that antibodies to κ- and λ-carrageenans are highly specific and precipitate only the homologous polysaccharides. Hence, it is possible to use antibodies as reagents of high sensitivity for detection and microdetermination of the corresponding structures in polysaccharides of the carrageenan group. For example, a serum containing anti-κ activity was used to demonstrate the presence of κ-carrageenan in gametophytes and its absence from sporophytes of the algae belonging to the family Gigartinaceae.[529] It was stated that 4-linked D-galactose 6-sulfate residues are the most important structural fragments of λ-carrageenan, and they determine the specificity of interaction with antibody. According to this observation, the immunological reaction made it possible to distinguish λ- [G2S-D2S,6S]$_n$ and ξ- [G2S-D2S]$_n$ carrageenans. A similar role of antigenic determinant in κ-carrageenan is fulfilled by 3,6-anhydro-D-galactose residues, but anti-ι-carrageenan was not highly specific and reacted with both ι- and κ-carrageenans.[530] Later on, monoclonal antibodies were obtained that are specific to κ-, ι-, and λ-structure, respectively.[531]

Antibodies were used in several investigations to elucidate the difference between carrageenans isolated from various algal species.[532–535] They turned out to be especially suitable reagents to study the localization of carrageenans in algal cells during the biosynthesis[536] or in algal tissues during the growth and development of algae.[531,537] An immunological reaction was also used for a highly sensitive immunoenzymatic procedure for quantitative determination of κ-carrageenan in foodstuffs.[538]

6. Polysaccharides of the Agar Group

Neutral regular agarose [G-LA]$_n$ is the most powerful gelling polysaccharide, and hence, high quality agars should have a high content of unsubstituted agarose. Such preparations were obtained from representatives of the orders Gelidiales (species of the genera *Gelidium*,[539–542] *Gelidiella*,[543,544] *Pterocladia*[124,130,545]) and Ahnfeltiales (the genus *Ahnfeltia*[546]). The galactans may be slightly methylated at C-6 of β-D-galactose and C-2 of 3,6-anhydro-α-L-galactose residues, but agarose from *G. acerosa* is considerably more methylated and contains both G6Me and LA2Me in amounts comparable with G and LA.[245] Methylation of agarose at position 6 was shown to increase the melting points of the corresponding gels.[547] Several red algal species contain totally substituted 2,6'-di-*O*-methyl-agaroses, and gels of these polysaccharides can be melted only under pressure, since their melting points are higher than the boiling point of water.[548–550]

Partially cyclized, methylated and sulfated agarans G,G6M-LA,L6S, known as porphyrans, are present in *Porphyra* and other representatives of the order Bangiales.[413,417,478,481,551–554] Alkaline treatment converts them to partially methylated agaroses. The term "masked repeating structure" seems to have been applied for the first time to the porphyran from *P. umbilicalis*.[99] Interestingly, the conchocelis phases of *P. leucosticta* and *B. atropurpurea* contain agarans, which are chemically distinct from porphyrans of generic phases of the algae, since they are sulfated at C-6 of L-galactose and practically devoid of 3,6-anhydrogalactose and methylated galactose residues.[555]

Recently, representatives of the genus *Gracilaria* became of the greatest importance as the source of agar because of their abundance in *Nature* and successful cultivation, particularly in Chile and Asian countries (see the reviews in refs. 556 and 557). The genus includes many species that differ in details of agar-like galactan structures. Usually, the polysaccharides contain some methyl groups, sulfate and pyruvate, and even branches, such as xylose or 4-*O*-methyl-L-galactose residues. These substituents diminish the gel-forming ability.[19] Sulfate is present mainly as L6S instead of LA, and this feature masks the regularity of molecules. Studies on biosynthesis of agar in *G. chilensis*[288,524,525,558,559] confirmed the common opinion[284] that L6S is the biogenetic precursor of LA, and hence, the unfinished biosynthetic process may be completed by alkaline treatment of algal biomass during the extraction of agar. Therefore, numerous investigations were performed in order to choose the appropriate *Gracilaria* species and elaborate the most effective conditions for alkaline modification giving the best yields of agar of high quality (in addition to papers cited in the reviews,[556,557] several later papers may be mentioned as examples[129,560–576]). No marked differences were observed in the yield, composition and gel-forming properties of agar preparations extracted from gametophytic and sporophytic forms of *Gracilaria verrucosa*, although juvenile plants contained agar of lower quality due to higher content of the biogenetic precursor of agarose.[128] A great similarity in agar composition of different generations was also found in representatives of the order Gelidiales.[540,577]

Polysaccharides belonging to the agar group according to their chemical structure were also found in algae of various other taxonomic positions.[179,182,578–580] For example, about 70 species of red algae belonging to different orders are mentioned in the review[581] on marine macroalgal research in New Zealand, and 50 of them apparently contain polysaccharides of the agar group. It should be noted that many of them are highly substituted derivatives of agaran or agarose and may be practically devoid of gelling properties. Structures of several complex agar-like polysaccharides were carefully elucidated, whereas those based mostly on the NMR spectrometric data, as already mentioned, may require confirmation by independent chemical methods.

The large order Ceramiales consists of several families, and the algae most investigated belong to the family Rhodomelaceae. A gel-forming galactan from *P. morrowii* was shown to be a hybrid of the usual agarose and agarose 6-sulfate, the latter structure being isolated after action of bacterial β-agarase followed by alkaline treatment[485] (Scheme 10). It is clear that sulfate at position 6 in D6S is stable to alkali and cannot be split under alkaline conditions. Agarose 6-sulfate was also isolated from *Polysiphonia nigrescens* by direct fractionation of the total extract containing partially cyclized agarans highly substituted on C-6 not only by sulfate but also by methyl groups and single stubs of β-D-xylose.[582] Similarly, substituted agarans were obtained from the algae of the same genus *Polysiphonia lanosa*,[260] *Polysiphonia strictissima*,[583] *Polysiphonia abscissoides*,[583] and *Polysiphonia atterima*[584] as well as from *Streblocladia glomerulata*[583] and *Pterosiphonia pennata*.[584] *Odonthalia corymbifera* was shown to contain substituted agarose highly methylated at C-6 and sulfated at C-4 of 3-linked β-D-galactose residues.[100] A complex agaran–agarose hybrid, similarly sulfated mainly at C-4, but partially methylated and sulfated also at other positions and having single xylopyranosyl stubs at C-6 was isolated from *Bostrychia montagnei*.[585] Several species of the genus *Laurencia* obviously contain similar highly substituted agaran–agarose hydrids.[357,580,586–588] Single β-D-xylopyranosyl branches at C-3 of 4-linked α-L-Gal$p$ together with specific sulfation pattern were found in a galactan sulfate from *L. nipponica*, partial reductive hydrolysis of the polysaccharide (Scheme 6) being used for the first time to obtain oligosaccharide fragments important for structural analysis.[181] The same position of xylosyl substituents was found in the xylogalactan sulfate from *Chondria macrocarpa*.[180] Structures of highly substituted agaran–agaroses were ascribed also to polysaccharides from *Cladhymenia oblongifolia*,[433] two species of the genus *Lophurella*,[440] two species of the genus *Lenormandia*,[580] *Bryocladia ericoides*,[580] and *Osmundaria colensoi*.[580] It is noteworthy that a gel-forming galactan sulfate from *R. larix* was formerly described as a methylated and sulfated agarose,[356,484] but subsequent reinvestigation of its structure indicated the occurrence of minor fraction having the carrageenan-type backbone.[589]

Algal species belonging to other families of the order Ceramiales also contain agar-like polysaccharides. Thus, 6-O-methyl-agarose with 10–20% of L6S was found in *Euptilota formosissima*,[583] and similar agaroses having considerable precursor content were isolated from *Ceramium rubrum*,[261,439] *Ceramium uncinatum*,[443] and *Griffithsia antarctica*[443] (Ceramiaceae). The most heavily sulfated agaroses were found in representatives of the family Delesseriaceae, where the polysaccharides extracted from *Myriogramme denticulata*[437] and *Hymenena palmata* f. *marginata*[447] contained D4S-L2S,3S as the main repeating units, while *H. palmata*, *Hymenena variolosa*, and *Acrosorium decumbens* contained D2S,4S,6S-L2S,3S and even totally sulfated repeating disaccharides.[447] Polysaccharides of the agar group were also found in algae belonging to other orders of Rhodophyta. Thus, heavily sulfated agaroses also containing pyruvate were isolated from two species of *Sarcodia* (the order Plocamiales).[590] A substituted agaran was obtained, together with the main neutral xylan, from *P. stenogona* (the order Palmariales).[591] Similarly, the main sulfated mannan is accompanied by an agar-like xylogalactan in *N. fastigiata* (the order Nemaliales).[592]

Agarans with the highest content of different substituents are characteristic of calcareous algae of the order Corallinales.[593,594] A nonconventional sulfated xylogalactan devoid of 3,6-anhydrogalactose was first isolated from *C. officinalis*.[164] The structure of this polysaccharide was later carefully investigated by chemical methods.[102,178] The presence of the typical agaran backbone was demonstrated, where single D-xylopyranose residues occupy positions 6 in 3-linked D-Gal; sulfate is also attached to the same position as well as to positions 2 and 3 of 4-linked L-Gal. Four isomeric mono-O-methylgalactoses were also found and their absolute configurations (D6M, L2M, L3M, and L4M) were determined. Similar sulfated and methylated xylogalactans were detected in several other species of the order Corallinales and isolated from *C. pilulifera*,[166,323,595] *B. cretacea*,[596] and *J. maximus*.[165] It is noteworthy that ^{13}C NMR spectra of native xylogalactans display sharp signals of carbon atoms of xylose residues, whereas the signals of the heavily substituted backbone were strongly broadened and superimposed and could not be interpreted (Fig. 3).

At the same time, the ^{13}C NMR spectra of desulfation and periodate oxidation products of native xylogalactans were sufficiently resolved to be completely assigned.[166] Combined chemical and spectral evidence made it possible to evaluate the following specific structural features for sulfated polysaccharides of corallinean red algae[594]:

(i) The absence of 3,6-anhydro-L-galactose and its biogenetic precursor L6S.
(ii) The majority of methyl and sulfate groups occupy positions 2 and 3 of α-L-Gal, and only small amount of them occupy position 6 of β-D-Gal.
(iii) The highest xylose content in agar-like polysaccharides. Single stubs of β-D-Xyl$p$ are linked to C-6 of β-D-Gal.

These conclusions were confirmed by a later study on the system of xylogalactans isolated from *Bossiella orbigniana*[597] and *Jania rubens*.[598] Several additional minor structural details were found, such as small amounts of the D enantiomer of 2-*O*-methylgalactose in the former polysaccharide. The latter xylogalactan contained some 3,6-anhydro-L-galactose, partly sulfated or methylated on C-2, replacing the L-Gal units, and side stubs of 2,3-di- and 3-*O*-methyl-D-galactose and of 3-*O*-methyl-L-galactose, which may be characteristic for this particular species.

7. Polysaccharides of the Carrageenan Group

Carrageenans are present in the algae belonging to the order Gigartinales. Three types of carrageenans, ι-, κ-, and λ-, are widely used as gelling, stabilizing, and viscosity-binding agents in the food industry, and in cosmetics and pharmaceutical formulations. Structures of these polysaccharides as [G4S-DA2S]$_n$, [D4S-DA]$_n$, and [G2S-D2S,6S]$_n$, respectively, were elucidated in several classical works of Rees and coworkers.[107,109,210,599] The first samples of κ- and λ-carrageenans were obtained by fractionation[6] of a total sulfated galactan from *C. crispus*, whereas ι-carrageenan was isolated from *Eucheuma spinosum* (the correct name is *E. denticulatum*).[107] It was shown in 1973 by several groups of authors that different generations of carrageenophytes may contain different polysaccharides,[600–603] the gametophytic plants producing κ/ι-type carrageenans, whereas the sporophytic plants produce λ-type carrageenans. Now it is known that this feature is characteristic for species of the families Gigartinaceae[604] and Phyllophoraceae.[605] Differences in polysaccharide composition between generic phases in other families of the order are negligible. In accordance with this observation, both generations of the cultivated tropical algae, *K. alvarezii* and *E. denticulatum* belonging to Solieriaceae, are used for industrial production of κ- and ι-carrageenans, respectively, but isolation of λ-carrageenan requires selection of the sporophytic stage of several species of the genera *Gigartina* and *Chondrus*.

The real samples of carrageenans of κ–ι-type contain both disaccharide repeating units in different proportions, while native polysaccharides additionally contain some biogenetic precursors G4S-D6S and G4S-D2S,6S (μ- and ν-units).[133,287,389,606,607] The nature and amount of different disaccharide repeating units in these carrageenans may be determined by using NMR spectra.[378] Alkaline treatment during the isolation process[120] made it possible to diminish the content of μ- and ν-units, giving rise to κ–ι-hybrids known as "κ-2 carrageenan."[608–610] It is evident that physicochemical properties of such hybrids depend not only on the ratio of different repeating units but also on the mode of their distribution along the polymer chains.[397,398,611] As already described, analysis of disaccharide distribution is possible by means of specific enzymatic hydrolysis, using κ- and ι-carrageenases.[400]

Many additional theoretically possible disaccharides, differing in amount and position of substituents and in the presence or absence of 3,6-anhydrogalactose, were found in hybrid κ–ι-type carrageenans. Thus, units of non-sulfated carrageenose G-DA (β-carrageenan) are present together with G4S-DA (κ-carrageenan) in undersulfated κ-fractions of polysaccharides from *T. crinitus*[306,407,612] and *F. lumbricalis*,[112,122,306,613,614] both polysaccharides retaining the gel-forming ability typical of κ-carrageenans. A KCl-soluble fraction of *T. crinitus* polysaccharides was transformed by treatment with alkali into an unique carrageenan composed of G2S,4S-DA repeating disaccharides.[615] Units G-D6S (γ-carrageenan, the biogenetic precursor of β-carragenan) were found in β/κ/γ-hybrid from *E. gelatinae*.[105] Units G-DA2S (α-carrageenan) were detected in carrageenans from Burmese and Thai samples of *Catenella nipae*, together with the ι-structure,[430] and units of G-D2S,6S (δ-carrageenan, the biogenetic precursor of α-carrageenan) were found in a separate sample of the same species.[104] Units of carrageenose 6'-sulfate (D6S-DA) were first detected in a κ-like polysaccharide from *P. nervosa*[306,407,408] and then found also in *Rissoella verruculosa* and named ω-carrageenan.[114] Its biogenetic precursor G6S-D6S (ψ-carrageenan) was found in the tetrasporophytic form of *Stenogramme interrupta*.[616] A polysaccharide from *Phacelocarpus peperocarpos* is composed predominantly of carrabiose 4',6'-disulfate.[424] Structure G2S-DA2S (θ-carrageenan) was first obtained as the result of alkaline modification of λ-carrageenan[291] and only recently was detected as the main repeating unit of a galactan from *Callophyllis hombroniana*.[617] It is interesting to mention that θ-carrageenan, in spite of the identity of its backbone with gelling κ- and ι-carrageenans, showed no conformational transition on cooling and heating and was practically devoid of gel-forming ability, evidently due to the presence of sulfate at C-2 of 3-linked D-Gal.[618]

The presence of methyl groups and 1-carboxyethylidene acetals leads to an even more complex composition of carrageenan chains. Thus, $4',6'$-O-(1-carboxyethylidene)carrabiose 2-sulfate and carrabiose 2-sulfate units were found in highly pyruvated carrageenans from the genus *Callophycus*,[196] whereas a galactan from *Sarconema filiforme* was shown to contain the same disaccharides together with carrabiose $2,4'$-disulfate units.[619] The presence of a pyruvated ι-carrageenan was detected in *Stenogramme interrupta*.[620] Survival of $4',6'$-O-(1-methoxycarbonylethylidene) group under partial reductive hydrolysis conditions made it possible to isolate two novel pyruvated carrabiitol derivatives from permethylated samples of both pyruvated β- and pyruvated α-carrageenans (the former being prepared by solvolytic desulfation of the latter, as isolated from *Callophycus tridentifer*).[621] Red algae of the genus *Erythroclonium* contain the same pyruvylated carrabiose 2-sulfate, units of ι- and α-carrageenan, and their $6'$-O-methylated counterparts.[423] Highly 6-O-methylated ι-carrageenans containing also 3-O-methylgalactose and pyruvate were found in two species of the genus *Rhabdonia*.[192] A nearly idealized, completely $6'$-O-methylated ι-carrageenan was isolated (after alkaline modification) from *Claviclonium ovatum*.[622]

Carrageenan [G-D] devoid of 3,6-anhydrogalactose and sulfate has not been found in Nature but was prepared by desulfation of λ-carrageenan.[109,409,419] It should be noted that commercial and natural samples of λ-carrageenan usually deviate from the ideal structure [G2S-D2S,6S]$_n$. Thus, the λ-polysaccharide from *C. crispus* is undersulfated and contains 2-sulfate in the 3-linked residues in only ~70% of the disaccharide units.[110] In contrast, the polysaccharide from tetrasporophytes of *G. skottsbergii* was shown to contain about 8% of disaccharide units having an extra 4-sulfate in 3-linked residues,[394] whereas the polysaccharide from *Gigartina decipiens* has about 15% of 3-linked residues oversulfated at C-6.[419] In spite of these deviations, the polysaccharides from the tetrasporic stages of *G. decipiens*,[419] *Iridaea undulosa*,[420] and *G. skottsbergii*[394] were used to assign the signals in NMR spectra of λ-carrageenan. The polysaccharide corresponding very closely to the idealized structure of λ-carrageenan was isolated from tetrasporic stage of *Gigartina lanceata*.[623] Other representatives of the genus *Gigartina* contain predominantly ξ-carrageenans partly pyruvated or sulfated at C-6 of 3-linked residues. The presence of pyruvate in λ-carrageenan-type polysaccharides was observed first in 1972,[195] and then pyruvated ξ-units GP,2S-D2S (π-carrageenan) were detected as the main components of a carrageenan from *P. middendorfii* (the sporophyte generation of *Gigartina papillata*)[113] and found also in carrageenans from the tetrasporic stages of *Gigartina clavifera*, *Gigartina alveata*,[624] *Gigartina chapmanii*,[623] and *Gigartina pistillata*.[625]

8. DL-Hybrids

According to the rule, carrageenans should be composed of D-galactose derivatives only. At the same time, small but appreciable amounts of L-galactose may often be found in algae containing carrageenans as the main polysaccharides.[96] It means that agar-type polysaccharides may coexist with carrageenans even in typical carragenophytes, such as *C. crispus*,[626] *G. skottsbergii*,[187,188] *I. undulosa*,[139] *K. alvarezii*,[627,628] and *Gymnogongrus torulosus*.[629,630] Carrageenans and agarans in the latter species were found in a ratio of 1:0.5, showing the highest amount of agaran-type structure for a carrageenophyte.[630] Stepwise extraction, careful fractionation, and chemical modifications enabled detection in the alga, besides "pure" κ/ι-carrageenans, also carrageenan-agaran complexes and possibly real DL-hybrids. The opposite situations, when carrageenan-type components coexist with the main agar-type polysaccharides in agarophytes, such as *Digenea simplex*,[631] *R. larix*,[589] or *Gloiopeltis furcata*,[189] are also well documented. Detailed analysis revealed small amounts of 4-linked D-galactose derivatives even in the representatives of Gelidiales[282] and Bangiales,[300,482] previously regarded as typical agarophytes. Galactans containing D- and L-galactose derivatives in a ratio deviating from unity were often found in species, formerly belonging to the old order Cryptonemiales, which was then transformed into several other groups (many old papers are cited in refs. 631 and 632). According to the new taxonomy, DL-hybrids were isolated recently from species belonging to Halymeniales (*Pachymenia lusoria*,[633] *Grateloupia indica*,[634] *Aeodes nitidissima*,[445] *Carpopeltis* sp.,[635] and *Halymenia durvillei*[636]), Gigartinales (*Kallymenia berggrenii*[637] and *Callophyllis variegata*[638]), Rhodymeniales (*Hymenocladia sanguinea*[436] and several species of the genera *Champia*[444] and *Lomentaria*[185,449]), Plocamiales (*Plocamium costatum*[639]), and Bonnemaisoniales (*Asparagopsis armata*[640]). Such polysaccharides usually consist of particularly complex molecules heavily substituted with sulfate, pyruvate, methyl groups, and monosaccharide branches.

The papers just cited contain only indirect evidence for the presence of agar and carrageenan fragments as different parts of the same hybrid polymeric molecule. It is clear that only isolation of oligosaccharides containing both structural types in one molecule can be regarded as an unambiguous argument for the hybrid structure of polymers. Such an attempt was made in the course of investigation of a polysaccharide isolated from *G. divaricata*.[253] The galactan contained D- and L-galactose in a ratio of 4:1 with minor amount of 3,6-anhydrogalactose. Acetolysis of the polysaccharide afforded disaccharides D-G (**28**) and G-L (**29**) and trisaccharides G-D-G (**30**) and G-L-G (**31**), which obviously can be equally formed from the molecular hybrid and from the mixture of agar and carrageenan (Scheme 7). Only one tetrasaccharide G-D-G-D (**32**) corresponding to a carrageenan-type structure was isolated. The necessary trisaccharides D-G-L or L-G-D and tetrasaccharides G-D-G-L, G-L-G-D, D-G-L-G, or L-G-D-G were not found, and hence, the problem of the hybrid nature of the polysaccharide remained open. A similar result was obtained by partial degradation of funoran from *G. furcata*.[189] Nevertheless, it is usually assumed that DL-hybrids do really exist, but agaran and carrageenan chains are arranged in blocks, giving rise to very low yield of hybrid oligosaccharides after partial depolymerization.[630]

9. Biological Activity of Sulfated Galactans

In addition to valuable physicochemical properties, sulfated polysaccharides of the red algae are promising biologically active polymers, and many structural works already cited aimed at isolation and characterization of new compounds having potential medical application. The most important types of biological activity of sulfated galactans are listed briefly in this section.

Carragenans are known as powerful agents causing inflammation, and so paw edema induced by carrageenans constitutes a classical model extensively used in anti-inflammatory drug development. It was shown in the comparative investigation of three commercial carrageenans as paw edema and pleurisy inductors, that ι- and λ-carrageenans have higher inflammatory potential than does κ-carrageenan, probably due to their higher sulfate content.[639]

As with heparin and other sulfated polysaccharides, red algal galactans are potent anticoagulants.[640–642] At the first approximation, this activity depends on the degree of sulfation. In accordance with this opinion, oversulfation of κ-carrageenan by chemical methods was shown to increase its anticoagulant activity.[643] At the same time, studies on many other sulfated polysaccharides isolated from different algae clearly demonstrated that structure–activity dependence in this case is much more complex.[644–647] Interesting data on the role of structural factors for anticoagulant activity were obtained by comparison of sulfated galactans isolated from *Botryocladia occidentalis*[648] and *Gelidium crinale*.[649] The former polysaccharide was shown to have a backbone $[G-D]_n$ containing approximately one-third of the total α-D-Gal as 2,3-disulfate and another one-third as 2-sulfate. The same type of structure was ascribed to the galactan from *G. crinale*, where only 15% of the total α-units are 2,3-disulfated and another 55% are 2-sulfated. These two galactans did not differ in thrombin inhibition mediated by antithrombin, but in assays with heparin cofactor II instead of antithrombin the activity of the polysaccharide from *G. crinale* was much lower. In contrast, when factor Xa was used instead of thrombin, the polysaccharide from *G. crinale* was the more potent anticoagulant. It is clear that different structural features are important for interaction of sulfated polysaccharides with different components of the multistage coagulation cascade. The authors explained the results on the basis of different sulfation patterns in two galactans. Unfortunately, no data on determination of the absolute configuration of galactose in the galactan from *G. crinale* were given in ref. 649. Being a member of the genus *Gelidium*, the alga should contain an agaran-like polysaccharide. This attribution may be confirmed by the ^{13}C NMR spectrum of the desulfated polysaccharide, given by the authors.[649] The spectrum differs from the known spectrum of [G-D]$_n$ but coincides well with the spectrum of unsulfated agaran [G-L]$_n$. Since a similar spectrum was obtained for the desulfated polysaccharide from *B. occidentalis* (being the member of Rhodymeniales, the alga should also contain an agar-like polysaccharide), it may be concluded that both polysaccharides have agaran-type backbones, and the attribution of *B. occidentalis* polysaccharide to carrageenans[648] is an experimental error of the authors.

Like many other sulfated polysaccharides of marine algae,[650,651] red algal galactans often demonstrate potent antiviral activity, the most interesting being their inhibitory action on *Herpes simplex*, dengue, and human immunodeficiency viruses. The inhibitory effect of negatively charged sulfated polysaccharides appears to be based mainly on their ability to interfere with the initial attachment of the virus to the target cells by interaction with the positively charged viral external proteins.[652–654] Antiviral polysaccharides were obtained from agarophytes *Acanthophora spicifera*,[655] *Gracilaria corticata*,[567,656] and *B. montagnei*,[657] from carrageenophytes *G. skottsbergii*,[645] *S. interrupta*,[658] and *Meristiella gelidium*,[659] as well as from *N. fastigiata*, containing a sulfated mannan,[660] and from several species of the genera *Schizymenia*[661–663] and *Grateloupia*[664,665] probably containing DL-hybrids. Some attempts were made to diminish the undesirable anticoagulant activity of potential anti-HIV preparations by partial radical depolymerization of commercial carrageenans followed by sulfation[666] or acylation with carboxylic acid anhydrides.[667] The high antiviral activity of red algal polysaccharides was explained by a reasonable hypothesis on the structural and conformational similarity of disaccharide repeating units of sulfated galactans and heparan sulfate, which serves as negatively charged cell-surface target for most of the viruses.[655]

Sulfated galactans may interact with numerous regulatory proteins affording inhibition of complement activation, of tumor cell adhesion to P-selectin, of several matrix-degrading enzymes, and so on.[668–670] These features depend strongly on the sulfate content and molecular mass distribution of the polysaccharides and may explain their action as anti-inflammatory substances,[671,672] macrophage activators[673] and antitumor agents.[674,675] The antitumor activity of galactans may often be enhanced by partial depolymerization through partial acid hydrolysis, action of free radicals,[676] radiolysis,[677] or treatment with specific enzymes.[678] Application of mixtures of low-molecular λ-carrageenan with 5-fluorouracil[679] and even a chemical conjugate of 5-fluorouracil with porphyran[680] was suggested for tumor-growth inhibition. Antitumor activity may be associated with the anti-angiogenic action of carrageenans.[681,682] At the same time, carrageenan-based hydrogels have been developed for the controlled delivery of platelet-derived growth factor as the angiogenesis-stimulating agent for bone-tissue engineering applications.[683]

Some sulfated galactans, or products of their partial depolymerization, exhibit antioxidant activity.[684,685] The chemistry of radical scavenging effects is not very clear, especially taking into account the fact that several products of chemical modifications of natural polysaccharides, such as oversulfated, phosphorylated, acetylated and phthaloylated derivatives, may demonstrate stronger antioxidant activity than the raw material.[685–688]

Carrageenans were shown to have a protective effect against endotoxins of gram-negative bacteria, since they can increase the resistance of mice to the toxic action of bacterial lipopolysaccharides.[689]

As in terrestrial ecosystems,[690] oligosaccharides produced by the action of pathogens on the algal cell-wall polysaccharides may serve as specific recognition signals activating the defense reactions of the host.[691,692] A very interesting example is the cell recognition in the *C. crispus*–*Acrochaete operculata* host–pathogen association, where the host gametophytes containing κ-carrageenan are resistant to the pathogen, whereas the sporophytic generation containing λ-carrageenan is susceptible to infection. The virulence of the green algal pathogen *A. operculata* was shown to be mediated by the recognition of the corresponding carrageenan oligosaccharides released from its red algal host: κ-carrageenan oligosaccharides inhibit *A. operculata* virulence, while λ-carrageenan oligosaccharides enhance it.[693] This finding may have a great importance in developing the disease control for maricultured algal crops.

V. Sulfated Mannans

The presence of sulfated mannans in the red algae was described for the first time in 1973. It was found that hydrolysis of a water-soluble mucilage from *N. vermiculare* gave mannose and xylose instead of the expected galactose.[86,87] Subsequent fractionation led to isolation of a neutral xylan (see before, Section III) and a sulfated polysaccharide containing, aside from D-mannose, also 3.1% of D-xylose and 15.5% of sulfate. The structure of this unusual polysaccharide was carefully investigated by desulfation, methylation, acetolysis, and alkaline degradation.[86,87,694–696] To facilitate the use of ^{13}C NMR spectroscopy for the structural analysis of the xylomannan, isomeric sulfates of methyl 3-O-methyl-α-D-mannopyranoside were synthesized and their NMR spectra were interpreted.[697] Using all of these approaches, the polysaccharide was shown to have a linear backbone of 3-linked α-D-mannopyranose residues sulfated at positions 6 or 4. Several backbone units may carry β-D-xylopyranose residues as single stubs at position 2.

Similar sulfated xylomannans were then found in other representatives of the order Nemaliales, namely, in *N. fastigiata*,[660,698–700] two species of the genus *Liagora*,[89,701] *S. hatei*,[702] and *Nemalion helmintoides*.[703] All of these polysaccharides have identical backbones of (1 → 3)-linked α-D-mannopyranose units but differ slightly in degree of sulfation and xylosylation and in substitution patterns. Thus, sulfate groups at positions 4 and 6 were found in the xylomannans from *Nemalion*[694,703] and *Liagora*,[89,701] at positions 2 and 6 in the xylomannan from *N. fastigiata*,[660] whereas the polysaccharide from *S. hatei*[702] was sulfated only at position 4. Single stubs of β-D-xylopyranose were located mostly at position 2, but the polysaccharide from *S. hatei*[702] contained xylose residues at positions 2, 4, and 6, while in the polysaccharide from *L. valida*[89] not only β-D-xylopyranose but also 3-O-methyl-β-D-xylopyranose and (1 → 4)-linked short chains of β-D-xylopyranose units were found attached to position 2 of the backbone.

Essentially, linear (1 → 3)-α-D-mannans obtained by desulfation of the native sulfated xylomannans are insoluble in water and even in alkali. Native polysaccharides often have "erratic" solubility. For example, the polysaccharide from *L. valida* was separated into two approximately equal fractions, soluble and insoluble, by dissolution in concentrated aqueous sodium chloride.[89] Surprisingly, both fractions scarcely differed in composition, and hence, their solution behavior may be explained only by an uneven distribution of sulfate groups along the main chain, leaving blocks of non-sulfated mannose residues for intermolecular association. Such association may be observed also in the interaction of sulfated mannans with accompanying xylans and galactans leading to some difficulties in isolation of pure polysaccharides.[660] It should be noted that sulfated galactans or xylogalactans may be present in the species of Nemaliales in minor amounts together with sulfated xylomannans. These galactans were shown to belong to the agar group of polysaccharides.[592,704]

There are two examples of sulfated xylomannans deviating from the structures just described. *G. rugosa* (as *G. squalida*), an alga of the order Nemaliales, was shown to contain a polysaccharide having a backbone built up of (1 → 3)-linked

β-D-mannopyranose residues.[88] Recently, the presence of sulfated xylomannans was detected for the first time in two species belonging to Ceramiales, namely, in *Chondrophycus papillosus* and *Chondrophycus flagelliferus*. In this case, the backbone of (1→4)-linked β-D-mannopyranose 2-sulfate units is partially substituted at positions 6 by single β-D-mannopyranose 2-sulfate or β-D-xylopyranose residues.[705]

Along with many other sulfated polysaccharides, sulfated xylomannans show various biological activities. They demonstrated moderate anticoagulant action[660] but showed considerable antiviral (especially antiherpetic) activity, which depends on the degree of sulfation and position of the sulfate group.[660,702,703] Thus, some sulfated xylomannans may be regarded as potentially important antiviral agents.[651,706]

VI. GLYCURONANS

Uronic acids are present in several types of red algal polysaccharides. First, glucuronic acid is known as a minor component of several sulfated galactans, especially DL-hybrids, where it usually forms single branches on the galactan backbone. The highest glucuronic acid content has been found in a sulfated glucuronogalactan from the gametophytic form of *Schizymenia dubyi*, a member of Gigartinales, where the ratio of galactose and glucuronic acid was as high as 1:0.75.[662,707,708] Second, glucuronic acid is a component of the backbones of extracellular polysaccharides produced by unicellular red algae of the order Porphyridiales (see later, Section VII). Third, glucuronic acid is the main component of a mucilage from the freshwater red alga *Batrachospermum* sp., which backbone appears to be built of alternating D-glucuronic acid and D-galactose residues.[709] Fourth, the most unusual case is the presence of alginic acids in the calcareous species of the family Corallinaceae.[594] Alginic acids are linear polymers built up of 1→4-linked residues of β-D-mannuronic and α-L-guluronic acids.[710] These polysaccharides were considered to be specific components of brown seaweeds and several bacteria, but the presence of a similar polysaccharide was surprisingly discovered[711] in 1982 in the coralline alga *Serraticardia maxima*. Later on, glycuronans were isolated from *C. pilulifera*, *B. cretacea*, *Clathromorphum nereostratum*, *Amphiroa fragilissima*, and *Corallina mediterranea*.[323,595] All of them were shown to be alginic acids similar to those from brown seaweeds and containing β-D-ManA and α-L-GulA in the ratio of 0.4–0.7. Thus, the calcareous red algae of the family Corallinaceae differ strongly from the other groups of red algae in their polysaccharide composition. They synthesize sulfated galactans of extremely complex structure (see earlier, Section IV.6) and contain alginic acids, which are absent from other red algae. It is very interesting to elucidate the origin of alginates, which may be produced either by algal cells or by some bacterial symbionts. Nevertheless, since alginic acids were invariably detected in numerous samples of coralline algae collected from various regions of the world's ocean,[594] the presence of alginic acids may be regarded as a systematic marker for the algae of the family Corallinaceae.

VII. HETEROPOLYSACCHARIDES OF UNICELLULAR AND FRESHWATER RED ALGAE. MISCELLANEOUS POLYSACCHARIDES

Red microalgae are regarded as a potential source of a wide range of biochemicals, including polysaccharides.[712] The most investigated representatives belong to the genera *Porphyridium* (the marine algae *Porphyridium* sp. and *Porphyridium cruentum*, the freshwater alga *Porphyridium aerugineum*), and *Rhodella*. The cells of these algae are encapsulated within a cell-wall polysaccharide complex, the external part of which ("soluble polysaccharide") dissolves in the medium, thus increasing its viscosity. Most of the capsule remains bound to the cell-wall and is designated as "bound polysaccharide."

The polysaccharide of *Porphyridium* sp. was shown to contain D-xylose, D- and L-galactose, D-glucose, D-glucuronic acid, several minor methylated sugars, and sulfate. It may be divided into fractions differing in charge and sugar composition.[713] Partial acid hydrolysis gave rise to a mixture of glucose 6-sulfate, galactose 6-sulfate, and galactose 3-sulfate[714] as well as to 3-*O*-(α-D-glucopyranosyluronic acid)-L-galactose. The same disaccharide building block was obtained also from polysaccharides of *P. cruentum*, *P. aerugineum*, and *Rhodella maculata* and characterized by MS and NMR spectroscopy.[715–718] These data did not confirm the previous report on the isolation of 3-*O*-(β-D-glucopyranosyluronic acid)-D-galactose, having the opposite configurations of glycosidic bond and galactose residue, from polysaccharides of *P. cruentum* and *P. aerugineum*.[719] At the same time, the extracellular polysaccharide of *P. cruentum* was shown to contain additionally 2-*O*-methyl-D-glucuronic acid, affording 3-*O*-(2-*O*-methyl-α-D-glucopyranosyluronic acid)-D-galactose and 3-*O*-(2-*O*-methyl-α-D-glucopyranosyluronic acid)-D-glucose upon partial acid hydrolysis.[715] The polysaccharide is covalently linked to the protein moiety of the complex.[720]

In an attempt to elucidate the molecular structure of the polysaccharide of *Porphyridium* sp., uronic acid degradation with lithium in ethylenediamine was used to obtain neutral oligosaccharide fragments[721] (Scheme 11). Two higher oligosaccharides (**44** and **45**) were obtained, demonstrating the presence of single xylosyl branches as well as of contiguous β-D-xylopyranose residues in the linear polysaccharide

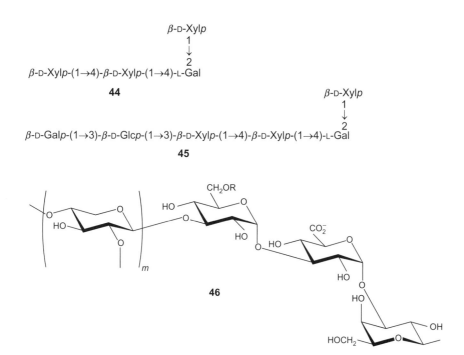

SCHEME 11. Two branched oligosaccharides, **44** and **45**, obtained after treatment of a polysaccharide from *Porphyridium* sp. with lithium in ethylenediamine,[721] and the structure **46** ($m=2$ or 3) of a backbone of this polymer (R = H, sulfate, terminal Gal, or terminal Xyl) proposed according to the data of partial acid hydrolysis.[722]

chains. It should be noted that the hexasaccharide fragment contained both D-Gal and L-Gal in one molecule. Unfortunately, no data were given on the presence of sulfate in the starting material and on the fate of sulfate groups under treatment with lithium in ethylenediamine. New information was later obtained by investigation of a similar polysaccharide, using different chemical methods of structural analysis in combination with NMR spectroscopy.[722] An important acidic tetrasaccharide fragment β-D-Xylp-(1 → 3)-α-D-Glcp-(1 → 3)-α-D-GlcpA-(1 → 3)-L-Gal was isolated demonstrating the sequence of four main monosaccharides in the backbone. Surprisingly, the authors were unable to distinguish between 2- and 4-linked xylose residues, and hence, alternative structures **46** (Scheme 11) with these two possibilities have been suggested for the linear polysaccharide backbone.[722] Sulfate groups, and nonreducing terminal Xyl and Gal residues, were attached at O-6 of Glc residue of the main chain.

Extracellular polysaccharides are produced by some species of the genus *Rhodella*, but their composition seems to be even more complex than that of *Porphyridium*. In addition to monosaccharide components just mentioned, rhamnose and 3-*O*-methyl-xylose were found in the mucilage of *R. maculata*.[723,724] An extremely complex proteoglycan was shown to be produced by *Rhodella grisea*.[725]

Extracellular polysaccharides of the red microalgae demonstrate high viscosity[726] and resistance to enzymatic digestion.[727] They may be obtained on a large scale using biotechnological methods[728] and may have a wide range of promising industrial and medicinal applications. Chemical modifications, such as oversulfation[729] and quaternization,[730] were used to obtain polymers demonstrating new rheological and biological properties. A strong inhibitory effect on *H. simplex* and *Varicella zoster* viruses was shown for the polysaccharide from *Porphyridium* sp.,[731] and its low-molecular fragments obtained by microwave degradation exerted antioxidant activity.[732]

Information on the polysaccharide composition of multicellular freshwater red algae is very scanty. An acidic polysaccharide isolated from *Batrachospermum* sp. was shown to contain D- and L-galactose, D-mannose, D-xylose, L-rhamnose, 3-*O*-methyl-L-rhamnose, 3-*O*-methyl-D-galactose, and D-glucuronic acid, but was free of sulfate. Partial acid hydrolysis afforded two preponderant acidic oligosaccharides, both containing galactose and glucuronic acid in 1:1 ratio, and hence, a repeating sequence of these two residues was suggested as a major structural feature of the polysaccharide.[709] Two species of red algae were adapted to both freshwater and marine environments to compare the influence of these different conditions on their polysaccharide composition. Galactans of freshwater *B. atropurpurea* were found to be similar to those of marine isolates, containing disaccharide repeating units of agarose and porphyran types. *Bostrychia moritziana* also contained galactans of the agar group typical of the order Ceramiales, but polymers of freshwater sample were composed of a complex mixture of repeating disaccharide units, including 2-*O*-methyl-agarose, 6'-*O*-methyl-agarose, and 2-*O*-methyl-porphyran. In contrast, polymers of marine isolates of *B. moritziana* contained only trace amounts of 2-*O*-methyl disaccharides, but increased amounts of 6'-*O*-methyl disaccharide. Thus, the adaptive response of this alga to changing osmotic and ionic conditions may include changes in the pattern of methyl ether substitution of its agar-like galactan.[733]

The small cortical and large medullary cells of *K. alvarezii* were screened by wet sieving after extraction of carrageenans. It was found that, besides cellulose, the medullary cells contained a galactoglycan, strongly associated with other insoluble components of the cell walls, whereas the cortical cells contained a similar insoluble mannoglycan.[734] A subsequent study[735] enabled isolation of three alkali-soluble cell-wall components, a linear $(1 \rightarrow 4)$-β-D-glucomannan, a β-glucan sulfate, containing

(1→4)- and (1→3)-linkages in 9:1 ratio and carrying sulfate on C-6 of 64% of 4-linked glucose residues, and a component rich in galactose. The unusual glucan sulfate was possibly covalently linked or tightly associated with other cell-wall components, and these linkages rendered it insoluble in the carrageenan extraction step.

A unique polysaccharide composition was detected in *Apophloea lyallii*, a red seaweed that is endemic to New Zealand, where it spends a large amount of time out of the water. The polysaccharides extracted from the alga with hot water were non-sulfated and were composed of D-xylose, 2-*O*-methyl-D-galactose, D-glucuronic acid, D- and L-galactose, 3-*O*-methyl-galactose, and D-glucose. The polysaccharides were fractionated by anion-exchange chromatography, and the main fraction was shown to be a unique 2-*O*-methylgalacto-glucurono-xylo-glycan with a repeating trisaccharide unit represented by [→4)-α-D-(2-*O*-Me-Gal$p$)-(1→4)-β-D-Glc$p$A-(1→3)-α-D-Xyl$p$-(1→]$_n$.[736]

P. decipiens, similar to other Palmariales, contains a neutral xylan. In addition, an acidic component, containing 4.8% of uronic acids, 2.8% of sulfate, and 18.9% of protein, was obtained after separation from the xylan. This polymer was homogeneous in polyacrylamide gel electrophoresis. It was composed of galactose and xylose in the molar ratio 8.2:1.0 and of galacturonic and glucuronic acid in the ratio 1.5:1.0. Alkaline cleavage of the protein moiety indicated that it was a glycoprotein with *O*-glycosidic linkages between the carbohydrate and protein parts of molecules.[737]

VIII. ALGAL SYSTEMATICS AND POLYSACCHARIDE COMPOSITION

Several thousands of red algal species are known, but only a very limited number of them (mainly species of *Kappaphycus*, *Eucheuma*, and *Gracilaria*) are used on a large scale for industrial production of phycocolloids.[738,739] In order to find new sources of raw material, it is highly desirable to elucidate the relationship between the taxonomic position of the algae and the chemical structures of their main polysaccharides. Correspondingly, knowledge of the polysaccharide structures may be used as a very important tool in the algal classification. Both approaches require the preliminary structural analysis of polysaccharides of numerous algal species, but this process, as described in Sections IV.2–IV.5, is rather tedious and time consuming. Consequently, early attempts to find taxonomy–structure relationships were based on some indirect evidence on the chemical nature of polysaccharides. For example, gel-forming ability of aqueous extracts was suggested as a possible taxonomic criterion.[740,741] This property is directly connected with the practical application of polysaccharides and can be determined without difficulties, but the results are of limited value for taxonomy, since the gel-forming properties of a given polysaccharide are determined by superposition of many different structural factors. Another route to characterize the polysaccharides by specific enzymolysis[503] has the obvious limitations connected with the practical unavailability of the necessary enzymes.

Nevertheless, it is very attractive to have a simple procedure for characterizing the polysaccharide composition of a great number of algal samples. FTIR diffuse reflectance spectroscopy, which was formerly suggested for identification of carrageenans,[336] was then extended to the characterization of phycocolloids in diverse algal samples. The method has several important advantages; it requires only several milligrams of dry material, needs no extraction or preparation of pellets or films, and, hence, rapidly gives information on the native composition of algal samples. At the same time, the method has several shortcomings, since it gives no direct evidence on the monosaccharide composition of polysaccharides or on the absolute configuration of monosaccharide constituents. Assignment of spectral bands may be done only tentatively, especially in the case of samples of unusual composition. The method was applied to determination of the cell-wall composition for 224 species of algae belonging to 28 families of the order Gigartinales, and the results were published, together with the most probable attribution of more than 30 IR spectral peaks to specific bonds in the molecules of sulfated galactans.[742]

Introduction of the reductive hydrolysis procedure made it possible to determine the monosaccharide composition of algal polysaccharides, including 3,6-anhydrogalactose and 2-O-methyl-3,6-anhydrogalactose content, in hydrolyzates of small samples of algal biomass before extraction of the polysaccharides. Moreover, a combination of total and partial reductive hydrolysis gave evidence on the absolute configuration of 3,6-anhydrogalactose and, hence, allowed distinction between representatives of agars and carrageenans as well as detecting the presence of DL-hybids.[213–215] Being applied to 44 species of red algae collected from Kamchatka coastal waters, the method demonstrated substantial differences in polysaccharide composition between representatives of different orders and families.[217] Reductive hydrolysis in combination with other methods of structural analysis of polysaccharides was successfully used to solve some problems in classification of the species belonging to the genus *Porphyra*[743] and to the family Gigartinaceae.[744] Thus, differences in the galactan structures in representatives of *Sarcothalia* and *Gigartina* indicate the presence of chemotaxonomically distinct groupings of species, which are aligned to those determined by *rbc*L sequence analysis.[745]

NMR spectroscopy gives valuable information on the structures of polysaccharides, but its application requires preliminary isolation of polysaccharide samples from

TABLE III
Specific Polysaccharides in Representatives of Several Orders of the Red Algae

Order	Specific Water-Soluble Polysaccharides
Bangiales	Partially methylated and sulfated agaran–agarose hybrids (porphyrans)
Nemaliales	Sulfated xylomannans and neutral xylans
Gelidiales	Typical agars
Corallinales	Sulfated and methylated xylogalactans of the agar group and alginic acids
Halymeniales	Complex sulfated galactans (DL-hybrids)
Gracilariales	Agars, including highly substituted agaran–agarose hybrids
Gigartinales	Typical carrageenans and DL-hybrids
Rhodymeniales	Complex sulfated galactans (DL-hybrids)
Palmariales	Neutral xylans and sulfated galactans of the agar group
Ceramiales	Highly modified galactans of the agar group and DL-hybrids

the algal biomass. The method has been widely used for characterization of red algal polysaccharides in connection with the taxonomic position of the species (see reviews refs. 581 and 746).

As follows from Table III, two orders, Nemaliales and Palmariales, containing mannans and xylans, respectively, as the main polysaccharides, can be distinguished from other orders simply according to the results of acid hydrolysis of the biomass (a good example is the transfer of *L. simplex* into *P. decipiens* after identification of xylan as the main cell-wall component of the alga[73]). Other orders contain galactans, and more-detailed investigation is necessary to distinguish between representatives of different orders, families or genera. It is clear that only a combination of several chemical and physicochemical methods of structural analysis can give the most reliable information on the structure of polysaccharides. Such detailed structural data were successfully used several times for resolution of some special taxonomic problems.[747–749]

The classification of red algae is now being reinvestigated according to the data of molecular genetics.[745,750,751] Without any doubt, information on the polysaccharide structures can also be used as valuable supplementary material for modern taxonomy of the Rhodophyta.

REFERENCES

1. H. H. Selby and R. L. Whistler, Agar, in R. L. Whistler and J. N. BeMiller, (Eds.), *Industrial Gums: Polysaccharides and Their Derivatives,* 3rd ed. Academic Press, San Diego, CA, 1993, pp. 87–103.
2. G. H. Therkelsen, Carrageenan, in R. L. Whistler and J. N. BeMiller, (Eds.), *Industrial Gums: Polysaccharides and Their Derivatives,* 3rd ed. Academic Press, San Diego, CA, 1993, pp. 145–180.

3. C. Araki, Agar-agar. III. Acetylation of the agar-like substance of *Gelidium amansii* L., *J. Chem. Soc. Jpn.*, 58 (1937) 1338–1350 (*Chem. Abstr.*, 32 (1938) 4172).
4. C. Araki, Structure of the agarose constituent of agar-agar, *Bull. Chem. Soc. Jpn.*, 29 (1956) 543–544.
5. T. Mori, Seaweed polysaccharides, *Adv. Carbohydr. Chem.*, 8 (1953) 315–350.
6. D. B. Smith and W. H. Cook, Fractionation of carrageenin, *Arch. Biochem. Biophys.*, 45 (1953) 232–233.
7. N. S. Anderson, T. C. S. Dolan, and D. A. Rees, Evidence for a common structural pattern in the polysaccharide sulphates of the Rhodophyceae, *Nature*, 205 (1965) 1060–1062.
8. D. A. Rees, Structure, conformation, and mechanism in the formation of polysaccharide gels and networks, *Adv. Carbohydr. Chem. Biochem.*, 24 (1969) 267–332.
9. E. L. Percival and R. H. McDowell, Chemistry and Enzymology of Marine Algal Polysaccharides, Academic Press, London, 1967, pp 73–98, 127–156.
10. S. V. Yarotsky, A. S. Shashkov, and A. I. Usov, Analysis of ^{13}C-NMR spectra of some red seaweed galactans, *Sov. J. Bioorg. Chem.*, 3 (1977) 840–841 (English translation from *Bioorg. Khim.*, 3 (1977) 1135–1137).
11. S. S. Bhattacharjee, W. Yaphe, and G. K. Hamer, ^{13}C-N.m.r. spectroscopic analysis of agar, κ-carrageenan and ι-carrageenan, *Carbohydr. Res.*, 60 (1978) C1–C3.
12. T. J. Painter, Algal polysaccharides, in G. O. Aspinall, (Ed.), *The Polysaccharides*, Vol. 2, Academic Press, New York, 1983, pp. 195–285.
13. J. S. Craigie, Cell walls, in K. M. Cole and R. G. Sheath, (Eds.), *Biology of the Red Algae*, Cambridge University Press, Cambridge, 1990, pp. 221–257.
14. B. Kloareg and R. S. Quatrano, Structure of the cell walls of marine algae and ecophysiological functions of the matrix polysaccharides, *Oceanogr. Mar. Biol. Annu. Rev.*, 26 (1988) 259–315.
15. A. I. Usov, Sulfated polysaccharides of red seaweeds, *Food Hydrocolloids*, 6 (1992) 9–23.
16. A. I. Usov, Structural analysis of red seaweed galactans of agar and carrageenan groups, *Food Hydrocolloids*, 12 (1998) 301–308.
17. M. Lahaye, Developments on gelling algal galactans, their structure and physico-chemistry, *J. Appl. Phycol.*, 13 (2001) 173–184.
18. V. L. Campo, D. F. Kawano, D. B. da Silva, Jr., and I. Carvalho, Carrageenans: Biological properties, chemical modifications and structural analysis—A review, *Carbohydr. Polym.*, 77 (2009) 167–180.
19. D. A. Rees, E. R. Morris, D. Thom, and J. K. Madden, Shapes and interactions of carbohydrate chains, in G. O. Aspinall, (Ed.), *The Polysaccharides*, Vol. 1, Academic Press, New York, 1982, pp. 195–290.
20. L. Piculell, Gelling carrageenans, in A. M. Stephen, (Ed.), *Food Polysaccharides and Their Applications*, Marcel Dekker, New York, 1995, pp. 205–244.
21. G. A. De Ruiter and B. Rudolph, Carrageenan biotechnology, *Trends Food Sci. Technol.*, 8 (1997) 389–395.
22. R. Armisen and F. Galatas, Agar, in G. O. Phillips and P. A. Williams, (Eds.), *Handbook of Hydrocolloids*, Woodhead Publishing Ltd., Cambridge, England, 2000, pp. 21–40.
23. A. Imeson, Carrageenan, in G. O. Phillips and P. A. Williams, (Eds.), *Handbook of Hydrocolloids*, Woodhead Publishing Ltd., Cambridge, England, 2000, pp. 87–102.
24. H. J. Bixler and K. D. Johndro, Philippine natural grade or semi-refined carrageenan, in G. O. Phillips and P. A. Williams, (Eds.), *Handbook of Hydrocolloids*, Woodhead Publishing Ltd., Cambridge, England, 2000, pp. 425–442.
25. C. A. Stortz, Carrageenans: Structural and conformational studies, in K. J. Yarema, (Ed.), *Handbook of Carbohydrate Engineering, Taylor and Francis*, Taylor and Francis, CRC Press, Boca Raton, 2005, pp. 211–245.
26. R. H. Reed, Solute accumulation and osmotic adjustment, in K. M. Cole and R. G. Sheath, (Eds.), *Biology of the Red Algae*, Cambridge, Cambridge University Press, 1990, pp. 147–170.
27. G. O. Kirst, Low MW carbohydrates and ions in Rhodophyceae: Quantitative measurement of floridoside and digeneaside, *Phytochemistry*, 19 (1980) 1107–1110.

28. S. Bondu, N. Kervarec, E. Deslandes, and R. Pichon, Separation of floridoside and isofloridosides by HPLC and complete ^{1}H and ^{13}C NMR spectral assignments for D-isofloridoside, *Carbohydr. Res.*, 342 (2007) 2470–2473.
29. U. Karsten, D. N. Thomas, G. Weykam, C. Daniel, and G. O. Kirst, A simple and rapid method for extraction and separation of low molecular weight carbohydrates using high-performance liquid chromatography, *Plant Physiol. Biochem.*, 29 (1991) 373–378.
30. C. Simon-Colin, M.-A. Bessières, and E. Deslandes, An alternative HPLC method for the quantification of floridoside in salt-stressed cultures of the red alga *Grateloupia doryphora*, *J. Appl. Phycol.*, 14 (2002) 123–127.
31. S. D. Ascêncio, A. Orsato, R. A. França, M. E. R. Duarte, and M. D. Noseda, Complete ^{1}H and ^{13}C NMR assignment of digeneaside, a low-molecular-mass carbohydrate produced by red seaweeds, *Carbohydr. Res.*, 341 (2006) 677–682.
32. C. Simon-Colin, N. Kervarec, R. Pichon, and E. Deslandes, Complete ^{1}H and ^{13}C spectral assignment of floridoside, *Carbohydr. Res.*, 337 (2002) 279–280.
33. C. Simon-Colin, F. Michaud, J.-M. Léger, and E. Deslandes, Crystal structure and chirality of natural floridoside, *Carbohydr. Res.*, 338 (2003) 2413–2416.
34. C. Vonthron-Sénécheau, J. Sopkova-de Oliveira Santos, I. Mussio, and A.-M. Rusig, X-ray structure of floridoside isolated from the marine red algae *Dilsea carnosa*, *Carbohydr. Res.*, 343 (2008) 2697–2698.
35. A. Claude, S. Bondu, F. Michaud, N. Bourgougnon, and E. Deslandes, X-ray structure of a sodium salt of digeneaside isolated from red alga *Ceramium botryocarpum*, *Carbohydr. Res.*, 344 (2009) 707–710.
36. B. J. D. Meeuse, M. Andries, and J. A. Wood, Floriden starch, *J. Exp. Bot.*, 11 (1960) 129–140.
37. V. C. Barry, T. G. Halsall, E. L. Hirst, and J. K. N. Jones, The polysaccharides of the Florideæ. Floridean starch, *J. Chem. Soc.* (1949) 1468–1470.
38. S. Peat, J. R. Turvey, and J. M. Evans, The structure of floridean starch. Part I. Linkage analysis by partial acid hydrolysis, *J. Chem. Soc.* (1959) 3223–3227.
39. S. Peat, J. R. Turvey, and J. M. Evans, The structure of floridean starch. Part II. Enzymic hydrolysis and other studies, *J. Chem. Soc.* (1959) 3341–3344.
40. H. Ozaki, M. Maeda, and K. Nisizawa, Floridean starch of calcareous red alga, *Joculator maximus*, *J. Biochem.*, 61 (1967) 497–503.
41. S. Yu, A. Blennow, M. Bojko, F. Madsen, C. E. Olsen, and S. B. Engelsen, Physico-chemical characterization of floridean starch of red algae, *Starch/Stärke*, 54 (2002) 66–74.
42. R. Viola, P. Nivall, and M. Pedersen, The unique features of starch metabolism in red algae, *Proc. R. Soc. Lond. B*, 268 (2001) 1417–1422.
43. N. J. Patron and P. J. Keeling, Common evolutionary origin of starch biosynthetic enzymes in green and red algae, *J. Phycol.*, 41 (2005) 1131–1141.
44. T. Shimonaga, S. Fujiwara, M. Kaneko, A. Izumo, S. Nihei, P. B. Francisco, Jr., A. Satoh, N. Fujita, Y. Nakamura, and M. Tsuzuki, Variation in storage α-polyglucans of red algae: Amylose and semi-amylopectin types in *Porphyridium* and glycogen type in *Cyanidium*, *Mar. Biotechnol.*, 9 (2007) 192–202.
45. T. Shimonaga, M. Konishi, Y. Oyama, S. Fujiwara, A. Satoh, N. Fujita, C. Colleoni, A. Buléon, J.-L. Putaux, S. G. Ball, A. Yokoyama, Y. Hara, Y. Nakamura, and M. Tsuzuki, Variation in storage α-glucan of the Porphyridiales (Rhodophyta), *Plant Cell Physiol.*, 49 (2008) 103–116.
46. I. N. Stadnichuk, L. R. Semenova, G. P. Smirnova, and A. I. Usov, A highly branched storage polyglucan in the thermoacidophilic red microalga *Galdieria maxima* cells, *Prikl. Biokhim. Mikrobiol.*, 43 (2007) 88–93 (in Russian).
47. B. J. D. Meeuse and B. N. Smith, A note on the amylolytic breakdown of some raw algal starches, *Planta*, 57 (1962) 624–635.

48. S. Yu, L. Kenne, and M. Pedersen, α-1,4-Glucan lyase, a new class of starch/glycogen degrading enzyme. I. Efficient purification and characterization from red seaweeds, *Biochim. Biophys. Acta*, 1156 (1993) 313–320.
49. M. A. Baute, R. Baute, and G. Deffieux, Fungal enzymic activity degrading 1,4-α-D-glucans to 1,5-D-anhydrofructose, *Phytochemistry*, 27 (1988) 3401–3403.
50. S. Kametani, Y. Shiga, and H. Akanuma, Hepatic production of 1,5-anhydrofructose and 1,5-anhydroglucitol in rat by the third glycogenolytic pathway, *Eur. J. Biochem.*, 242 (1996) 832–838.
51. I. Lundt and S. Yu, 1,5-Anhydro-D-fructose: Biocatalytic and chemical synthetic methods for the preparation, transformation and derivatization, *Carbohydr. Res.*, 345 (2010) 181–190.
52. R. H. Marchessault and P. R. Sundararajan, Cellulose, in G. O. Aspinall, (Ed.), *The Polysaccharides*, Vol. 2, Academic Press, New York, 1983, pp. 11–95.
53. A. G. Ross, Some typical analyses of red seaweeds, *J. Sci. Food Agric.*, 4 (1953) 333–335.
54. L. S. Mukai, J. S. Craigie, and R. G. Brown, Chemical composition and structure of the cell walls of the conchocelis and thallus phases of *Porphyra tenera* (Rhodophyceae), *J. Phycol.*, 17 (1981) 192–198.
55. J. N. C. Whyte and J. R. Englar, Polysaccharides of the red alga *Rhodymenia pertusa*. II. Cell-wall glucan; proton magnetic resonance studies on permethylated polysaccharides, *Can. J. Chem.*, 49 (1971) 1302–1305.
56. R. Toffanin, S. H. Knutsen, C. Bertocchi, R. Rizzo, and E. Murano, Detection of cellulose in the cell wall of some red algae by ^{13}C NMR spectroscopy, *Carbohydr. Res.*, 262 (1994) 167–171.
57. M. Koyama, J. Sugiyama, and T. Itoh, Systematic survey on crystalline features of algal celluloses, *Cellulose*, 4 (1997) 147–160.
58. R. H. Newman and T. C. Davidson, Crystalline forms and cross-sectional dimensions of cellulose microfibrils in the Florideophyceae (Rhodophyta), *Bot. Mar.*, 47 (2004) 490–495.
59. I. Tsekos, N. Orologas, and W. Herth, Cellulose microfibril assembly and orientation in some bangiophyte red algae: Relationship between synthesizing terminal complexes and microfibril structure, shape, and dimensions, *Phycologia*, 38 (1999) 217–224.
60. E. Roberts and A. W. Roberts, A cellulose synthase (CESA) gene from the red alga *Porphyra yezoensis* (Rhodophyta), *J. Phycol.*, 45 (2009) 203–212.
61. I. Tsekos, The sites of cellulose synthesis in algae: Diversity and evolution of cellulose-synthesizing enzyme complexes, *J. Phycol.*, 35 (1999) 635–655.
62. J. K. N. Jones, The structure of the mannan present in *Porphyra umbilicalis*, *J. Chem. Soc.* (1950) 3292–3295.
63. A. I. Usov, S. V. Yarotsky, and M. L. Esteves, Polysaccharides of algae. 23. Polysaccharides of the red alga *Bangia fuscopurpurea* (Dillw.) Lyngb., *Sov. J. Bioorg. Chem.*, 4 (1978) 51–57 (English translation from *Bioorg. Khim.*, 4 (1978) 66–73).
64. M. R. Gretz, J. M. Aronson, and M. R. Sommerfeld, Cellulose in the cell walls of the Bangiophyceae (Rhodophyta), *Science*, 207 (1980) 779–781.
65. M. R. Gretz, M. R. Sommerfeld, and J. M. Aronson, Cell wall composition of the generic phase of *Bangia atropurpurea* (Rhodophyta), *Bot. Mar.*, 25 (1982) 529–535.
66. M. R. Gretz, J. M. Aronson, and M. R. Sommerfeld, Taxonomic significance of cellulosic cell walls in the Bangiales (Rhodophyta), *Phytochemistry*, 23 (1984) 2513–2514.
67. M. R. Gretz, J. M. Aronson, and M. R. Sommerfeld, Cell wall composition of the conchocelis phases of *Bangia atropurpurea* and *Porphyra leucosticta* (Rhodophyta), *Bot. Mar.*, 29 (1986) 91–96.
68. B. H. Howard, Hydrolysis of the soluble pentosans of wheat flour and *Rhodymenia palmata* by ruminal micro-organisms, *Biochem. J.*, 67 (1957) 643–651.
69. H. Björndal, K. E. Eriksson, P. J. Garegg, B. Lindberg, and B. Swan, Studies on the xylan from the red seaweed *Rhodymenia palmata*, *Acta Chem. Scand.*, 19 (1965) 2309–2315.

70. J. R. Turvey and E. L. Williams, The structures of some xylans from red algae, *Phytochemistry*, 9 (1970) 2383–2388.
71. A. I. Usov, S. V. Yarotsky, A. S. Shashkov, and V. P. Tishchenko, Polysaccharides of algae. 22. Polysaccharide composition of *Rhodymenia stenogona* Perest. and use of carbon-13 nmr spectroscopy for determining the structure of xylans, *Sov. J. Bioorg. Chem.*, 4 (1978) 44–50 (English translation from *Bioorg. Khim.*, 4 (1978) 57–65).
72. P. Kovač, J. Hirsch, A. S. Shashkov, A. I. Usov, and S. V. Yarotsky, ^{13}C-N.M.R. spectra of xylo-oligosaccharides and their application to the elucidation of xylan structures, *Carbohydr. Res.*, 85 (1980) 177–185.
73. N. M. Adams, R. H. Furneaux, I. J. Miller, and L. A. Whitehouse, Xylan from *Leptosarca simplex* and carrageenans from *Iridea*, *Cenacrum* and *Nemastoma* species from the Subantarctic Islands of New Zealand, *Bot. Mar.*, 31 (1988) 9–14.
74. J. R. Jerez, B. Matsuhiro, and C. C. Urzúa, Chemical modifications of the xylan from *Palmaria decipiens*, *Carbohydr. Polym.*, 32 (1997) 155–159.
75. M. C. Matulewicz, A. S. Cerezo, R. M. Jarret, and N. Syn, High resolution ^{13}C-NMR spectroscopy of "mixed linkage" xylans, *Int. J. Biol. Macromol.*, 14 (1992) 29–32.
76. M. Lahaye, C. Rondeau-Mouro, E. Deniaud, and A. Buléon, Solid-state ^{13}C NMR spectroscopy studies of xylans in the cell wall of *Palmaria palmata* (L. Kuntze, Rhodophyta), *Carbohydr. Res.*, 338 (2003) 1559–1569.
77. K. C. Morgan, J. L. C. Wright, and F. J. Simpson, Review of chemical constituents of the red alga *Palmaria palmata* (Dulse), *Econ. Bot.*, 34 (1980) 27–50.
78. O. Marrion, A. Schwertz, J. Fleurence, J. L. Guéant, and C. Willaume, Improvement of the digestibility of the proteins of the red alga *Palmaria palmata* by physical processes and fermentation, *Nahrung/Food*, 47 (2003) 339–344.
79. E. Deniaud, J. Fleurence, and M. Lahaye, Preparation and chemical characterization of cell wall fractions enriched in structural proteins from *Palmaria palmata* (Rhodophyta), *Bot. Mar.*, 46 (2003) 366–377.
80. Y. Joubert and J. Fleurence, Simultaneous extraction of proteins and DNA by an enzymatic treatment of the cell wall of *Palmaria palmata* (Rhodophyta), *J. Appl. Phycol.*, 20 (2008) 55–61.
81. E. Deniaud, J. Fleurence, and M. Lahaye, Interactions of the mix-linked β-(1,3)/β-(1,4)-D-xylans in the cell walls of *Palmaria palmata* (Rhodophyta), *J. Phycol.*, 39 (2003) 74–82.
82. E. Deniaud, B. Quemener, J. Fleurence, and M. Lahaye, Structural studies of the mix-linked β-(1→3)/β-(1→4)-D-xylans from the cell wall of *Palmaria palmata* (Rhodophyta), *Int. J. Biol. Macromol.*, 33 (2003) 9–18.
83. A. S. Cerezo, A. Lezerovich, R. Labriola, and D. A. Rees, A xylan from the red seaweed *Chaetangium fastigiatum*, *Carbohydr. Res.*, 19 (1971) 289–296.
84. A. S. Cerezo, The fine structure of *Chaetangium fastigiatum* xylan: Studies of the sequence and configuration of the (1→3) linkages, *Carbohydr. Res.*, 22 (1972) 209–211.
85. J. R. Nunn, H. Parolis, and I. Russell, Polysaccharides of the red alga *Chaetangium erinaceum*. Part I. Isolation and characterization of the water-soluble xylan, *Carbohydr. Res.*, 26 (1973) 169–180.
86. A. I. Usov, K. S. Adamyants, S. V. Yarotsky, A. A. Anoshina, and N. K. Kochetkov, The isolation of a sulphated mannan and a neutral xylan from the red seaweed *Nemalion vermiculare* Sur, *Carbohydr. Res.*, 26 (1973) 282–283.
87. A. I. Usov, K. S. Adamyants, S. V. Yarotsky, A. A. Anoshina, and N. K. Kochetkov, Polysaccharides of algae. 14. Isolation of sulfated mannan and neutral xylan from the red alga *Nemalion vermiculare*, *Zh. Obshch. Khim.*, 44 (1974) 416–420 (in Russian).
88. A. I. Usov, S. V. Yarotsky, and M. L. Estevez, Polysaccharides of algae. 32. Polysaccharides of the red seaweed *Galaxaura squalida* Kjellm, *Sov. J. Bioorg. Chem.*, 7 (1981) 692–699 (English translation from *Bioorg. Khim.*, 7 (1981) 1261–1270).

89. A. I. Usov and I. M. Dobkina, Polysaccharides of algae. 43. Neutral xylan and sulfated xylomannan from the red seaweed *Liagora valida*, *Sov. J. Bioorg. Chem.*, 17 (1991) 596–603 (English translation from *Bioorg. Khim.*, 17 (1991) 1051–1058).
90. Y. Fukuyama, A. D. Kolender, M. Nishioka, H. Nonami, M. C. Matulewicz, R. Erra-Balsells, and A. S. Cerezo, Matrix-assisted ultraviolet laser desorption/ionization time-of-flight mass spectrometry of β-(1→3), β-(1→4)-xylans from Nothogenia fastigiata using nor-harmane as matrix, *Rapid Commun. Mass Spectrom.*, 19 (2005) 349–358.
91. W. Nerinckx, A. Broberg, J.Ø. Duus, P. Ntarima, L. A. S. Parolis, H. Parolis, and M. Claeyssens, Hydrolysis of *Nothogenia erinacea* xylan by xylanases from families 10 and 11, *Carbohydr. Res.*, 339 (2004) 1047–1060.
92. P. Mandal, C. A. Pujol, E. B. Damonte, T. Ghosh, and B. Ray, Xylans from *Scinaia hatei*: Structural features, sulfation and anti-HSV activity, *Int. J. Biol. Macromol.*, 46 (2010) 173–178.
93. C. J. Lawson and D. A. Rees, An enzyme for the metabolic control of polysaccharide conformation and function, *Nature (London)*, 227 (1970) 392–393.
94. D. A. Rees, Estimation of the relative amounts of isomeric sulphate esters in some sulphated polysaccharides, *J. Chem. Soc.* (1961) 5168–5171.
95. M. Duckworth and W. Yaphe, The structure of agar. Part I. Fractionation of a complex mixture of polysaccharides, *Carbohydr. Res.*, 16 (1971) 189–197.
96. C. A. Stortz and A. S. Cerezo, Novel findings in carrageenans, agaroids and "hybrid" red seaweed galactans, *Curr. Top. Phytochem.*, 4 (2000) 121–134.
97. C. Araki and S. Hirase, Studies on the chemical constitution of agar-agar. XXI. Re-investigation of methylated agarose of *Gelidium amansii*, *Bull. Chem. Soc. Jpn.*, 33 (1960) 291–295.
98. G. Lewis, N. Stanley, and G. Guist, Commercial production and applications of algal hydrocolloids, in C. Lembi, (Ed.), *Algae and Human Affairs*, University of Washington, Seattle, 1988, pp. 206–232.
99. N. S. Anderson and D. A. Rees, Porphyran: A polysaccharide with a masked repeating structure, *J. Chem. Soc.* (1965) 5880–5887.
100. A. I. Usov and E. G. Kozlova, Polysaccharides of algae. 20. Studies on odonthalan, a sulfated polysaccharide from the red alga *Odonthalia corymbifera* (Gmel.) J.Ag., *Sov. J. Bioorg. Chem.*, 1 (1975) 693–699 (English translation from *Bioorg. Khim.*, 1 (1975) 912–918).
101. T. J. Painter, The polysaccharides of *Furcellaria fastigiata*. I. Isolation and partial mercaptolysis of a gel-fraction, *Can. J. Chem.*, 38 (1960) 112–118.
102. M. R. Cases, C. A. Stortz, and A. S. Cerezo, Structure of the 'corallinans'—Sulfated xylogalactans from *Corallina officinalis*, *Int. J. Biol. Macromol.*, 16 (1994) 93–97.
103. S. H. Knutsen, D. E. Myslabodski, B. Larsen, and A. I. Usov, A modified system of nomenclature for red algal galactans, *Bot. Mar.*, 37 (1994) 163–169.
104. E. Zablakis and G. A. Santos, The carrageenan of *Catenella nipae* Zanard., a marine red alga, *Bot. Mar.*, 29 (1986) 319–322.
105. C. W. Greer and W. Yaphe, Characterization of hybrid (*beta-kappa-gamma*) carrageenan from *Eucheuma gelatinae* J. Agardh (Rhodophyta, Solieriaceae) using carrageenase, infrared and ^{13}C-nuclear magnetic resonance spectroscopy, *Bot. Mar.*, 27 (1984) 473–478.
106. Monograph Nr. 1, Carrageenan. FMC Corporation, Marine Colloids Division, 1977, p. 6.
107. N. S. Anderson, T. C. S. Dolan, and D. A. Rees, Carrageenans. Part 7. Polysaccharides from *Eucheuma spinosum* and *Eucheuma cottonii*. The covalent structure of *iota*-carrageenan, *J. Chem. Soc. Perkin Trans.*, 1 (1973) 2173–2176.
108. A. N. O'Neill, Derivatives of 4-O-β-D-galactopyranosyl-3,6-anhydro-D-galactose from *kappa*-carrageenin, *J. Am. Chem. Soc.*, 77 (1955) 6324–6326.
109. T. C. S. Dolan and D. A. Rees, The carrageenans. Part 2. The positions of the glycosidic linkages and the sulphate esters in *lambda*-carrageenan, *J. Chem. Soc.* (1965) 3534–3539.

110. N. S. Anderson, T. C. S. Dolan, C. J. Lawson, A. Penman, and D. A. Rees, Carrageenans. Part 5. The masked repeating structure of kappa- and mu-carrageenans, *Carbohydr. Res.*, 7 (1968) 468–473.
111. D. J. Stancioff and N. F. Stanley, Infrared and chemical studies on algal polysaccharides, *Proc. Int. Seaweed Symp.*, 6 (1969) 595–609.
112. A. Penman and D. A. Rees, Carrageenans. Part 9. Methylation analysis of galactan sulphates from *Furcellaria fastigiata, Gigartina canaliculata, Gigartina chamissoi, Gigartina atropurpurea, Ahnfeltia durvillaei, Gymnogongrus furcellatus, Eucheuma isiforme, Eucheuma uncinatum, Aghardiella tenera, Pachymenia hymantophora,* and *Gloiopeltis cervicornis.* Structure of xi-carrageenan, *J. Chem. Soc. Perkin Trans.*, 1 (1973) 2182–2187.
113. V. DiNinno, E. L. McCandless, and R. A. Bell, Pyruvic acid derivative of a carrageenan from a marine red algae (*Petrocelis* species), *Carbohydr. Res.*, 71 (1979) C1–C4.
114. J. Mollion, S. Moreau, and D. Christiaen, Isolation of a new type of carrageenan from *Rissoella verruculosa* (Bert.) J. Ag. (Rhodophyta, Gigartinales), *Bot. Mar.*, 29 (1986) 549–552.
115. J. N. BeMiller, Agar. Preparation of agar and agarose; methanolysis; mercaptolysis, *Methods, Carbohydr. Chem.*, 5 (1965) 65–69.
116. T. J. Painter, Kappa-Carrageenan. Isolation of κ-carrageenan from red algae, *Methods Carbohydr. Chem.*, 5 (1965) 98–100.
117. J. S. Craigie and C. Leigh, Carrageenans and agars, in J. A. Hellebust and J. S. Craigie, (Eds.), *Handbook of Phycological Methods. Physiological and Biochemical Methods,* Cambridge University Press, Cambridge, 1978, pp. 109–131.
118. M. Lahaye, C. Rochas, and W. Yaphe, A new procedure for determining the heterogeneity of agar polymers in the cell walls of *Gracilaria* spp. (Gracilariaceae, Rhodophyta), *Can. J. Bot.*, 64 (1986) 579–585.
119. S. H. Knutsen, E. Murano, M. D'Amato, R. Toffanin, R. Rizzo, and S. Paoletti, Modified procedures for extraction and analysis of carrageenan applied to the red alga *Hypnea musciformis*, *J. Appl. Phycol.*, 7 (1995) 565–576.
120. R. A. Hoffmann, M. J. Gidley, D. Cooke, and W. J. Frith, Effect of isolation procedures on the molecular composition and physical properties of *Eucheuma cottonii* carrageenan, *Food Hydrocolloids*, 9 (1995) 281–289.
121. C. Araki, Agar-agar. I. Chemical changes in the process of preparation of agar-agar, *J. Chem. Soc. Jpn.*, 58 (1937) 1085–1088 (*Chem. Abstr.*, 32 (1938) 4172).
122. S. H. Knutsen and H. Grasdalen, Characterization of water-extractable polysaccharides from Norwegian *Furcellaria lumbricalis* (Huds.) Lamour. (Gigartinales, Rhodophyceae) by IR and NMR spectroscopy, *Bot. Mar.*, 30 (1987) 497–505.
123. S. Hjerten, A new method for preparation of agarose for gel electrophoresis, *Biochim. Biophys. Acta*, 62 (1962) 445–449.
124. M. I. Errea and M. C. Matulewicz, Cold water-soluble polysaccharides from tetrasporic *Pterocladia capillacea*, *Phytochemistry*, 37 (1994) 1075–1078.
125. K. Izumi, A new method for fractionation of agar, *Agr. Biol. Chem.*, 34 (1970) 1739–1740.
126. S. Hjerten, Some new methods for the preparation of agarose, *J. Chromatogr.*, 61 (1971) 73–80.
127. M. Duckworth, K. C. Hong, and W. Yaphe, The agar polysaccharides of *Gracilaria* species, *Carbohydr. Res.*, 18 (1971) 1–9.
128. A. I. Usov, E. G. Ivanova, and V. F. Makienko, Polysaccharides of algae. 29. Agar composition in different generations of *Gracilaria verrucosa* (Huds.) Papenf., *Sov. J. Bioorg. Chem.*, 5 (1979) 1219–1224 (English translation from *Bioorg. Khim.*, 5 (1979) 1647–1653).
129. A. I. Usov and E. G. Ivanova, Polysaccharides of algae. 42. Composition and properties of agars from the Black Sea red seaweeds *Gracilaria verrucosa* (Huds.) Papenf. f. *procerima* and *Gracilaria dura* (Ag.) J. Ag., *Sov. J. Bioorg. Chem.*, 16 (1990) 885–890 (English translation from *Bioorg. Khim.*, 16 (1990) 1545–1551).

130. M. I. Errea and M. C. Matulewicz, Hot water-soluble polysaccharides from tetrasporic *Pterocladia capillacea*, *Phytochemistry*, 42 (1996) 1071–1073.
131. A. S. Cerezo, The carrageenan system of *Gigartina scottsbergii* S. et G. Part I. Studies on a fraction of κ-carrageenan, *J. Chem. Soc. C* (1967) 992–997.
132. A. S. Cerezo, The carrageenan system of *Gigartina scottsbergii* S. et G. Part II. Analysis of the system and studies on the structure of the fraction precipitated at 0.3-0.4 M potassium chloride, *J. Chem. Soc. C*. (1967) 2491–2495.
133. C. A. Stortz and A. S. Cerezo, The systems of carrageenans from cystocarpic and tetrasporic stages from *Iridaea undulosa*: Fractionation with potassium chloride and methylation analysis of the fractions, *Carbohydr. Res.*, 242 (1993) 217–227.
134. A. J. Pernas, O. Smidsrød, B. Larsen, and A. Haug, Chemical heterogeneity of carrageenans as shown by fractional precipitation with potassium chloride, *Acta Chem. Scand.*, 21 (1967) 98–110.
135. O. Smidsrød, B. Larsen, A. J. Pernas, and A. Haug, The effect of alkali treatment on the chemical heterogeneity and physical properties of some carrageenans, *Acta Chem. Scand.*, 21 (1967) 2585–2598.
136. M. C. Matulewicz and A. S. Cerezo, A rapid turbidimetric method for the determination of the precipitation curves of carrageenans, *J. Sci. Food Agric.*, 26 (1975) 243–250.
137. J. Mollion, H. Morvan, F. Bellanger, and S. Moreau, Carbon-13 NMR study of heterogeneity in the carrageenan system from *Rissoella verruculosa*, *Phytochemistry*, 27 (1988) 2023–2026.
138. N. Caram-Lelham, R. L. Cleland, and L. O. Sundeloef, Temperature and salt optimization of κ-carrageenan fractionation by DEAE-cellulose, *Int. J. Biol. Macromol.*, 16 (1994) 71–75.
139. C. A. Stortz, M. R. Cases, and A. S. Cerezo, The system of agaroids and carrageenans from the soluble fraction of the tetrasporic stage of the red seaweed *Iridaea undulosa*, *Carbohydr. Polym.*, 34 (1997) 61–65.
140. M. A. Roberts, H.-J. Zhong, J. Prodolliet, and D. M. Goodall, Separation of high-molecular-mass carrageenan polysaccharides by capillary electrophoresis with laser-induced fluorescence detection, *J. Chromatogr. A*, 817 (1998) 353–366.
141. C. Rochas and M. Lahaye, Average molecular weight and molecular weight distribution of agarose and agarose-type polysaccharides, *Carbohydr. Polym.*, 10 (1989) 289–298.
142. G. Sworn, W. M. Marrs, and R. J. Hart, Characterisation of carrageenans by high-performance size-exclusion chromatography using a LiChrospher 1000 DIOL column, *J. Chromatogr.*, 403 (1987) 307–311.
143. D. Slootmaekers, J. A. P. P. Dijk, F. A. Varkevisser, C. J. Bloys van Treslong, and H. Reynaers, Molecular characterization of κ- and λ-carrageenan by gel permeation chromatography, light scattering, sedimentation analysis and osmometry, *Biophys. Chem.*, 41 (1991) 51–59.
144. V. Spichtig and S. Austin, Determination of the low molecular weight fraction of food-grade carrageenans, *J. Chromatogr. B*, 861 (2008) 81–87.
145. T. Hjerde, O. Smidsrød, and B. E. Christensen, Analysis of the conformational properties of κ- and ι-carrageenan by size-exclusion chromatography combined with low angle laser light scattering, *Biopolymers*, 49 (1999) 71–80.
146. C. Viebke, J. Borgström, and L. Piculell, Characterisation of kappa- and iota-carrageenan coils and helices by MALLS/GPC, *Carbohydr. Polym.*, 27 (1995) 145–154.
147. M. A. Roberts and B. Quemener, Measurement of carrageenans in food: Challenges, progress, and trends in analysis, *Trends Food Sci. Technol.*, 10 (1999) 169–181.
148. W. Yaphe and G. P. Arsenault, Improved resorcinol reagent for the determination of fructose, and of 3,6-anhydrogalactose in polysaccharides, *Anal. Biochem.*, 13 (1965) 143–148.
149. W. Anderson and W. Bowtle, Determination of iota-carrageenan with 2-thiobarbituric acid, *Analyst*, 99 (1974) 178–183.

150. E. M. Hoffmann and N. Bauknecht, A dye binding assay for the quantification of soluble and cell-bound acidic polysaccharides produced by red algae, *Anal. Biochem.*, 267 (1999) 245–251.
151. H. S. Soedjak, Colorimetric determination of carrageenans and other anionic hydrocolloids with methylene blue, *Anal. Chem.*, 66 (1994) 4514–4518.
152. M. F. Chaplin, Monosaccharides, in M. F. Chaplin and J. F. Kennedy, (Eds.), *Carbohydrate Analysis. A Practical Approach,* IRL Press, Oxford, Washington, DC, 1986, pp. 1–36.
153. E. Lau and A. Bacic, Capillary gas chromatography of partially methylated alditol acetates on a high-polarity, cross-linked, fused-silica BPX 70 column, *J. Chromatogr.*, 637 (1993) 100–103.
154. C. A. Stortz, M. C. Matulewicz, and A. S. Cerezo, Separation and identification of O-acetyl-O-methyl-galactononitriles by gas-liquid chromatography and mass spectrometry, *Carbohydr. Res.*, 111 (1982) 31–39.
155. H. Björndal, C. G. Hellerqvist, B. Lindberg, and S. Svensson, Gas-liquid chromatography and mass spectrometry in methylation analysis of polysaccharides, *Angew. Chem., Int. Ed. Engl.*, 9 (1970) 610–619.
156. G. J. Gerwig, J. P. Kamerling, and J. F. G. Vliegenthart, Determination of the D- and L-configuration of neutral monosaccharides by high-resolution capillary G.L.C., *Carbohydr. Res.*, 62 (1978) 349–357.
157. K. Leontein, B. Lindberg, and J. Lönngren, Assignment of absolute configuration of sugars by g.l.c. of their acetylated glycosides formed from chiral alcohols, *Carbohydr. Res.*, 62 (1978) 359–362.
158. R. Takano, K. Kamei-Hayashi, S. Hara, and S. Hirase, Assignment of the absolute configuration of partially methylated galactoses by combined gas-liquid chromatography–mass spectrometry, *Biosci. Biotech. Biochem.*, 57 (1993) 1195–1197.
159. M. R. Cases, A. S. Cerezo, and C. A. Stortz, Separation and quantitation of enantiomeric galactoses and their mono-O-methyl ethers as their diastereomeric acetylated 1-deoxy-1-(2-hydroxypropyla-mino)alditols, *Carbohydr. Res.*, 269 (1995) 333–341.
160. R. J. Sturgeon, Enzymic determination of D-galactose, *Methods Carbohydr. Chem.*, 8 (1980) 131–133.
161. R. J. Sturgeon, The enzymic determination of the enantiomers of galactose, *Carbohydr. Res.*, 200 (1990) 499–501.
162. N. W. Pirie, The preparation of heptaacetyl-dl-galactose by the acetolysis of agar, *Biochem. J.*, 30 (1936) 369–373.
163. C. Araki, Agar-agar VII L-Galactose and its derivatives, *J. Chem. Soc. Jpn.*, 59 (1938) 424–432 (*C. A.*, 35 (1941) 7946).
164. J. R. Turvey and P. R. Simpson, Polysaccharides from *Corallina officinalis*, *Proc. Int. Seaweed Symp.*, 5 (1966) 323–328.
165. R. Takano, J. Hayashi, K. Hayashi, S. Hara, and S. Hirase, Structure of a water-soluble polysaccharide sulfate from the red seaweed *Joculator maximus* Manza, *Bot. Mar.*, 39 (1996) 95–102.
166. A. I. Usov, M. I. Bilan, and A. S. Shashkov, Structure of a sulfated xylogalactan from the calcareous red alga *Corallina pilulifera* P. et R. (Rhodophyta, Corallinaceae), *Carbohydr. Res.*, 303 (1997) 93–102.
167. J. R. Nunn and M. M. von Holdt, Red-seaweed polysaccharides. Part II. *Porphyra capensis* and the separation of D- and L-galactose by crystallization, *J. Chem. Soc.* (1957) 1094–1097.
168. S. Hirase and C. Araki, Isolation of 6-*O*-methyl-D-galactose from the agar of *Ceramium boydenii*, *Bull. Chem. Soc. Jpn.*, 34 (1961) 1048.
169. C. Araki, K. Arai, and S. Hirase, Studies on the chemical constitution of agar-agar. XXIII. Isolation of D-xylose, 6-*O*-methyl-D-galactose, 4-*O*-methyl-L-galactose and *O*-methylpentose, *Bull. Chem. Soc. Jpn.*, 40 (1967) 959–962.
170. C. Araki, Some recent studies on the polysaccharides of agarophytes, *Proc. Int. Seaweed Symp.*, 5 (1966) 3–17.
171. J. R. Nunn and H. Parolis, A polysaccharide from *Aeodes orbitosa*, *Carbohydr. Res.*, 6 (1968) 1–11.
172. A. J. Farrant, J. R. Nunn, and H. Parolis, Sulphated polysaccharides of the Grateloupiaceae family. Part VI. A polysaccharide from *Pachymenia carnosa*, *Carbohydr. Res.*, 19 (1971) 161–168.

173. M. Lahaye, J. F. Revol, C. Rochas, J. McLachlan, and W. Yaphe, The chemical structure of *Gracilaria crassissima* (P. et H. Crouan in Schramm *et* Maze) and *G. tikvahiae* McLaughlin (Gigartinales, Rhodophyta) cell-wall polysaccharides, *Bot. Mar.*, 31 (1988) 491–501.
174. K. T. Bird, Evidence of 4-*O*-methyl-α-L-galactose in the agar of *Gracilaria verrucosa* strain G-16, *J. Appl. Phycol.*, 2 (1990) 383–384.
175. Y. Karamanos, M. Ondarza, F. Bellanger, and D. Christiaen, The linkage of 4-*O*-methyl-L-galactose in the agar polymers from *Gracilaria verrucosa*, *Carbohydr. Res.*, 187 (1989) 93–101.
176. J. S. Craigie and A. Jurgens, Structure of agars from *Gracilaria tikvahiae* Rhodophyta: Location of 4-*O*-methyl-L-galactose and sulphate, *Carbohydr. Polym.*, 11 (1989) 265–278.
177. S. Hirase, C. Araki, and K. Watanabe, Component sugars of the polysaccharide of the red seaweed *Grateloupia elliptica*, *Bull. Chem. Soc. Jpn.*, 40 (1967) 1445–1448.
178. M. R. Cases, C. A. Stortz, and A. S. Cerezo, Methylated, sulphated xylogalactans from the red seaweed *Corallina officinalis*, *Phytochemistry*, 31 (1992) 3897–3900.
179. A. I. Usov, E. G. Ivanova, and A. S. Shashkov, Polysaccharides of algae. 33. Isolation and ^{13}C-NMR spectral study of some new gel-forming polysaccharides from Japan Sea red seaweeds, *Bot. Mar.*, 26 (1983) 285–294.
180. R. H. Furneaux and T. T. Stevenson, The xylogalactan sulfate from *Chondria macrocarpa* (Ceramiales, Rhodophyta), *Hydrobiologia*, 204/205 (1990) 615–620.
181. A. I. Usov and M. Ya. Elashvili, Polysaccharides of algae. 44. Investigation of sulfated galactan from *Laurencia nipponica* Yamada (Rhodophyta, Rhodomelaceae) using partial reductive hydrolysis, *Bot. Mar.*, 34 (1991) 553–560.
182. R. Falshaw, R. H. Furneaux, and D. E. Stevenson, Agars from nine species of red seaweed in the genus *Curdiea* (Gracilariaceae, Rhodophyta), *Carbohydr. Res.*, 308 (1998) 107–115.
183. C. Araki, Agar-agar. II. Agar-like substance of *Gelidium amansii* L., *J. Chem. Soc. Jpn.*, 58 (1937) 1214–1234 (*C. A.*, 32 (1938) 4172).
184. J. R. Nunn, H. Parolis, and I. Russell, Sulphated polysaccharides of the Solieriaceae family. Part II. The acidic components of the polysaccharide from the red alga *Anatheca dentata*, *Carbohydr. Res.*, 29 (1973) 281–289.
185. R. Takano, Y. Nose, K. Hayashi, S. Hara, and S. Hirase, Agarose-carrageenan hybrid polysaccharide from *Lomentaria catenata*, *Phytochemistry*, 37 (1994) 1615–1619.
186. J. S. Craigie and H. Rivero-Carro, Agarocolloids from carrageenophytes, Abstr. XIVth Int. Seaweed Symp., University of Western Brittany: Brest, France, 1992, p. 71.
187. M. Ciancia, M. C. Matulewicz, and A. S. Cerezo, L-Galactose containing galactans from the carrageenophyte *Gigartina scottsbergii*, *Phytochemistry*, 34 (1993) 1541–1543.
188. M. Ciancia, M. C. Matulewicz, and A. S. Cerezo, A L-galactose-containing carrageenan from cystocarpic *Gigartina scottsbergii*, *Phytochemistry*, 45 (1997) 1009–1013.
189. R. Takano, H. Iwane-Sakata, K. Hayashi, S. Hara, and S. Hirase, Concurrence of agaroid and carrageenan chains in funoran from the red seaweed *Gloiopeltis furcata* Post. et Ruprecht (Cryptonemiales, Rhodophyta), *Carbohydr. Polym.*, 35 (1998) 81–87.
190. C. Bellion, G. Brigand, J. C. Prome, D. Welti, and S. Bociek, Identification et caracterisation des precurseurs biologiques des carraghenanes par spectroscopie de RMN du ^{13}C, *Carbohydr. Res.*, 119 (1983) 31–48.
191. C. M. Ibanez and B. Matsuhiro, Structural studies on the soluble polysaccharide from *Iridaea membranacea*, *Carbohydr. Res.*, 146 (1986) 327–334.
192. A. Chiovitti, M.-L. Liao, G. T. Kraft, S. L. A. Munro, D. J. Craik, and A. Bacic, Cell wall polysaccharides from Australian red algae of the family Solieriaceae (Gigartinales, Rhodophyta): Highly methylated carrageenans from the genus *Rhabdonia*, *Bot. Mar.*, 39 (1996) 47–59.

193. D. W. Renn, G. A. Santos, L. E. Dumont, C. A. Parent, N. F. Stanley, D. J. Stancioff, and K. B. Guiseley, Beta-carrageenan—Isolation and characterization, *Carbohydr. Polym.*, 22 (1993) 247–252.
194. S. Hirase, Studies on the chemical constitution of agar-agar. XIX. Pyruvic acid as a constituent of agar-agar (Part I). Identification and estimation of pyruvic acid in the hydrolysate of agar, *Bull. Chem. Soc. Jpn.*, 30 (1957) 68–70.
195. S. Hirase and K. Watanabe, The presence of pyruvate residues in λ-carrageenan and a similar polysaccharide, *Bull. Inst. Chem. Res. Kyoto Univ.*, 50 (1972) 332–336.
196. A. Chiovitti, A. Bacic, D. J. Craik, S. L. A. Munro, G. T. Kraft, and M.-L. Liao, Cell-wall polysaccharides from Australian red algae of the family Solieriaceae (Gigartinales, Rhodophyta): Novel, highly pyruvated carrageenans from the genus *Callophycus*, *Carbohydr. Res.*, 299 (1997) 229–243.
197. W. M. Nelson, G. A. Knight, R. Falshaw, R. H. Furneaux, A. Falshaw, and S. M. Lynds, Characterisation of the enigmatic red alga *Gelidium alanii* (Gelidiales) from northern New Zealand—Morphology, distribution, agar chemistry, *J. Appl. Phycol.*, 6 (1994) 497–507.
198. M. Duckworth and W. Yaphe, Definitive assay for pyruvic acid in agar and other algal polysaccharides, *Chem. Ind.*, 23 (1970) 747–748.
199. W. Yaphe, Colorimetric determination of 3,6-anhydrogalactose and galactose in marine algal polysaccharides, *Anal. Chem.*, 32 (1960) 1327–1330.
200. W. Yaphe, Colorimetric determination of 3,6-anhydrogalactose with the indolyl-3-acetic acid reagent, *Nature*, 197 (1963) 488–489.
201. B. Matsuhiro and A. B. Zanlungo, Colorimetric determination of 3,6-anhydrogalactose in polysaccharides from red seaweeds, *Carbohydr. Res.*, 118 (1983) 276–279.
202. C. Araki and S. Hirase, Chemical constitution of agar-agar. XV. Exhaustive mercaptolysis of agaragar, *Bull. Chem. Soc. Jpn.*, 26 (1953) 463–467 (*Chem. Abstr.*, 49 (1955) 9516).
203. A. N. O'Neill, 3,6-Anhydro-D-galactose as a constituent of κ-carrageenin, *J. Am. Chem. Soc.*, 77 (1955) 2837–2839.
204. C. Araki, Chemical studies on agar-agar. XIV. Investigation of component sugars of agar-agar by methanolysis. 1. Separation of 3,6-anhydro-L-galactose dimethylacetal, *J. Chem. Soc. Jpn.*, 65 (1944) 725–730 (*Chem. Abstr.*, 41 (1947) 3496).
205. N. K. Kochetkov, A. I. Usov, and L. I. Miroshnikova, Polysaccharides of algae. 4. Fractionation and methanolysis of a sulfated polysaccharide from *Laingia pacifica* Yamada, *Zh. Obshch. Khim.*, 40 (1970) 2469–2473 (in Russian).
206. A. I. Usov, R. A. Lotov, and N. K. Kochetkov, Polysaccharides of algae. 7. Preliminary investigation of polysaccharides of the red alga *Rhodomela larix* (Turn.) C.Ag., *Zh. Obshch. Khim.*, 41 (1971) 1154–1160 (in Russian).
207. B. Quemener, M. Lahaye, and F. Metro, Assessment of methanolysis for the determination of composite sugars of gelling carrageenans and agarose by HPLC, *Carbohydr. Res.*, 266 (1995) 53–64.
208. Y. Hama, H. Nakagawa, T. Sumi, X. Xia, and K. Yamaguchi, Anhydrous mercaptolysis of agar: An efficient preparation of 3,6-anhydro-L-galactose diethyl dithioacetal, *Carbohydr. Res.*, 318 (1999) 154–156.
209. A. L. Clingman, J. R. Nunn, and A. M. Stephen, Red-seaweed polysaccharides. Part I. *Gracilaria confervoides*, *J. Chem. Soc.* (1957) 197–203.
210. N. S. Anderson, T. C. S. Dolan, and D. A. Rees, Carrageenans. Part III. Oxidative hydrolysis of methylated κ-carrageenan and evidence for a masked repeating structure, *J. Chem. Soc. C* (1968) 596–601.
211. M. Errea, M. Ciancia, M. Matulewicz, and A. Cerezo, Separation and quantitation of enantiomeric 3,6-anhydrogalactoses by conversion to the corresponding diastereomeric acetylated *sec*-butyl 3,6-anhydrogalactonates, *Carbohydr. Res.*, 311 (1998) 235–238.

212. P. J. Garegg, B. Lindberg, P. Konradsson, and I. Kvarnström, Hydrolysis of glycosides under reducing conditions, *Carbohydr. Res.*, 176 (1988) 145–148.
213. A. I. Usov, A new chemical tool for characterization and partial depolymerization of red algal galactans, *Hydrobiologia*, 260/261 (1993) 641–645.
214. A. I. Usov and M. Ya Elashvili, Quantitative determination of 3,6-anhydrogalactose derivatives and partial fragmentation of the red algal galactans under reductive hydrolysis conditions, *Bioorg. Khim.*, 17 (1991) 839–848 (in Russian).
215. T. T. Stevenson and R. H. Furneaux, Chemical methods for the analysis of sulphated galactans from red algae, *Carbohydr. Res.*, 210 (1991) 277–298.
216. C. N. Jol, T. G. Neiss, B. Penninkhof, B. Rudolph, and G. A. De Ruiter, A novel high-performance anion-exchange chromatographic method for the analysis of carrageenans and agars containing 3,6-anhydrogalactose, *Anal. Biochem.*, 268 (1999) 213–222.
217. A. I. Usov and N. G. Klochkova, Polysaccharides of algae. 45. Polysaccharide composition of red seaweeds from Kamchatka coastal waters (Northwestern Pacific) studied by reductive hydrolysis of biomass, *Bot. Mar.*, 35 (1992) 371–378.
218. B. Soerbo, Sulfate: Turbidimetric and nephelometric methods, *Methods Enzymol.*, 143 (1987) 3–6.
219. D. M. Sullivan, Sulfate determination: Ion chromatography, *Methods Enzymol.*, 143 (1987) 7–11.
220. K. S. Dodgson and R. G. Price, A note on the determination of the ester sulphate content of sulphated polysaccharides, *Biochem. J.*, 84 (1962) 106–110.
221. J. S. Craigie, Z. C. Wen, and J. P. van der Meer, Interspecific, intraspecific and nutritionally-determined variations in the composition of agars from *Gracilaria* spp, *Bot. Mar.*, 27 (1984) 55–61.
222. C. A. Stortz and A. S. Cerezo, Specific fragmentation of carrageenans, *Carbohydr. Res.*, 166 (1987) 317–323.
223. C. A. Stortz and A. S. Cerezo, Room temperature, low-field ^{13}C-n.m.r. spectra of degraded kappa/iota carrageenans, *Int. J. Biol. Macromol.*, 13 (1991) 101–104.
224. M. Ciancia, M. C. Matulewicz, C. A. Stortz, and A. S. Cerezo, Room temperature, low-field ^{13}C-n.m.r. spectra of degraded carrageenans. Part II. On the specificity of the autohydrolysis reaction in kappa/iota and mu/nu structures, *Int. J. Biol. Macromol.*, 13 (1991) 337–340.
225. M. D. Noseda and A. S. Cerezo, Room temperature, low-field ^{13}C-n.m.r. spectra of degraded carrageenans: Part III. Autohydrolysis of a lambda carrageenan and of its alkali-treated derivative, *Int. J. Biol. Macromol.*, 15 (1993) 177–181.
226. M. Ciancia, Y. Sato, H. Nonami, A. S. Cerezo, R. Erra-Balsells, and M. C. Matulewicz, Autohydrolysis of a partially cyclized mu/nu-carrageenan and structural elucidation of the oligosaccharides by chemical analysis, NMR spectroscopy and UV-MALDI mass spectrometry, *ARKIVOC*, Xii (2005) 319–331.
227. C. Araki, Chemical studies on agar-agar. XIII. 1. Separation of agarobiose from the agar-agar-like substance of *Gelidium amansii* by partial hydrolysis, *J. Chem. Soc. Jpn.*, 65 (1944) 533–538 (*Chem. Abstr.*, 42 (1948) 1210).
228. B. Kazlowski, C. L. Pan, and Y. T. Ko, Separation and quantification of neoagaro- and agaro-oligosaccharide products generated from agarose digestion by β-agarase and HCl in liquid chromatography systems, *Cabohydr. Res.*, 343 (2008) 2443–2450.
229. M. K. Fatema, H. Nonami, D. R. B. Ducatti, A. G. Gonçalves, M. E. R. Duarte, M. D. Noseda, A. S. Cerezo, R. Erra-Balsells, and M. C. Matulewicz, Matrix-assisted laser desorption/ionization time-of-flight (MALDI-TOF) mass spectrometry analysis of oligosaccharides and oligosaccharide alditols obtained by hydrolysis of agaroses and carrageenans, two important types of red seaweed polysaccharides, *Carbohydr. Res.*, 345 (2010) 275–283.
230. H. Chen, X. Yan, P. Zhu, and J. Lin, Antioxidant activity and hepatoprotective potential of agaro-oligosaccharides in *vitro* and in *vivo*, *Nutri. J.*, 5 (2006) 31.

231. A. Penman and D. A. Rees, Carrageenans. Part XI. Mild oxidative hydrolysis of κ- and ι-carrageenans and the characterisation of oligosaccharide sulphates, *J. Chem. Soc. Perkin Trans.*, 1 (1973) 2191–2196.
232. A. I. Usov and E. G. Ivanova, Identification of 3,6-anhydrogalactose derivatives as *N*-phenylosotriazoles, *Sov. J. Bioorg. Chem.*, 7 (1981) 752–757 (English translation from *Bioorg. Khim.*, 7 (1981) 1372–1378).
233. A. I. Usov and I. M. Dobkina, Identification of various cleavage products of red seaweed galactans in the form of pyridylamino derivatives, *Sov. J. Bioorg. Chem.*, 11 (1985) 596–603(English translation from *Bioorg. Khim.*, 11 (1985) 1110–1118).
234. C. Araki and S. Hirase, Studies on the chemical constitution of agar-agar. XVII. Isolation of crystalline agarobiose dimethylacetal by partial methanolysis of agar-agar, *Bull. Chem. Soc. Jpn.*, 27 (1954) 109–112.
235. C. Araki and S. Hirase, Partial methanolysis of the mucilage of *Chondrus ocellatus* Holmes, *Bull. Chem. Soc. Jpn.*, 29 (1956) 770–775.
236. S. Hirase, C. Araki, and T. Ito, Isolation of agarobiose derivative from the mucilage of *Gloiopeltis furcata*, *Bull. Chem. Soc. Jpn.*, 31 (1958) 428–431.
237. T. J. Painter, Confirmatory evidence for the structure of carrobiose, *J. Chem. Soc.* (1964) 1396–1400.
238. S. Hirase, Studies on the chemical constitution of agar-agar. XIX. Pyruvic acid as a constituent of agar-agar (Part 2). Isolation of a pyruvic acid-linking disaccharide derivative from the methanolysis products of agar, *Bull. Chem. Soc. Jpn.*, 30 (1957) 70–75.
239. S. Hirase, Studies on the chemical constitution of agar-agar. XIX. Pyruvic acid as a constituent of agar-agar (Part 3). Structure of the pyruvic acid-linking disaccharide derivative isolated from the methanolysis products of agar, *Bull. Chem. Soc. Jpn.*, 30 (1957) 75–79.
240. N. S. Anderson and D. A. Rees, The repeating structure of some polysaccharide sulphates from red seaweeds, *Proc. Int. Seaweed Symp.*, 5 (1966) 243–249.
241. C. J. Lawson, D. A. Rees, D. J. Stancioff, and N. F. Stanley, Carrageenans. Part VIII. Repeating structures of galactan sulphates from *Furcellaria fastigiata, Gigartina canaliculata, Gigartina chamissoi, Gigartina atropurpurea, Ahnfeltia durvillaei, Gymnogongrus furcellatus, Eucheuma cottonii, Eucheuma spinosum, Eucheuma isiforme, Eucheuma uncinatum, Aghardhiella tenera, Pachymenia hymantophora*, and *Gloiopeltis cervicornis*, *J. Chem. Soc. Perkin Trans.*, 1 (1973) 2177–2182.
242. J. N. C. Whyte, S. P. C. Hosford, and J. R. Englar, Assignment of agar or carrageenan structures to red algal polysaccharides, *Carbohydr. Res.*, 140 (1985) 336–341.
243. A. I. Usov and M. Ya. Elashvili, Specific fragmentation of red algal galactans under reducing conditions, *F. E. C. S. Fifth Int. Conf. Chem. Biotechnol. Biol. Active Nat. Prod. Conf. Proc. Bulgarian Acad. Sci.*, 2 (1989) 346–350.
244. R. Falshaw and R. H. Furneaux, The structural analysis of disaccharides from red algal galactans by methylation and reductive partial-hydrolysis, *Carbohydr. Res.*, 269 (1995) 183–189.
245. A. I. Usov and E. G. Ivanova, Polysaccharides of algae. 46. Studies on agar from the red seaweed *Gelidiella acerosa*, *Sov. J. Bioorg. Chem.*, 18 (1992) 588–595 (English translation from Bioorg. Khim. 18 (1992) 1108–1116).
246. A. G. Gonçalves, D. R. B. Ducatti, M. E. R. Duarte, and M. D. Noseda, Sulfated and pyruvylated disaccharide alditols obtained from a red seaweed galactan: ESIMS and NMR approaches, *Carbohydr. Res.*, 337 (2002) 2443–2453.
247. A. G. Gonçalves, D. R. B. Ducatti, R. G. Paranha, M. E. R. Duarte, and M. D. Noseda, Positional isomers of sulfated oligosaccharides obtained from agarans and carrageenans: Preparation and capillary electrophoresis separation, *Carbohydr. Res.*, 340 (2005) 2123–2134.
248. K. Morgan and A. N. O'Neill, Degradative studies on λ-carrageenin, *Can. J. Chem.*, 37 (1959) 1201–1209.

249. C. J. Lawson and D. A. Rees, Carrageenans. Part VI. Reinvestigation of the acetolysis products of λ-carrageenan, revision of the structure of 'α-1,3-galactotriose', and a further example of the reverse specificities of glycoside hydrolysis and acetolysis, *J. Chem. Soc. C* (1968) 1301–1304.
250. N. K. Kochetkov, A. I. Usov, and M. A. Rechter, Polysaccharides of algae. 8. Acetolysis of λ-polysaccharide from *Tichocarpus crinitus* (Gmel.) Rupr., *Zh. Obshch. Khim.*, 41 (1971) 1160–1165(in Russian).
251. A. J. Farrant, J. R. Nunn, and H. Parolis, Sulphated polysaccharides of the Grateloupiaceae family. Part VII. Investigation of the acetolysis products of a partially desulphated sample of the polysaccharide of *Pachymenia carnosa*, *Carbohydr. Res.*, 25 (1972) 283–292.
252. A. J. R. Allsobrook, J. R. Nunn, and H. Parolis, Investigation of the acetolysis products of the sulphated polysaccharide of *Aeodes ulvoidea*, *Carbohydr. Res.*, 40 (1975) 337–344.
253. A. I. Usov and V. V. Barbakadze, Polysaccharides of algae. 27. Partial acetolysis of the sulfated galactan from the red seaweed *Grateloupia divaricata* Okam, *Sov. J. Bioorg. Chem.*, 4 (1978) 814–820(English translation from *Bioorg. Khim.*, 4 (1978) 1107–1115).
254. J. R. Nunn, H. Parolis, and I. Russell, Sulphated polysaccharides of the Solieriaceae family. Part I. A polysaccharide from *Anatheca dentata*, *Carbohydr. Res.*, 20 (1971) 205–215.
255. A. J. R. Allsobrook, J. R. Nunn, and H. Parolis, The linkage of 4-*O*-methyl-L-galactose in the sulphated polysaccharide of *Aeodes ulvoidea*, *Carbohydr. Res.*, 36 (1974) 139–145.
256. J. R. Turvey, Sulfates of the simple sugars, *Adv. Carbohydr. Chem.*, 20 (1965) 183–218.
257. T. J. Painter, Isolation of sulphated sugars from some algal polysaccharides, *Chem. Ind.* (1959) 1488.
258. J. R. Turvey and D. A. Rees, Isolation of L-galactose 6-sulphate from a seaweed polysaccharide, *Nature*, 189 (1961) 831–832.
259. N. K. Kochetkov, A. I. Usov, L. I. Miroshnikova, and O. S. Chizhov, Polysaccharides of algae. 12. Partial hydrolysis of a polysaccharide from *Laingia pacifica* Yamada, *Zh. Obshch. Khim.*, 43 (1973) 1832–1839 (in Russian).
260. J. F. Batey and J. R. Turvey, The galactan sulphate of the red alga *Polysiphonia lanosa*, *Carbohydr. Res.*, 43 (1975) 133–143.
261. J. R. Turvey and E. L. Williams, The agar-type polysaccharide from the red alga *Ceramium rubrum*, *Carbohydr. Res.*, 49 (1976) 419–425.
262. D. A. Rees, A note on the characterization of carbohydrate sulphates by acid hydrolysis, *Biochem. J.*, 88 (1963) 343–345.
263. R. Takano, Desulfation of sulfated carbohydrates, *Trends Glycosci. Glycotechnol.*, 14 (2002) 343–351.
264. A. Karlsson and S. K. Singh, Acid hydrolysis of sulphated polysaccharides. Desulphation and the effect of molecular mass, *Carbohydr. Polym.*, 38 (1999) 7–15.
265. T. G. Kantor and M. Schubert, A method for the desulfation of chondroitin sulfate, *J. Am. Chem. Soc.*, 79 (1957) 152–153.
266. N. K. Kochetkov, A. I. Usov, and L. I. Miroshnikova, Polysaccharides of algae. 1. Water-soluble polysaccharides of the red seaweed *Laingia pacifica*, *Zh. Obshch. Khim.*, 37 (1967) 792–796 (in Russian).
267. A. I. Usov, K. S. Adamyants, L. I. Miroshnikova, A. A. Shaposhnikova, and N. K. Kochetkov, Solvolytic desulphation of sulphated carbohydrates, *Carbohydr. Res.*, 18 (1971) 336–338.
268. A. I. Usov and K. S. Adamyants, Solvolytic desulfation of dextran sulfate, *Sov. J. Bioorg. Chem.*, 1 (1975) 506–510 (English translation from *Bioorg. Khim.*, 1 (1975) 659–664).
269. K. Nagasawa, Y. Inoue, and T. Kamata, Solvolytic desulfation of glycosaminoglycuronan sulfates with dimethyl sulfoxide containing water or methanol, *Carbohydr. Res.*, 58 (1977) 47–55.
270. K. Nagasawa, Y. Inoue, and T. Tokuyasu, An improved method for the preparation of chondroitin by solvolytic desulfation of chondroitin sulfates, *J. Biochem.*, 86 (1979) 1323–1329.
271. I. J. Miller and J. W. Blunt, Desulfation of algal galactans, *Carbohydr. Res.*, 309 (1998) 39–43.

272. D. A. Navarro, M. L. Flores, and C. A. Stortz, Microwave-assisted desulfation of sulfated polysaccharides, *Carbohydr. Polym.*, 69 (2007) 742–747.
273. T. Toida, K. Sato, N. Sakamoto, S. Sakai, S. Hosoyama, and R. J. Linhardt, Solvolytic depolymerization of chondroitin and dermatan sulfates, *Carbohydr. Res.*, 344 (2009) 888–893.
274. A. I. Usov, L. I. Miroshnikova, and N. K. Kochetkov, Polysaccharides of algae. 9. Solvolytic desulfation of a polysaccharide from the red seaweed *Laingia pacifica* Yamada, *Zh. Obshch. Khim.*, 42 (1972) 945–949 (in Russian).
275. R. Takano, M. Matsuo, K. Kamei-Hayashi, S. Hara, and S. Hirase, A novel regioselective desulfation method specific to carbohydrate 6-sulfate using silylating reagents, *Biosci. Biotech. Biochem.*, 56 (1992) 1577–1580.
276. R. Takano, T. Kanda, K. Hayashi, K. Yoshida, and S. Hara, Desulfation of sulfated carbohydrates mediated by silylating reagents, *J. Carbohydr. Chem.*, 14 (1995) 885–888.
277. M. Matsuo, R. Takano, K. Kamei-Hayashi, and S. Hara, A novel regioselective desulfation of polysaccharide sulfates: Specific 6-O-desulfation with *N*, *O*-bis(trimethylsilyl)acetamide, *Carbohydr. Res.*, 241 (1993) 209–215.
278. R. Takano, K. Hayashi, S. Hara, and S. Hirase, Funoran from the red seaweed, *Gloiopeltis complanata*: Polysaccharides with sulphated agarose structure and their precursor structure, *Carbohydr. Polym.*, 27 (1995) 305–311.
279. R. Takano, Z. Ye, T.-V. Ta, K. Hayashi, Y. Kariya, and S. Hara, Specific 6-O-desulfation of heparin, *Carbohydr. Lett.*, 3 (1998) 71–77.
280. A. A. Kolender and M. C. Matulewicz, Sulfated polysaccharides from the red seaweed *Georgiella confluens*, *Carbohydr. Res.*, 337 (2002) 57–68.
281. M. I. Errea and M. C. Matulewicz, Unusual structures in the polysaccharides from the red seaweed *Pterocladiella capillacea* (Gelidiaceae, Gelidiales), *Carbohydr. Res.*, 338 (2003) 943–953.
282. A. A. Kolender and M. C. Matulewicz, Desulfation of sulfated galactans with chlorotrimethylsilane. Characterization of β-carrageenan by ^{1}H-NMR spectroscopy, *Carbohydr. Res.*, 339 (2004) 1619–1629.
283. R. S. Tipson, Sulfonic esters of carbohydrates, *Adv. Carbohydr. Chem.*, 8 (1953) 107–216.
284. D. A. Rees, Enzymic synthesis of 3:6-anhydro-L-galactose within porphyran from L-galactose 6-sulphate units, *Biochem. J.*, 81 (1961) 347–352.
285. K. F. Wong and J. S. Craigie, Sulfohydrolase activity and carrageenan biosynthesis in *Chondrus crispus* (Rhodophyceae), *Plant Physiol.*, 61 (1978) 663–666.
286. M. Zinoun, M. Diouris, P. Potin, J. Y. Floc'h, and E. Deslandes, Evidence of sulfohydrolase activity in the red alga *Callipeltis jubata*, *Bot. Mar.*, 40 (1997) 49–53.
287. J. T. Aguilan, J. E. Broom, J. A. Hemmingson, F. M. Dayrit, M. N. E. Montaño, M. C. A. Dancel, M. R. Niñonuevo, and R. H. Furneaux, Structural analysis of carrageenan from farmed varieties of Philippine seaweed, *Bot. Mar.*, 46 (2003) 179–192.
288. J. A. Hemmingson and R. H. Furneaux, Variation of native agar gel strength in light deprived *Gracilaria chilensis* Bird, McLachlan et Oliveira, *Bot. Mar.*, 46 (2003) 307–314.
289. R. D. Villanueva, L. Hilliou, and I. Sousa-Pinto, Postharvest culture in the dark: An eco-friendly alternative to alkali treatment for enhancing the gel quality of κ/ι-hybrid carrageenan from *Chondrus crispus* (Gigartinales, Rhodophyta), *Bioresour. Technol.*, 100 (2009) 2633–2638.
290. R. E. Rincones, S. Yu, and M. Pedersen, Effect of dark treatment on the starch degradation and the agar quality of cultivated *Gracilariopsis lemaneiformis* (Rhodophyta, Gracilariales) from Venezuela, *Hydrobiologia*, 260/261 (1993) 633–640.
291. M. Ciancia, M. D. Noseda, M. C. Matilewicz, and A. S. Cerezo, Alkali-modification of carrageenans: Mechanism and kinetics in the kappa/iota, mu/nu and lambda-series, *Carbohydr. Polym.*, 20 (1993) 95–98.

292. M. D. Noseda and A. S. Cerezo, Alkali modification of carrageenans-II. The cyclization of model compounds containing non-sulfated β-D-galactose units, *Carbohydr. Polym.*, 26 (1995) 1–3.
293. M. Ciancia, M. C. Matulewicz, and A. S. Cerezo, Alkaline modification of carrageenans. Part III. Use of mild alkaline media and high ionic strengths, *Carbohydr. Polym.*, 32 (1997) 293–295.
294. A. G. Viana, M. D. Noseda, M. E. R. Duarte, and A. S. Cerezo, Alkali modification of carrageenans. Part V. The iota-nu hybrid carrageenan from *Eucheuma denticulatum* and its cyclization to iota-carrageenan, *Carbohydr. Polym.*, 58 (2004) 455–460.
295. M. D. Noseda, A. G. Viana, M. E. R. Duarte, and A. S. Cerezo, Alkali modification of carrageenans. Part IV. Porphyrans as model compounds, *Carbohydr. Polym.*, 42 (2000) 301–305.
296. R. M. G. Zibetti, M. D. Noseda, A. S. Cerezo, and M. E. R. Duarte, The system of galactans from *Cryptonemia crenulata* (Halymeniaceae, Halymeniales) and the structure of two major fractions. Kinetic studies on the alkaline cyclization of the unusual diad G2S→D(L)6S, *Carbohydr. Res.*, 340 (2005) 711–722.
297. R. G. M. Zibetti, M. E. R. Duarte, M. D. Noseda, F. G. Colodi, D. R. B. Ducatti, L. G. Ferreira, M. A. Cardoso, and A. S. Cerezo, Galactans from *Cryptonemia* species. Part II: Studies on the system of galactans of *Cryptonemia seminervis* (Halymeniales) and on the structure of major fractions, *Carbohydr. Res.*, 344 (2009) 2364–2374.
298. D. Jouanneau, M. Guibet, P. Boulenguer, J. Mazoyer, M. Smietana, and W. Helbert, New insights into the structure of hybrid κ-/μ-carrageenan and its alkaline conversion, *Food Hydrocolloids*, 24 (2010) 452–461.
299. D. A. Navarro and C. A. Stortz, Microwave-assisted alkaline modification of red seaweed galactans, *Carbohydr. Polym.*, 62 (2005) 187–191.
300. D. A. Navarro and C. A. Stortz, Determination of the configuration of 3,6-anhydrogalactose and cyclizable α-galactose 6-sulfate units in red seaweed galactans, *Carbohydr. Res.*, 338 (2003) 2111–2118.
301. M. Jaseja, R. N. Rej, F. Sauriol, and A. S. Perlin, Novel regio- and stereoselective modifications of heparin in alkaline solution. Nuclear magnetic resonance spectroscopic evidence, *Can. J. Chem.*, 67 (1989) 1449–1456.
302. E. Percival and J. K. Wold, The acid polysaccharide from the green seaweed *Ulva lactuca*. Part II. The site of the ester sulphate, *J. Chem. Soc.* (1963) 5459–5468.
303. C. Araki, Agar-agar. V. D-Galactose and its derivatives, *J. Chem. Soc. Jpn.*, 58 (1937) 1362–1383 (*Chem. Abstr.*, 32 (1938) 4172).
304. E. G. V. Percival and J. C. Somerville, Acetylation and methylation of agar-agar and the isolation of 2,4,6-trimethyl-α-d-galactose by hydrolysis, *J. Chem. Soc.* (1937) 1615–1619.
305. E. T. Dewar and E. G. V. Percival, The polysaccharides of carrageen. II. The *Gigartina stellata* polysaccharide, *J. Chem. Soc.* (1947) 1622–1626.
306. A. I. Usov and V. S. Arkhipova, Polysaccharides of algae. 30. Methylation of κ-carrageenan type polysaccharides of the red seaweeds *Tichocarpus crinitus* (Gmel.) Rupr., *Furcellaria fastigiata* (Huds.) Lam. and *Phyllophora nervosa* (De Cand.) Grev., *Sov. J. Bioorg. Chem.*, 7 (1981) 222–226 (English translation from *Bioorg. Khim.*, 7 (1981) 385–390).
307. C. A. Stortz and A. S. Cerezo, The potassium chloride—Soluble carrageenans of the red seaweed *Iridaea undulosa* B, *Carbohydr. Res.*, 145 (1986) 219–235.
308. M. C. Matulewicz, M. Ciancia, M. D. Noseda, and A. S. Cerezo, Methylation analysis of carrageenans from tetrasporic and cystocarpic stages of *Gigartina scottsbergii*, *Phytochemistry*, 29 (1990) 3407–3410.
309. N. K. Kochetkov, A. I. Usov, and L. I. Miroshnikova, Polysaccharides of algae. 11. Methylation analysis of a polysaccharide from *Laingia pacifica* Yamada, *Zh. Obshch. Khim.*, 42 (1972) 2309–2315.
310. A. S. Cerezo, The carrageenan system of *Gigartina scottsbergii* S. et G. Part IV. Methylation analysis of a partly desulphated derivative, *Carbohydr. Res.*, 36 (1974) 201–204.

311. M. R. Cases, C. A. Stortz, and A. S. Cerezo, Separation and identification of partially ethylated galactoses as their acetylated aldononitriles and alditols by capillary gas chromatography and mass spectrometry, *J. Chromatogr.*, 662 (1994) 293–299.
312. I. Ciucanu, Per-O-methylation reaction for structural analysis of carbohydrates by mass spectrometry, *Anal. Chim. Acta*, 576 (2006) 147–155.
313. S. Hakomori, Rapid permethylation of glycolipids and polysaccharides, catalyzed by methylsulfinyl carbanion in dimethyl sulfoxide, *J. Biochem.*, 55 (1964) 205–208.
314. I. Ciucanu and F. Kerek, A simple and rapid method for the permethylation of carbohydrates, *Carbohydr. Res.*, 131 (1984) 209–217.
315. I. Ciucanu and R. Carpita, Per-O-methylation of neutral carbohydrates directly from aqueous samples for gas chromatography and mass spectrometry analysis, *Anal. Chim. Acta*, 585 (2007) 81–85.
316. I. J. Miller and J. W. Blunt, Mobility of sulfate ester during structural determination of red algal galactans, *Bot. Mar.*, 45 (2002) 559–565.
317. D. A. Rees, F. B. Williamson, S. A. Frangou, and E. R. Morris, Fragmentation and modification of ι-carrageenan and characterization of the polysaccharide order-disorder transition in solution, *Eur. J. Biochem.*, 122 (1982) 71–79.
318. A. I. Usov, M. A. Rechter, and N. K. Kochetkov, Polysaccharides of algae. 3. Isolation and preliminary investigation of a λ-polysaccharide from *Tichocarpus crinitus* (Gmel.) Rupr., *Zh. Obshch. Khim.*, 39 (1969) 905–911 (in Russian).
319. V. V. Barbakadze and A. I. Usov, Polysaccharides of algae. 26. Methylation and periodate oxidation of a polysaccharide from the red seaweed *Grateloupia divaricata* Okam., *Sov. J. Bioorg. Chem.*, 4 (1978) 808–813(English translation from *Bioorg. Khim.*, 4 (1978) 1100–1106).
320. G. O. Aspinall, Chemical characterization and structure determination of polysaccharides, in O. Aspinall, (Ed.), *The Polysaccharides*, Vol. 1, Academic Press, New York, 1982, pp. 35–131.
321. D. Grant and A. Holt, The properties of some sulphated derivatives of D-glucose and D-galactose, *J. Chem. Soc.* (1960) 5026–5031.
322. J. R. Turvey, M. J. Clancy, and T. P. Williams, Sulphates of monosaccharides and derivatives. Part II. Periodate oxidation, *J. Chem. Soc.* (1961) 1692–1697.
323. A. I. Usov and M. I. Bilan, Polysaccharides of algae. 49. Isolation of alginate, sulfated xylogalactan, and floridean starch from calcareous red alga *Bossiella cretacea* (P. et R.) Johansen (Rhodophyta, Corallinaceae), *Russ. J. Bioorg. Chem.*, 22 (1996) 106–112 (English translation from *Bioorg. Khim.*, 22 (1996) 126–133).
324. A. I. Usov, L. I. Miroshnikova, and N. K. Kochetkov, Polysaccharides of algae. 10. Periodate oxidation of a polysaccharide from *Laingia pacifica* Yamada, *Zh. Obshch. Khim.*, 42 (1972) 1623–1630 (in Russian).
325. W. B. Neely, Infrared spectra of carbohydrates, *Adv. Carbohydr. Chem.*, 12 (1957) 13–33.
326. H. Spedding, Infrared spectroscopy and carbohydrate chemistry, *Adv. Carbohydr. Chem.*, 19 (1964) 23–50.
327. M. Mathlouthi and J. L. Koenig, Vibrational spectra of carbohydrates, *Adv. Carbohydr. Chem. Biochem.*, 44 (1986) 7–89.
328. A. G. Lloyd, K. S. Dodgson, R. G. Price, and F. A. Rose, Infrared spectra of sulphate esters. I. Polysaccharide sulphates, *Biochim. Biophys. Acta*, 46 (1961) 108–115.
329. A. G. Lloyd and K. S. Dodgson, Infrared studies on sulphate esters. II. Monosaccharide sulphates, *Biochim. Biophys. Acta*, 46 (1961) 116–120.
330. N. S. Anderson, T. C. S. Dolan, A. Penman, D. A. Rees, G. P. Mueller, D. J. Stancioff, and N. F. Stanley, Carrageenans. Part IV. Variations in the structure and gel properties of κ-carrageenan, and the characterisation of sulphate esters by infrared spectroscopy, *J. Chem. Soc. C* (1968) 602–606.
331. C. Rochas, M. Lahaye, and W. Yaphe, Sulfate content of carrageenan and agar determined by infrared spectroscopy, *Bot. Mar.*, 29 (1986) 335–340.

332. B. Matsuhiro and P. Rivas, Second-derivative Fourier transform infrared spectra of seaweed galactans, *J. Appl. Phycol.*, 5 (1993) 45–51.
333. M. Seccal, J. P. Huvenne, P. Legrand, B. Sombret, J. C. Mollet, A. Mouradi-Givernaud, and M. C. Verdus, Direct structural identification of polysaccharides from red algae by FTIR microspectrometry. 1. Localization of agar in *Gracilaria verrucosa* sections, *Microchim. Acta*, 112 (1993) 1–10.
334. M. Sekkal, P. Legrand, J. P. Huvenne, and M. C. Verdus, The use of FTIR microspectroscopy as a new tool for the identification *in situ* of polygalactans in red seaweeds, *J. Mol. Struct.*, 294 (1993) 227–230.
335. I. Fournet, E. Ar Gall, E. Deslandes, J.-P. Huvenne, B. Sombret, and J. Y. Floc'h, *In situ* measurements of cell wall components in the red alga *Solieria chordalis* (Solieriaceae, Rhodophyta) by FTIR microspectrometry, *Bot. Mar.*, 40 (1997) 45–48.
336. T. Chopin and E. Whalen, A new and rapid method for carrageenan identification by FT IR diffuse reflectance spectroscopy directly on dried, ground algal material, *Carbohydr. Res.*, 246 (1993) 51–59.
337. M. Cerná, A. S. Barros, A. Nunes, S. M. Rocha, I. Delgadillo, J. Copiková, and M. A. Coimbra, Use of FT-IR spectroscopy as a tool for the analysis of polysaccharide food additives, *Carbohydr. Polym.*, 51 (2003) 383–389.
338. S. P. Jacobsson and A. Hagman, Chemical composition analysis of carrageenan by infrared spectroscopy using partial least-squares and neural networks, *Anal. Chim. Acta*, 284 (1993) 137–147.
339. J. Prado-Fernández, J. A. Rodriguez-Vásquez, E. Tojo, and J. M. Andrade, Quantitation of κ-, ι- and λ-carrageenans by mid-infrared spectroscopy and PLS regression, *Anal. Chim. Acta*, 480 (2003) 23–37.
340. E. Tojo and J. Prado, Chemical composition of carrageenan blends determined by IR spectroscopy combined with a PLS multivariate calibration method, *Carbohydr. Res.*, 338 (2003) 1309–1312.
341. T. Turquois, S. Acquistapace, F. Arce Vera, and D. H. Welti, Composition of carrageenan blends inferred from ^{13}C-NMR and infrared spectroscopic analysis, *Carbohydr. Polym.*, 31 (1996) 269–278.
342. B. Matsuhiro, Vibrational spectroscopy of seaweed galactans, *Hydrobiologia*, 326/327 (1996) 481–489.
343. L. Pereira, A. M. Amado, A. T. Critchley, F. van de Velde, and P. J. A. Ribeiro-Claro, Identification of selected seaweed polysaccharides (phycocolloids) by vibrational spectroscopy (FTIR-ATR and FT-Raman), *Food Hydrocolloids*, 23 (2009) 1903–1909.
344. T. Malfait, H. Van Dael, and F. van Cauwelaert, Molecular structure of carrageenans and kappa oligomers: A Raman spectroscopic study, *Int. J. Biol. Macromol.*, 11 (1989) 259–264.
345. M. Sekkal and P. Legrand, A spectroscopic investigation of the carrageenans and agar in the 1500–100 cm^{-1} spectral range, *Spectrochim. Acta*, 49A (1993) 209–221.
346. L. Pereira and J. F. Mesquita, Carrageenophytes of occidental Portuguese coast: 1-spectroscopic analysis in eight carrageenophytes from Buarcos bay, *Biomol. Eng.*, 20 (2003) 217–222.
347. L. Pereira, A. Sousa, H. Coelho, A. M. Amado, and P. J. A. Ribeiro-Claro, Use of FTIR, FT-Raman and ^{13}C-NMR spectroscopy for identification of some seaweed phycocolloids, *Biomol. Eng.*, 20 (2003) 223–228.
348. M. Dyrby, R. V. Petersen, J. Larsen, B. Rudolf, L. Nørgaard, and S. B. Engelsen, Towards on-line monitoring of the composition of commercial carrageenan powders, *Carbohydr. Polym.*, 57 (2004) 337–348.
349. N. K. Kochetkov and O. S. Chizhov, Mass spectrometry of carbohydrate derivatives, *Adv. Carbohydr. Chem.*, 21 (1966) 39–94.
350. J. Lönngren and S. Svensson, Mass spectrometry in structural analysis of natural carbohydrates, *Adv. Carbohydr. Chem. Biochem.*, 29 (1974) 41–106.
351. A. Dell, F.a.b.-Mass spectrometry of carbohydrates, *Adv. Carbohydr. Chem. Biochem.*, 45 (1987) 19–72.
352. O. S. Chizhov, B. M. Zolotarev, A. I. Usov, M. A. Rechter, and N. K. Kochetkov, Mass-spectrometric characterization of 3,6-anhydro-galactose derivatives, *Carbohydr. Res.*, 16 (1971) 29–38.

353. A. I. Usov and N. K. Kochetkov, Polysaccharides of algae. 2. Polysaccharides of the red seaweed *Odonthalia corymbifera* (Gmel.) J. Ag. Isolation of 6-*O*-methyl-D-galactose, *Zh. Obshch. Khim.*, 38 (1968) 234–238(in Russian).
354. D. C. De Jongh and K. Biemann, Mass spectra of O-isopropylidene derivatives of pentoses and hexoses, *J. Am. Chem. Soc.*, 86 (1964) 67–74.
355. V. L. DiNinno, R. A. Bell, and E. L. McCandless, Identification of 4,6-*O*-(1-carboxyethylidene)-D-galactose by combined gas chromatography mass spectrometry, *Biomed. Mass Spectrom.*, 5 (1978) 671–673.
356. A. I. Usov and E. G. Ivanova, Polysaccharides of algae. 19. Partial methanolysis of sulfated polysaccharides of the red alga *Rhodomela larix* (Turn.) C.Ag., *Sov. J. Bioorg. Chem.*, 1 (1975) 511–515 (English translation from *Bioorg. Khim.*, 1 (1975) 665–671).
357. A. I. Usov and M. Ya, Elashvili, Polysaccharides of algae. 51. Partial reductive hydrolysis of a sulfated galactan from the red alga *Laurencia coronopus* J.Ag. (Rhodophyta, Rhodomelaceae), *Russ. J. Bioorg. Chem.*, 23 (1997) 468–473 (English translation from *Bioorg. Khim.*, 23 (1997) 505–511).
358. J. Zaia, Mass spectrometry of oligosaccharides, *Mass Spectrom. Rev.*, 23 (2004) 161–227.
359. D. J. Harvey, Matrix-assisted laser desorption/ionization mass spectrometry of carbohydrates, *Mass Spectrom. Rev.*, 18 (1999) 349–451.
360. D. J. Harvey, Analysis of carbohydrates and glycoconjugates by matrix-assisted laser desorption/ionization mass spectrometry: An update covering the period 1999–2000, *Mass Spectrom. Rev.*, 25 (2006) 595–662.
361. D. J. Harvey, Analysis of carbohydrates and glycoconjugates by matrix-assisted laser desorption/ionization mass spectrometry: An update covering the period 2001–2002, *Mass Spectrom. Rev.*, 27 (2008) 125–201.
362. L. Chi, J. Amster, and R. J. Linhardt, Mass spectrometry for the analysis of highly charged sulfated carbohydrates, *Curr. Anal. Chem.*, 1 (2005) 223–240.
363. D. Ekeberg, S. H. Knutsen, and M. Sletmoen, Negative-ion electrospray ionization—Mass spectrometry (ESI-MS) as a tool for analyzing structural heterogeneity in *kappa*-carrageenan oligosaccharides, *Carbohydr. Res.*, 334 (2001) 49–59.
364. A. Antonopoulos, P. Favetta, W. Helbert, and M. Lafosse, Isolation of κ-carrageenan oligosaccharides using ion-pair liquid chromatography—Characterisation by electrospray ionisation mass spectrometry in positive-ion mode, *Carbohydr. Res.*, 339 (2004) 1301–1309.
365. A. Antonopoulos, P. Favetta, W. Helbert, and M. Lafosse, On-line liquid chromatography electrospray ionization mass spectrometry for the characterization of κ- and ι-carrageenans. Application to the hybrid ι-/ν-carrageenans, *Anal. Chem.*, 77 (2005) 4125–4136.
366. A. Antonopoulos, P. Favetta, W. Helbert, and M. Lafosse, On-line liquid chromatography-electrospray ionization mass spectrometry for κ-carrageenan oligosaccharides with a porous graphitic carbon column, *J. Chromatogr. A*, 1147 (2007) 37–41.
367. G. Yu, X. Zhao, B. Yang, S. Ren, H. Guan, Y. Zhang, A. M. Lawsom, and W. Chai, Sequence determination of sulfated carrageenan-derived oligosaccharides by high-sensitivity negative-ion electrospray tandem mass spectrometry, *Anal. Chem.*, 78 (2006) 8499–8505.
368. J. Li, F. Han, X. Lu, X. Fu, C. Ma, Y. Chu, and W. Yu, A simple method of preparing diverse neoagaro-oligosaccharides with *β*-agarase, *Carbohydr. Res.*, 342 (2007) 1030–1033.
369. Y. Fukuyama, M. Ciancia, H. Nonami, A. S. Cerezo, R. Erra-Balsells, and M. C. Matulewicz, Matrix-assisted ultraviolet laser-desorption ionization and electrospray-ionization time-of-flight mass spectrometry of sulfated neocarrabiose oligosaccharides, *Carbohydr. Res.*, 337 (2002) 1553–1562.
370. R. Erra-Balsells and H. Nonami, UV-MALDI-TOF MS analysis of carbohydrates. Reviewing comparative studies performed using *nor*-harmane and classical UV-MALDI matrices, *Environ. Control Biol.*, 46 (2008) 65–90.

371. K. Ohara, J.-C. Jacquinet, D. Jouanneau, W. Helbert, M. Smietana, and J.-J. Vasseur, Matrix-assisted laser desorption-ionization mass spectrometric analysis of polysulfated-derived oligosaccharides using pyrenemethylguanidine, *J. Am. Soc. Mass Spectrom.*, 20 (2009) 131–137.
372. B. Coxon, Proton magnetic resonance spectroscopy: Part I, *Adv. Carbohydr. Chem. Biochem.*, 27 (1972) 7–84.
373. L. D. Hall, Solutions to the hidden-resonance problem in proton nuclear magnetic resonance spectroscopy, *Adv Carbohydr. Chem. Biochem.*, 29 (1974) 11–40.
374. P. A. J. Gorin, Carbon-13 nuclear magnetic resonance spectroscopy of polysaccharides, *Adv. Carbohydr. Chem. Biochem.*, 38 (1981) 13–104.
375. K. Bock and C. Pedersen, Carbon-13 nuclear magnetic resonance spectroscopy of monosaccharides, *Adv. Carbohydr. Chem. Biochem.*, 41 (1983) 27–66.
376. K. Bock, C. Pedersen, and H. Pedersen, Carbon-13 nuclear magnetic resonance data for oligosaccharides, *Adv. Carbohydr. Chem. Biochem.*, 42 (1984) 193–225.
377. A. I. Usov, NMR spectroscopy of red seaweed polysaccharides: Agars, carrageenans, and xylans, *Bot. Mar.*, 27 (1984) 189–202.
378. F. van de Velde, S. H. Knutsen, A. I. Usov, H. S. Rollema, and A. S. Cerezo, ^{1}H and ^{13}C high resolution NMR spectroscopy of carrageenans: Application in research and industry, *Trends Food Sci. Technol.*, 13 (2002) 73–92.
379. K. Izumi, N.m.r. spectra of some 3,6-anhydro-D-galactose derivatives, *Carbohydr. Res.*, 27 (1973) 278–281.
380. E. B. Rathbone, A. M. Stephen, and K. G. R. Pachler, P.m.r. spectroscopy of monomethyl ethers of D-galactopyranose and its derivatives, *Carbohydr. Res.*, 20 (1971) 357–367.
381. M. J. Harris and J. R. Turvey, Sulphates of monosaccharides and derivatives. Part IX. The conformations of glycoside sulphates in solution determined by n.m.r. spectroscopy, *Carbohydr. Res.*, 15 (1970) 57–63.
382. P. A. J. Gorin and T. Ishikawa, Configuration of pyruvic acid ketals, 4,6-O-linked to D-galactose units, in bacterial and algal polysaccharides, *Can. J. Chem.*, 45 (1967) 521–532.
383. K. Izumi, Structural analysis of agar-type polysaccharides by NMR spectroscopy, *Biochim. Biophys. Acta*, 320 (1973) 311–317.
384. K. S. Young, S. S. Bhattacharjee, and W. Yaphe, Enzymic cleavage of the α-linkages in agarose, to yield agarooligosaccharides, *Carbohydr. Res.*, 66 (1978) 207–212.
385. S. S. Bhattacharjee, W. Yaphe, and N. M. R. Enzymic, Spectroscopic analysis of agar-type polysaccharides, in H. A. Hoppe, T. Levring, and Y. Tanaka, (Eds.), *Marine algae in Pharmaceutical Sciences*, Vol. 1, Walter de Gruyter, Berlin, 1979, pp. 645–655.
386. D. Welty, Carrageenans. Part 12. The 300 MHz proton magnetic resonance spectra of methyl β-D-galactopyranoside, methyl 3,6-anhydro-α-D-galactopyranoside, agarose, κ-carrageenan, and segments of ι-carrageenan and agarose sulphate, *J. Chem. Res.* (1977) (S) 312–313, (M) 3566–3587.
387. E. Murano, R. Toffanin, F. Zanetti, S. H. Knutsen, S. Paoletti, and R. Rizzo, Chemical and macromolecular characterisation of agar polymers from *Gracilaria dura* (C.Agardh) J.Agardh (Gracilariaceae, Rhodophyta), *Carbohydr. Polym.*, 18 (1992) 171–178.
388. S. H. Knutsen, D. E. Myslabodski, and H. Grasdalen, Characterization of carrageenan fractions from Norwegian *Furcellaria lumbricalis* (Huds.) Lamour. by ^{1}H-n.m.r. spectroscopy, *Carbohydr. Res.*, 206 (1990) 367–372.
389. M. Ciancia, M. C. Matulewicz, P. Finch, and A. S. Cerezo, Determination of the structures of cystocarpic carrageenans from *Gigartina scottsbergii* by methylation analysis and NMR spectroscopy, *Carbohydr. Res.*, 238 (1993) 241–248.
390. E. Tojo and J. Prado, A simple ^{1}H NMR method for the quantification of carrageenans in blends, *Carbohydr. Polym.*, 53 (2003) 325–329.

391. S. H. Knutsen and H. Grasdalen, The use of neocarrabiose oligosaccharides with different length and sulphate substitution as model compounds for ^{1}H-NMR spectroscopy, *Carbohyd. Res.*, 229 (1992) 233–244.
392. S. H. Knutsen, M. Sletmoen, T. Kristensen, T. Barbeyron, B. Kloareg, and P. Potin, A rapid method for the separation and analysis of carrageenan oligosaccharides released by *iota*- and *kappa*-carrageenase, *Carbohyd. Res.*, 331 (2001) 101–106.
393. D. Jouanneau, P. Boulenguer, J. Mazoyer, and W. Helbert, Complete assignment of ^{1}H and ^{13}C NMR spectra of standard neo-ι-carrabiose oligosaccharides, *Carbohyd. Res.*, 345 (2010) 547–551.
394. M. Guibet, N. Kervarec, S. Génicot, Y. Chevolot, and W. Helbert, Complete assignment of ^{1}H and ^{13}C NMR spectra of *Gigartina scottsbergii* λ-carrageenan using carrabiose oligosaccharides prepared by enzymatic hydrolysis, *Carbohyd. Res.*, 341 (2006) 1859–1869.
395. G. Yu, H. Guan, A. S. Ioanoviciu, S. A. Sikkander, C. Thanawiroon, J. K. Tobacman, T. Toida, and R. J. Linhardt, Structural studies on κ-carrageenan derived oligosaccharides, *Carbohyd. Res.*, 337 (2002) 433–440.
396. M. Bosco, A. Segre, S. Miertus, A. Cesàro, and S. Paoletti, The disordered conformation of κ-carrageenan in solution as determined by NMR experiments and molecular modeling, *Carbohyd. Res.*, 340 (2005) 943–958.
397. F. van de Velde, H. A. Peppelman, H. S. Rollema, and R. H. Tromp, On the structure of κ/ι-hybrid carrageenans, *Carbohyd. Res.*, 331 (2001) 271–283.
398. F. van de Velde, A. S. Antipova, H. S. Rollema, T. V. Burova, N. V. Grinberg, L. Pereira, P. M. Gilsenan, R. H. Tromp, B. Rudolf, and V. Ya, Grinberg, The structure of κ/ι-hybrid carrageenans II. Coil-helix transition as a function of chain composition, *Carbohyd. Res.*, 340 (2005) 1113–1129.
399. S. H. Knutsen and H. Grasdalen, Analysis of carrageenans by enzymic degradation, gel filtration and ^{1}H NMR spectroscopy, *Carbohyd. Polym.*, 19 (1992) 199–210.
400. M. Guibet, P. Boulenguer, J. Mazoyer, N. Kervarec, A. Antonopoulos, M. Lafosse, and W. Helbert, Composition and distribution of carrabiose moieties in hybrid κ-/ι-carrageenans using carrageenases, *Biomacromolecules*, 9 (2008) 408–415.
401. A. S. Shashkov, A. I. Usov, and S. V. Yarotsky, Carbon-13 NMR spectra of methyl 3,6-anhydro-α-D-galactopyranoside and its methyl ethers, *Sov. J. Bioorg. Chem.*, 3 (1977) 35–38 (English translation from *Bioorg. Khim.*, 3 (1977) 46–49).
402. A. I. Usov and S. V. Yarotsky, Synthesis of methyl 3-*O*-methyl- and 3,6-di-*O*-methyl-β-D-galactopyranosides, *Sov. J. Bioorg. Chem.*, 3 (1977) 554–557 (English translation from *Bioorg. Khim.*, 3 (1977) 746–751).
403. A. I. Usov, V. V. Barbakadze, S. V. Yarotsky, and A. S. Shashkov, Polysaccharides of algae. 28. Application of ^{13}C-NMR spectroscopy for structural studies of galactan from the red seaweed *Grateloupia divaricata* Okam., *Sov. J. Bioorg. Chem.*, 4 (1978) 1085–1089 (English translation from *Bioorg. Khim.*, 4 (1978) 1507–1512).
404. A. S. Shashkov, A. I. Usov, Yu.A. Knirel, B. A. Dmitriev, and N. K. Kochetkov, Determination of the absolute configuration of sugars in oligo- and polysaccharides using glycosylation effects in ^{13}C-NMR spectra, *Sov. J. Bioorg. Chem.*, 7 (1981) 746–751(English translation from *Bioorg. Khim.*, 7 (1981) 1364–1371).
405. A. S. Shashkov, G. M. Lipkind, Yu.A. Knirel, and N. K. Kochetkov, Stereochemical factors determining the effects of glycosylation on the carbon-13 chemical shifts in carbohydrates, *Magn. Reson. Chem.*, 26 (1988) 735–747.
406. A. S. Shashkov, A. I. Usov, and S. V. Yarotsky, Polysaccharides of algae. 24. Application of ^{13}C-NMR spectroscopy for structural analysis of polysaccharides of the agar group, *Sov. J. Bioorg. Chem.*, 4 (1978) 57–63 (English translation from *Bioorg. Khim.*, 4 (1978) 74–81).

407. S. V. Yarotsky, A. S. Shashkov, and A. I. Usov, Polysaccharides of algae. 25. Application of ^{13}C-NMR spectroscopy for structural analysis of κ-carrageenan group polysaccharides,, *Sov. J. Bioorg. Chem.*, 4 (1978) 537–542 (English translation from *Bioorg. Khim.*, 4 (1978) 745–751).
408. A. I. Usov and A. S. Shashkov, Polysaccharides of algae. 34. Detection of iota-carrageenan in *Phyllophora brodiaei* (Turn.) J.Ag. (Rhodophyta) using ^{13}C-NMR spectroscopy, *Bot. Mar.*, 28 (1985) 367–373.
409. A. I. Usov, S. V. Yarotsky, and A. S. Shashkov, ^{13}C-NMR spectroscopy of red algal galactans, *Biopolymers*, 19 (1980) 977–990.
410. W. Voelter, E. Breitmaier, E. B. Rathbone, and A. M. Stephen, Influence of methylation on carbon-13 chemical shifts of galactose derivatives, *Tetrahedron*, 29 (1973) 3845–3848.
411. P. J. Garegg, B. Lindberg, and I. Kvarnstrom, Preparation and N.M.R. studies of pyruvic acid and related acetals of pyranosides: Configuration at the acetal carbon atom, *Carbohydr. Res.*, 77 (1979) 71–78.
412. P. A. J. Gorin, M. Mazurek, H. S. Duarte, M. Iacomini, and J. H. Duarte, Properties of ^{13}C-N.M.R. spectra of *O*-(1-carboxyethylidene) derivatives of methyl β-D-galactopyranoside: Models for determination of pyruvic acid acetal structures in polysaccharides, *Carbohydr. Res.*, 100 (1982) 1–15.
413. D. J. Brasch, H. M. Chang, C. T. Chuah, and L. D. Melton, The galactan sulfate from the edible, red alga *Porphyra columbina*, *Carbohydr. Res.*, 97 (1981) 113–125.
414. C. Rochas, M. Rinaudo, and M. Vincendon, Spectroscopic characterization and conformation of oligo kappa carrageenans, *Int. J. Biol. Macromol.*, 5 (1983) 111–115.
415. C. W. Greer, C. Rochas, and W. Yaphe, Iota-carrageenan oligosaccharides as model compounds for structural analysis of iota-carrageenan by ^{13}C-NMR spectroscopy, *Bot. Mar.*, 28 (1985) 9–14.
416. C. Rochas, M. Lahaye, W. Yaphe, and M. T. P. Viet, ^{13}C-N.M.R. spectroscopic investigation of agarose oligomers, *Carbohydr. Res.*, 148 (1986) 199–207.
417. M. Lahaye, W. Yaphe, and C. Rochas, ^{13}C-N.m.r.-spectral analysis of sulfated and desulfated polysaccharides of the agar type, *Carbohydr. Res.*, 143 (1985) 240–245.
418. M. Lahaye, W. Yaphe, M. T. P. Viet, and C. Rochas, ^{13}C-N.M.R. spectroscopic investigation of methylated and charged agarose oligosaccharides and polysaccharides, *Carbohydr. Res.*, 190 (1989) 249–265.
419. R. Falshaw and R. H. Furneaux, Carrageenan from the tetrasporic stage of *Gigartina decipiens* (Gigartinaceae, Rhodophyta), *Carbohydr. Res.*, 252 (1994) 171–182.
420. C. A. Stortz, B. E. Bacon, R. Cherniak, and A. S. Cerezo, High-field NMR spectroscopy of cystocarpic and tetrasporic carrageenans from *Iridaea undulosa*, *Carbohydr. Res.*, 261 (1994) 317–326.
421. M. Ji, M. Lahaye, and W. Yaphe, Structure of agar from *Gracilaria* spp. (Rhodophyta) collected in the People's Republic of China, *Bot. Mar.*, 28 (1985) 521–528.
422. M. Lahaye and W. Yaphe, The chemical structure of agar from Gracilaria compressa (C.Agardh) Greville, G. cervicornis (Turner) J. Agardh, G. damaecornis J. Agardh and G. domingensis Sonder ex Kutzing (Gigartinales, Rhodophyta), *Bot. Mar.*, 32 (1989) 369–377.
423. A. Chiovitti, A. Bacic, D. J. Craik, S. L. A. Munro, G. T. Kraft, and M.-L. Liao, Carrageenans with complex substitution patterns from red algae of the genus *Erythroclonium*, *Carbohydr. Res.*, 305 (1998) 243–252.
424. M.-L. Liao, A. Chiovitti, S. L. A. Munro, D. J. Craik, G. T. Kraft, and A. Bacic, Sulfated galactans from Australian specimens of the red alga *Phacelocarpus peperocarpos* (Gigartinales, Rhodophyta), *Carbohydr. Res.*, 296 (1996) 237–247.
425. G. M. Lipkind, A. S. Shashkov, Yu.A. Knirel, E. V. Vinogradov, and N. K. Kochetkov, A computer-assisted structural analysis of regular polysaccharides on the basis of carbon-13 NMR data, *Carbohydr. Res.*, 175 (1988) 59–75.

426. N. K. Kochetkov, E. V. Vinogradov, Yu.A. Knirel, A. S. Shashkov, and G. M. Lipkind, A programme SCAN for the ^{13}C NMR-based structural analysis of linear polysaccharides using personal computer, *Bioorg. Khim.*, 18 (1992) 116–125 (in Russian).
427. P.-E. Jansson, L. Kenne, and G. Widmalm, Computer-assisted structural analysis of polysaccharides with an extended version of CASPER using proton and carbon-13 NMR data, *Carbohydr. Res.*, 188 (1989) 169–191.
428. R. Stenutz, P.-E. Jansson, and G. Widmalm, Computer-assisted structural analysis of oligo- and polysaccharides: An extension of CASPER to multibranched structures, *Carbohydr. Res.*, 306 (1998) 11–17.
429. C. A. Stortz and A. S. Cerezo, The ^{13}C NMR spectroscopy of carrageenans: Calculation of chemical shifts and computer-aided structural determination, *Carbohydr. Polym.*, 18 (1992) 237–242.
430. R. Falshaw, R. H. Furneaux, H. Wong, M.-L. Liao, A. Bacic, and S. Chandrkrachang, Structural analysis of carrageenans from Burmese and Thai samples of *Catenella nipae* Zanardini, *Carbohydr. Res.*, 285 (1996) 81–98.
431. I. J. Miller and J. W. Blunt, New ^{13}C NMR methods for determining the structure of algal polysaccharides. Part 1. The effect of substitution on the chemical shifts of simple diad galactans, *Bot. Mar.*, 43 (2000) 239–250.
432. I. J. Miller and J. W. Blunt, New ^{13}C NMR methods for determining the structure of algal polysaccharides. Part 2. Galactans consisting of mixed diads, *Bot. Mar.*, 43 (2000) 251–261.
433. I. J. Miller and J. W. Blunt, New ^{13}C NMR methods for determining the structure of algal polysaccharides. Part 3. The structure of the polysaccharide from *Cladhymenia oblongifolia*, *Bot. Mar.*, 43 (2000) 263–271.
434. I. J. Miller, The structure of a pyruvylated carrageenan extracted from *Stenogramme interrupta* as determined by ^{13}C NMR spectroscopy, *Bot. Mar.*, 41 (1998) 305–315.
435. I. J. Miller, Further evaluation of the structure of the polysaccharide from *Plocamium costatum* with the use of set theory, *Hydrobiologia*, 398/399 (1999) 385–389.
436. I. J. Miller, The structure of the polysaccharide from *Hymenocladia sanguinea* through ^{13}C NMR spectroscopy, *Bot. Mar.*, 44 (2001) 245–251.
437. I. J. Miller, Evaluation of the structure of the polysaccharide from *Myriogramme denticulata* as determined by ^{13}C NMR spectroscopy, *Bot. Mar.*, 44 (2001) 253–259.
438. I. J. Miller, The structure of the carrageenan extracted from the tetrasporophytic form of *Stenogramme interrupta* as determined by ^{13}C NMR spectroscopy, *Bot. Mar.*, 44 (2001) 583–587.
439. I. J. Miller and J. W. Blunt, Evaluation of the structure of the polysaccharides from *Chondria macrocarpa* and *Ceramium rubrum* as determined by ^{13}C NMR spectroscopy, *Bot. Mar.*, 45 (2002) 1–8.
440. I. J. Miller, The structures of the polysaccharides from two *Lophurella* species as determined by ^{13}C NMR spectroscopy, *Bot. Mar.*, 45 (2002) 373–379.
441. I. J. Miller, Taxonomic implications of the chemical structures of galactans from two species of the genus *Trematocarpus*, *Bot. Mar.*, 45 (2002) 432–437.
442. I. J. Miller, The chemical structure of galactans from some New Zealand red algae, *Bot. Mar.*, 46 (2003) 572–577.
443. I. J. Miller, Evaluation of the structures of polysaccharides from two New Zealand members of the Ceramiaceae, *Bot. Mar.*, 46 (2003) 378–385.
444. I. J. Miller and J. W. Blunt, The structure of sulfated galactans from selected species of New Zealand *Champia*, *Bot. Mar.*, 48 (2005) 127–136.
445. I. J. Miller and J. W. Blunt, The structure of the galactan from *Aeodes nitidissima* (Halymeniales, Rhodophyta), *Bot. Mar.*, 48 (2005) 137–142.
446. I. J. Miller, The structure of the polysaccharide from *Grateloupia intestinalis* in New Zealand, *Bot. Mar.*, 48 (2005) 143–147.

447. I. J. Miller, Evaluation of the structures of polysaccharides from three taxa in the genus *Hymenena* and from *Acrosorium decumbens* (Rhodophyta, Delesseriaceae), *Bot. Mar.*, 48 (2005) 148–156.
448. I. J. Miller, The structure of polysaccharides from selected New Zealand species of *Grateloupia*, *Bot. Mar.*, 48 (2005) 157–166.
449. I. J. Miller and J. W. Blunt, The application of ^{13}C NMR spectroscopy to the red algal polysaccharides of selected New Zealand species of *Lomentaria*, *Bot. Mar.*, 48 (2005) 167–174.
450. A. I. Usov, Comments on the papers by Ian J. Miller devoted to carbon-13 NMR spectroscopy of red algal galactans, *Bot. Mar.*, 48 (2005) 178–180.
451. I. J. Miller, Response to "Comments on the papers by Ian J. Miller devoted to carbon-13 NMR spectroscopy of red algal galactans" by Anatolii I. Usov, *Bot. Mar.*, 48 (2005) 178–180181–185.
452. G. Michel, P. Nyval-Collen, T. Barbeyron, M. Czjzek, and W. Helbert, Bioconversion of red seaweed galactans: A focus on bacterial agarases and carrageenases, *Appl. Microbiol. Biotechnol.*, 71 (2006) 23–33.
453. C. Araki and K. Arai, Studies on the chemical constitution of agar-agar. XVIII. Isolation of a new crystalline disaccharide by enzymatic hydrolysis of agar-agar, *Bull. Chem. Soc. Jpn.*, 29 (1956) 339–345.
454. W. Yaphe, The use of agarase from *Pseudomonas atlantica* in the identification of agar in marine algae (Rhodophyceae), *Can. J. Microbiol.*, 3 (1957) 987–993.
455. W. Yaphe, The purification and properties of an agarase from a marine bacterium, *Pseudomonas atlantica*, *Proc. Int. Seaweed Symp.*, 5 (1966) 333–335.
456. L. M. Morrice, M. W. McLean, F. B. Williamson, and W. F. Long, β-Agarases I and II from *Pseudomonas atlantica*. Purifications and some properties, *Eur. J. Biochem.*, 135 (1983) 553–558.
457. J. R. Turvey and J. Christison, The hydrolysis of algal galactans by enzymes from a *Cytophaga* species, *Biochem. J.*, 105 (1967) 311–316.
458. M. Duckworth and J. R. Turvey, Specificity studies on the agarase from a *Cytophaga* species, *Proc. Int. Seaweed Symp.*, 6 (1969) 435–442.
459. M. Duckworth and J. R. Turvey, An extracellular agarase from a *Cytophaga* species, *Biochem. J.*, 113 (1969) 139–142.
460. A. R. Sampietro and M. A. Vattuone de Sampietro, Characterization of the agarolytic system of *Agarbacterium pastinator*, *Biochim. Biophys. Acta*, 244 (1971) 65–76.
461. M. A. Vattuone, E. A. de Flores, and A. R. Sampietro, Isolation of neoagarobiose and neoagarotetraose from agarose digested by *Pseudomonas elongata*, *Carbohydr. Res.*, 39 (1975) 164–167.
462. M. Malmqvist, Degradation of agarose gels and solutions by bacterial agarase, *Carbohydr. Res.*, 62 (1978) 337–348.
463. T. Aoki, T. Araki, and M. Kitamikado, Purification and characterization of a novel β-agarase from *Vibrio* sp. AP-2, *Eur. J. Biochem.*, 187 (1990) 461–465.
464. I. Yamaura, T. Matsumoto, M. Funatsu, H. Shigeiri, and T. Shibata, Purification and some properties of agarase from *Pseudomonas* sp. Pt-5, *Agric. Biol. Chem.*, 55 (1991) 2531–2536.
465. A. E. Jaffray, R. J. Anderson, and V. E. Coyne, Investigation of bacterial epiphytes of the agar-producing red seaweed *Gracilaria gracilis* (Stackhouse) Steentoft, Irvine *et* Farnham from Saldanha Bay, South Africa and Luderitz, Namibia, *Bot. Mar.*, 40 (1997) 569–576.
466. V. Parro and R. P. Mellado, Effect of glucose on agarase overproduction by *Streptomyces*, *Gene*, 145 (1994) 49–55.
467. M. Polne-Fuller, A multinucleated marine amoeba which digests seaweed, *J. Protozool.*, 34 (1987) 159–165.
468. M. Polne-Fuller, A. Rogerson, H. Amano, and A. Gibor, Digestion of seaweeds by the marine amoeba *Trichosphaerium*, *Hydrobiologia*, 204/205 (1990) 409–413.
469. A. I. Usov, M. D. Martynova, and N. K. Kochetkov, Detection of agarase in molluscs of the genus *Littorina*, *Dokl. Akad. Nauk SSSR*, 194 (1970) 455–457 (in Russian).

470. J. L. Gomez-Pinchetti and G. Garcia Reina, Enzymes from marine phycophages that degrade cell-walls of seaweeds, *Mar. Biol.*, 116 (1993) 553–558.
471. J. H. Erasmus, P. A. Cook, and V. E. Coyne, The role of bacteria in the digestion of seaweed by the abalone *Haliotis midae*, *Aquaculture*, 155 (1997) 377–386.
472. G. L. Skea, D. O. Mountfort, and K. D. Clements, Gut carbohydrases from the New Zealand marine herbivorous fishes *Kyphosus sydneyanus* (Kyphosidae), *Aplodactylus arctidens* (Aplodactylidae) and *Odax pullus* (Labridae), *Comp. Biochem. Physiol. B*, 140 (2005) 259–269.
473. A. I. Usov and L. I. Miroshnikova, Isolation of agarase from *Littorina mandshurica* by affinity chromatography on Biogel A, *Carbohydr. Res.*, 43 (1975) 204–207.
474. M. Malmqvist, Purification and characterization of two different agarose-degrading enzymes, *Biochim. Biophys. Acta*, 537 (1978) 31–43.
475. P. Potin, C. Richard, C. Rochas, and B. Kloareg, Purification and characterization of the α-agarase from *Alteromonas agarlyticus* (Cataldi) comb. nov., strain GJ1B, *Eur. J. Biochem.*, 214 (1993) 599–607.
476. C. Araki and K. Arai, Studies on the chemical constitution of agar-agar. XX. Isolation of a tetrasaccharide by enzymatic hydrolysis of agar-agar, *Bull. Chem. Soc. Jpn.*, 30 (1957) 287–293.
477. J. R. Turvey and J. Christison, The enzymic degradation of porphyran, *Biochem. J.*, 105 (1967) 317–321.
478. M. Duckworth and J. R. Turvey, The action of bacterial agarase on agarose, porphyran and alkali-treated porphyran, *Biochem. J.*, 113 (1969) 687–692.
479. M. Duckworth and J. R. Turvey, The specificity of an agarase from a *Cytophaga* species, *Biochem. J.*, 113 (1969) 693–696.
480. L. M. Morrice, M. W. McLean, W. F. Long, and F. B. Williamson, β-Agarases I and II from *Pseudomonas atlantica*. Substrate specificities, *Eur. J. Biochem.*, 137 (1983) 149–154.
481. L. M. Morrice, M. W. McLean, W. F. Long, and F. B. Williamson, Porphyran primary structure. An investigation using β-agarase I from *Pseudomonas atlantica* and ^{13}C-NMR spectroscopy, *Eur. J. Biochem.*, 133 (1983) 673–684.
482. G. Correc, J.-H. Hehemann, M. Czjzek, and W. Helbert, Structural analysis of the degradation products of porphyran digested by *Zobellia galactanivorans* β-porphyranase A, *Carbohydr. Polym.*, 83 (2011) 277–283.
483. M. Jam, D. Flament, J. Allouch, P. Potin, L. Thion, B. Kloareg, M. Czjzek, W. Helbert, G. Michel, and T. Barbeyron, The endo-β-agarases AgaA and AgaB from the marine bacterium *Zobellia galactanivorans*: Two paralogue enzymes with different molecular organizations and catalytic behaviours, *Biochem. J.*, 385 (2005) 703–713.
484. A. I. Usov and E. G. Ivanova, Polysaccharides of algae. 31. Enzymic hydrolysis of agar-like polysaccharide from the red seaweed *Rhodomela larix* (Turn.) C. Ag, *Sov. J. Bioorg. Chem.*, 7 (1981) 572–579 (English translation from *Bioorg. Khim.*, 7 (1981) 1060–1068).
485. A. I. Usov and E. G. Ivanova, Polysaccharides of algae. 37. Characterization of hybrid structure of substituted agarose from *Polysiphonia morrowii* (Rhodophyta, Rhodomelaceae) using β-agarase and ^{13}C-NMR spectroscopy, *Bot. Mar.*, 30 (1987) 365–370.
486. M. Duckworth and W. Yaphe, The structure of agar. Part II. The use of a bacterial agarase to elucidate structural features of the charged polysaccharides in agar, *Carbohydr. Res.*, 16 (1971) 435–445.
487. D. F. Day, M. Gomersall, and W. Yaphe, A *p*-nitrophenyl α-galactoside hydrolase from *Pseudomonas atlantica*. Localization of the enzyme, *Can. J. Microbiol.*, 21 (1975) 1476–1483.
488. D. F. Day and W. Yaphe, Enzymatic hydrolysis of agar: Purification and characterization of neoagarobiose hydrolase and *p*-nitrophenyl α-galactoside hydrolase, *Can. J. Microbiol.*, 21 (1975) 1512–1518.
489. D. Groleau and W. Yaphe, Enzymatic hydrolysis of agar: Purification and characterization of β-neoagarotetraose hydrolase from *Pseudomonas atlantica*, *Can. J. Microbiol.*, 23 (1977) 672–679.

490. C. Rochas, P. Potin, and B. Kloareg, NMR spectroscopic investigation of agarose oligomers produced by an α-agarase, *Carbohydr. Res.*, 253 (1994) 69–77.
491. T. Araki, Z. Lu, and T. Morishita, Optimization of parameters for isolation of protoplasts from *Gracilaria verrucosa* (Rhodophyta), *J. Mar. Biotachnol.*, 6 (1998) 193–197.
492. Y. Aoki and Y. Kamei, Preparation of recombinant polysaccharide-degrading enzymes from the marine bacterium, *Pseudomonas* sp. ND137 for the production of protoplasts of *Porphyra yezoensis*, *Eur. J. Phycol.*, 41 (2006) 321–328.
493. Y. Yoshizawa, A. Ametani, J. Tsunehiro, K. Nomura, M. Itoh, F. Fukui, and S. Kaminogawa, Macrophage stimulation activity of the polysaccharide fraction from a marine algae (*Porphyra yezoensis*): Structure-function relationships and improved solubility, *Biosci. Biotechnol. Biochem.*, 59 (1995) 1933–1937.
494. R. Kobayashi, M. Takisada, T. Suzuki, K. Kirimura, and S. Usami, Neoagarobiose as a novel moisturizer with whitening effect, *Biosci. Biotechnol. Biochem.*, 61 (1997) 162–163.
495. B. Hu, Q. Gong, Y. Wang, Y. Ma, J. Li, and W. Yu, Prebiotic effects of neoagaro-oligosaccharides prepared by enzymatic hydrolysis of agarose, *Anaerobe*, 12 (2006) 260–266.
496. W. Yaphe and B. Baxter, The enzymic hydrolysis of carrageenin, *Appl. Microbiol.*, 3 (1955) 380–383.
497. G. Sarwar, H. Oda, T. Sakata, and D. Kakimoto, Potentiality of artificial sea water salts for the production of carrageenase by a marine *Cytophaga* sp., *Microbiol. Immunol.*, 29 (1985) 405–411.
498. C. Bellion, G. K. Hamer, and W. Yaphe, The degradation of *Eucheuma spinosum* and *Eucheuma cottonii* carrageenans by ι-carrageenases and κ-carrageenases from marine bacteria, *Can. J. Microbiol.*, 28 (1982) 874–880.
499. J. L. Gomez-Pinchetti, M. Bjoerk, M. Pedersen, and G. Garcia Reina, Factors affecting protoplast yield of the carrageenophyte *Solieria filiformis* (Gigartinales, Rhodophyta), *Plant Cell Rep.*, 12 (1993) 541–545.
500. L. V. Benitez and J. M. Macaranas, Partial purification of a carrageenase from the tropical sea urchin *Diadema setosum*, *Proc. Int. Seaweed Symp.*, 9 (1979) 353–359.
501. J. Weigl and W. Yaphe, The enzymic hydrolysis of carrageenan by *Pseudomonas carrageenovora*: Purification of a κ-carrageenase, *Can. J. Microbiol.*, 12 (1966) 939–947.
502. K. H. Johnston and E. L. McCandless, Enzymic hydrolysis of the potassium chloride soluble fraction of carrageenan. Properties of λ-carrageenases from *Pseudomonas carrageenovora*, *Can. J. Microbiol.*, 19 (1973) 779–788.
503. W. Yaphe, The determination of κ-carrageenin as a factor in the classification of Rhodophyceae, *Can. J. Bot.*, 37 (1959) 751–757.
504. J. Weigl, J. R. Turvey, and W. Yaphe, The enzymic hydrolysis of κ-carrageenan with κ-carrageenase from *Pseudomonas carrageenovora*, *Proc. Int. Seaweed Symp.*, 5 (1966) 329–332.
505. M. W. McLean and F. B. Williamson, κ-Carrageenase from *Pseudomonas carrageenovora*, *Eur. J. Biochem.*, 93 (1979) 553–558.
506. J. Weigl and W. Yaphe, Glycosulfatase of *Pseudomonas carrageenovora*: Desulfation of disaccharide from κ-carrageenan, *Can. J. Microbiol.*, 12 (1966) 874–876.
507. M. W. McLean and F. B. Williamson, Glycosulphatase from *Pseudomonas carrageenovora*. Purification and some properties, *Eur. J. Biochem.*, 101 (1979) 497–505.
508. M. W. McLean and F. B. Williamson, Neocarratetraose 4-O-monosulphate β-hydrolase from *Pseudomonas carrageenovora*, *Eur. J. Biochem.*, 113 (1981) 447–456.
509. M. W. McLean and F. B. Williamson, Enzymes from *Pseudomonas carrageenovora*. Application to studies of carrageenan structure, *Proc. Int. Seaweed Symp.*, 10 (1981) 479–484.
510. C. W. Greer and W. Yaphe, Purification and properties of iota-carrageenase from a marine bacterium, *Can. J. Microbiol.*, 30 (1984) 1500–1506.
511. C. Rochas and A. Heyraud, Acid and enzymic hydrolysis of kappa carrageenan, *Polym. Bull.*, 5 (1981) 81–86.

512. C. Bellion, G. K. Hamer, and W. Yaphe, Analysis of kappa-iota, hybrid carrageenans with kappa-carrageenase iota-carrageenase and ^{13}C N.M.R, *Proc. Int. Seaweed Symp.*, 10 (1981) 379–384.
513. C. W. Greer and W. Yaphe, Enzymatic analysis of carrageenans: Structure of carrageenan from *Eucheuma nudum*, *Hydrobiologia*, 11(6/117), (1984) 563567.
514. C. W. Greer and W. Yaphe, Hybrid (iota-nu-kappa) carrageenan from *Eucheuma nudum* (Rhodophyta, Gigartinales) identified using iota- and kappa-carrageenases and ^{13}C-nuclear magnetic resonance spectroscopy, *Bot. Mar.*, 27 (1984) 479–484.
515. C. W. Greer and W. Yaphe, Characterization of hydrid (beta-kappa-gamma) carrageenan from Eucheuma gelatinae J. Agardh (Rhodophyta, Solieriaceae) using carrageenases, infrared and 13C-nuclear magnetic resonance spectroscopy, *Bot. Mar.*, 27 (1984) 473–478.
516. C. W. Greer, I. Shomer, M. E. Goldstein, and W. Yaphe, Analysis of carrageenan from *Hypnea musciformis* by using κ- and ι-carrageenases and ^{13}C-N.M.R. spectroscopy, *Carbohydr. Res.*, 129 (1984) 189–196.
517. K. Oestgaard, B. F. Wangen, S. H. Knutsen, and I. M. Aasen, Large-scale production and purification of κ-carrageenase from *Pseudomonas carrageenovora* for application in seaweed biotechnology, *Enzyme Microb. Technol.*, 15 (1993) 326–333.
518. G. Michel, L. Chantalat, E. Duee, T. Barbeyron, B. Henrissat, B. Kloareg, and O. Dideberg, The κ-carrageenase of *P. carrageenovora* features a tunnel-shaped active site: A novel insight in the evolution of clan-B glycoside hydrolases, *Structure*, 9 (2001) 513–525.
519. G. Michel, W. Helbert, R. Kahn, O. Dideberg, and B. Kloareg, The structural bases of the processive degradation of ι-carrageenan, a main cell wall polysaccharide of red algae, *J. Mol. Biol.*, 334 (2003) 421–433.
520. P. Nyvall Collen, M. Lemoine, R. Daniellou, J.-P. Guégan, S. Paoletti, and W. Helbert, Enzymatic degradation of κ-carrageenan in aqueous solution, *Biomacromolecules*, 10 (2009) 1757–1767.
521. D. Jouanneau, P. Boulenguer, J. Mazoyer, and W. Helbert, Hybridity of carrageenans water- and alkali-extracted from *Chondracanthus chamissoi*, *Mazzaella laminarioides*, *Sarcothalia crispata* and *S. radula*, *J. Appl. Phycol.*, 23 (2011) 105–114.
522. D. Jouanneau, P. Boulenguer, J. Mazzoyer, and W. Helbert, Enzymatic degradation of hybrid ι-/ν-carrageenan by *Alteromonas fortis* ι-carrageenase, *Carbohydr. Res.*, 345 (2010) 934–940.
523. D. A. Rees, Enzymic desulphation of porphyran, *Biochem. J.*, 80 (1961) 449–453.
524. S. Genicot-Joncour, A. Poinas, O. Richard, P. Potin, B. Rudolph, B. Kloareg, and W. Helbert, The cyclization of the 3,6-anhydro-galactose ring of ι-carrageenan is catalyzed by two D-galactose-2,6-sulfurylases in the red alga *Chondrus crispus*, *Plant Physiol.*, 151 (2009) 1609–1616.
525. J. A. Hemmingson, R. H. Furneaux, and H. Wong, In vivo conversion of 6-*O*-sulfo-L-galactopyranosyl residues into 3,6-anhydro-L-galactopyranosyl residues in *Gracilaria chilensis* Bird, McLachlan *et* Oliveira, *Carbohydr. Res.*, 296 (1996) 285–292.
526. S. R. Hanson, M. D. Best, and C.-H. Wong, Sulfatases: Structure, mechanism, biological activity, inhibition, and synthetic utility, *Angew. Chem. Int. Ed.*, 43 (2004) 5736–5763.
527. K. H. Johnston and E. L. McCandless, The immunologic response of rabbits to carrageenans, sulfated galactans extracted from marine algae, *J. Immunol.*, 101 (1968) 556–562.
528. K. H. Johnston and E. L. McCandless, Immunochemistry of carrageenans, *Proc. Int. Seaweed Symp.*, 6 (1969) 507–519.
529. S. P. C. Hosford and E. L. McCandless, Immunochemistry of carrageenans from gametophytes and sporophytes of certain red algae, *Can. J. Bot.*, 53 (1975) 2835–2841.
530. V. DiNinno and E. L. McCandless, Anti-carrageenans, *Immunochemistry*, 15 (1978) 273–274.
531. V. Vreeland, E. Zablackis, and W. M. Laetsch, Monoclonal antibodies as molecular markers for the intracellular and cell-wall distribution of carrageenan epitopes in *Kappaphycus* (Rhodophyta) during tissue development, *J. Phycol.*, 28 (1992) 328–342.

532. V. DiNinno and E. L. McCandless, The chemistry and immunochemistry of carrageenans from *Eucheuma* and related algal species, *Carbohydr. Res.*, 66 (1978) 85–93.
533. V. DiNinno and E. L. McCandless, The immunochemistry of λ-type carrageenans from certain red algae, *Carbohydr. Res.*, 67 (1978) 235–241.
534. M. J. Evelegh, C. M. Vollmer, and E. L. McCandless, Differentiation of lambda carrageenans from Rhodophyta with immunological and spectroscopic techniques, *J. Phycol.*, 14 (1978) 89–91.
535. V. L. DiNinno and E. L. McCandless, Immunochemistry of kappa-type carrageenans from certain red algae, *Carbohydr. Res.*, 72 (1979) 157–163.
536. M. R. Gretz, Y. Wu, J. Scott, and V. Vreeland, Ultrastructural immunogold localization of carrageenan in the Golgi complex of *Agardhiella subulata*, *J. Phycol.*, 27(Suppl.), (1991) 26.
537. E. M. Gordon-Mills and E. L. McCandless, Carrageenans in the cell walls of *Chondrus crispus* Stack. (Rhodophyceae, Gigartinales). I. Localization with fluorescent antibody, *Phycologia*, 14 (1975) 275–281.
538. S. Arakawa, H. Ishihara, O. Nishio, and S. Isomura, A sandwich enzyme-linked immunosorbent assay for kappa-carrageenan determination, *J. Sci. Food Agric.*, 57 (1991) 135–140.
539. K. Izumi, Chemical heterogeneity of the agar from *Gelidium amansii*, *Carbohydr. Res.*, 17 (1971) 227–230.
540. A. C. Onraët and B. L. Robertson, Seasonal variation in yield and properties of agar from sporophytic and gametophytic phases of *Onikusa prostoides* (Turner) Akatsuka (Gelidiaceae, Rhodophyta), *Bot. Mar.*, 30 (1987) 491–495.
541. A. Mouradi-Givernaud, T. Givernaud, H. Morvan, and J. Cosson, Agar from *Gelidium latifolium* (Rhodophyceae, Gelidiales): Biochemical composition and seasonal variations, *Bot. Mar.*, 35 (1992) 153–159.
542. M. R. Vignon, E. Morgan, and C. Rochas, *Gelidium sesquipedale* (Gelidiales, Rhodophyta). I. Soluble polymers, *Bot. Mar.*, 37 (1994) 325–329.
543. K. Prasad, A. K. Siddhanta, M. Ganesan, B. K. Ramavat, B. Jha, and P. K. Ghosh, Agars of *Gelidiella acerosa* of west and southeast coast of India, *Bioresour. Technol.*, 98 (2007) 1907–1915.
544. M. Y. Roleda, E. T. Ganzon-Fortes, N. E. Montaño, and F. N. de los Reyes, Temporal variation in the biomass, quantity and quality of agar from *Gelidiella acerosa* (Forsskål) Feldmann *et* Hamel (Rhodophyta: Gelidiales) from Cape Bolinao, NW Philippines, *Bot. Mar.*, 40 (1997) 487–495.
545. A. Chiovitti, L. J. McManus, G. T. Kraft, A. Bacic, and M.-L. Liao, Extraction and characterization of agar from Australian *Pterocladia lucida*, *J. Appl. Phycol.*, 16 (2004) 41–48.
546. K. Truus, R. Tuvikene, M. Vaher, T. Kailas, P. Toomik, and T. Pehk, Structural and compositional characteristics of gelling galactan from the red alga Ahnfeltia tobuchiensis (Ahnfeltiales, the Sea of Japan, *Carbohydr. Polym.*, 63 (2006) 130–135.
547. K. B. Guiseley, The relationship between methoxyl content and gelling temperature of agarose, *Carbohydr. Res.*, 13 (1970) 247–256.
548. R. H. Furneaux, I. J. Miller, and T. T. Stevenson, Agaroids from New Zealand members of the Gracilariaceae (Gracilariales, Rhodophyta)—A novel dimetylated agar, *Hydrobiologia*, 204/205 (1990) 645–654.
549. R. Takano, K. Hayashi, and S. Hara, Highly methylated agars with a high gel-melting point from the red seaweed, *Gracilaria eucheumoides*, *Phytochemistry*, 40 (1995) 487–490.
550. M. Tako, M. Higa, K. Medoruma, and Y. Nakasone, A highly methylated agar from red seaweed, *Gracilaria arcuata*, *Bot. Mar.*, 42 (1999) 513–517.
551. D. A. Rees and E. Convay, The structure and biosynthesis of porphyran: A comparison of some samples, *Biochem. J.*, 84 (1962) 411–416.
552. B. Baldan, P. Andolfo, F. Culoso, G. Tripodi, and P. Mariani, Polysaccharide localization in the cell wall of *Porphyra leucosticta* (Bangiophyceae, Rhodophyta) during the life cycle, *Bot. Mar.*, 38 (1995) 31–36.

553. Q. Zhang, N. Li, X. Liu, Z. Zhao, Z. Li, and Z. Xu, The structure of a sulfated galactan from *Porphyra haitanensis* and its in vivo antioxidant activity, *Carbohydr. Res.*, 339 (2004) 105–111.
554. Q. Zhang, H. Qi, T. Zhao, E. Deslandes, N. M. Ismaeli, F. Molloy, and A. T. Critchley, Chemical characteristics of a polysaccharide from *Porphyra capensis* (Rhodophyta), *Carbohydr. Res.*, 340 (2005) 2447–2450.
555. M. R. Gretz, E. L. McCandless, J. M. Aronson, and M. R. Sommerfeld, The galactan sulphates of the conchocelis phases of *Porphyra leucosticta* and *Bangia atropurpurea* (Rhodophyta), *J. Exp. Bot.*, 34 (1983) 705–711.
556. R. Armisen, World-wide use and importance of *Gracilaria*, *J. Appl. Phycol.*, 7 (1995) 231–243.
557. E. Murano, Chemical structure and quality of agars from *Gracilaria*, *J. Appl. Phycol.*, 7 (1995) 245–254.
558. J. A. Hemmingson and R. H. Furneaux, Biosynthetic activity and galactan composition in different regions of the thallus of *Gracilaria chilensis* Bird, McLachlan *et* Oliveira, *Bot. Mar.*, 40 (1997) 351–357.
559. J. A. Hemmingson and R. H. Furneaux, Manipulation of galactan biosynthesis in *Gracilaria chilensis* Bird, McLachlan *et* Oliveira by light deprivation, *Bot. Mar.*, 43 (2000) 285–289.
560. A. A. Lapshina, E. G. Ivanova, E. A. Titlyanov, and A. I. Usov, Agar from the unattached form of *Gracilaria verrucosa* (Huds.) Papenf. of Primorski territory, *Bioorg. Khim.*, 17 (1991) 1494–1499 (in Russian).
561. R. D. Villanueva, C. V. Pagba, and N. E. Montaño, Optimized agar extraction from *Gracilaria eucheumoides* Harvey, *Bot. Mar.*, 40 (1997) 369–372.
562. R. Falshaw, R. H. Furneaux, T. D. Pickering, and D. E. Stevenson, Agars from three Fijian *Gracilaria* species, *Bot. Mar.*, 42 (1999) 51–59.
563. R. D. Villanueva, N. E. Montaño, J. B. Romero, A. K. A. Aliganga, and E. P. Enriquez, Seasonal variations in the yield, gelling properties, and chemical composition of agars from *Gracilaria eucheumoides* and *Gelidiella acerosa* (Rhodophyta) from the Philippines, *Bot. Mar.*, 42 (1999) 175–182.
564. H. P. Calumpong, A. Maypa, M. Magbanua, and P. Suarez, Biomass and agar assessment of three species of *Gracilaria* from Nagros Island, central Philippines, *Hydrobiologia*, 398/399 (1999) 173–182.
565. K. G. Araño, G. C. Trono, Jr., N. E. Montaño, A. Q. Hurtado, and R. D. Villanueva, Growth, agar yield and quality of selected agarophyte species from the Philippines, *Bot. Mar.*, 43 (2000) 517–524.
566. E. Marinho-Soriano, Agar polysaccharides from *Gracilaria* species (Rhodophyta, Gracilariaceae), *J. Biotechnol.*, 89 (2001) 81–84.
567. S. Mazumder, P. K. Ghosal, C. A. Pujol, M. A. Carlucci, E. B. Damonte, and B. Ray, Isolation, chemical investigation and antiviral activity of polysaccharides from *Gracilaria corticata* (Gracilariaceae, Rhodophyta), *Int. J. Biol. Macromol.*, 31 (2002) 87–95.
568. M. R. S. Melo, J. P. A. Feitosa, A. L. P. Freitas, and R. C. M. de Paula, Isolation and characterization of soluble sulfated polysaccharide from the red seaweed *Gracilaria cornea*, *Carbohydr. Polym.*, 49 (2002) 491–498.
569. J. B. Romero, M. N. E. Montaño, F. E. Merca, R. D. Villanueva, M.-L. Liao, and A. Bacic, Seasonal variations in the composition and gel quality of agar from *Gracilaria edulis* in the Philippines, *Bot. Mar.*, 50 (2007) 191–194.
570. M. Guimarães, A. G. Viana, M. E. R. Duarte, S. D. Ascêncio, E. M. Plastino, and M. D. Noseda, Low-molecular-mass carbohydrates and soluble polysaccharides of green and red morphs of *Gracilaria domingensis* (Gracilariales, Phodophyta), *Bot. Mar.*, 50 (2007) 314–317.
571. H. Andriamanantoanina, G. Chambat, and M. Rinaudo, Fractionation of extracted Madagascan *Gracilaria corticata* polysaccharides: Structure and properties, *Carbohydr. Polym.*, 68 (2007) 77–88.

572. R. Meena, A. K. Siddhanta, K. Prasad, B. K. Ramavat, K. Eswaran, S. Thiruppathi, M. Ganesan, V. A. Mantri, and P. V. Subba Rao, Preparation, characterization and benchmarking of agarose from *Gracilaria dura* of Indian waters, *Carbohydr. Polym.*, 69 (2007) 179–188.
573. R. Meena, K. Prasad, M. Ganesan, and A. K. Siddhanta, Superior quality agar from *Gracilaria* species (Gracilariales, Rhodophyta) collected from the Gulf of Mannar, India, *J. Appl. Phycol.*, 20 (2008) 397–402.
574. J. S. Maciel, L. S. Chaves, B. W. S. Souza, D. I. A. Teixeira, A. L. P. Freitas, J. P. A. Feitosa, and R. C. M. de Paula, Structural characterization of cold extracted fraction of soluble sulfated polysaccharide from red seaweed *Gracilaria birdiae*, *Carbohydr. Polym.*, 71 (2008) 559–565.
575. V. Kumar and R. Fotedar, Agar extraction process for *Gracilaria cliftonii* (Withell, Millar, & Kraft, 1994), *Carbohydr. Polym.*, 78 (2009) 813–819.
576. R. D. Villanueva, A. M. M. Sousa, M. P. Gonçalves, M. Nilsson, and L. Hilliou, Production and properties of agar from the invasive marine alga, *Gracilaria vermiculophylla* (Gracilariales, Rhodophyta), *J. Appl. Phycol.*, 22 (2010) 211–220.
577. N. Y. Roleda, E. T. Ganzon-Fortes, and N. E. Montaño, Agar from vegetative and tetrasporic *Gelidiella acerosa* (Gelidiales, Rhodophyta), *Bot. Mar.*, 40 (1997) 501–506.
578. A. I. Usov, Polysaccharides of algae. 13. Monosaccharide composition of polysaccharides of several red algae from the Japan Sea, *Zh. Obshch. Khim.*, 44 (1974) 191–196.
579. I. J. Miller and R. H. Furneaux, Agars from New Zealand red algae in the family Gelidiaceae: A structural study, *N. Z. J. Sci.*, 25 (1982) 15–18.
580. I. J. Miller, R. Falshaw, and R. H. Furneaux, The chemical structures of polysaccharides from New Zealand members of Rhodomelaceae, *Bot. Mar.*, 36 (1993) 203–208.
581. C. L. Hurd, W. A. Nelson, R. Falshaw, and K. F. Neill, History, current status and future of marine macroalgal research in New Zealand: Taxonomy, ecology, physiology and human uses, *Phycol. Res.*, 52 (2004) 80–106.
582. H. Prado, M. Ciancia, and M. C. Matulewicz, Agarans from the red seaweed *Polysiphonia nigrescens* (Rhodomelaceae, Ceramiales), *Carbohydr. Res.*, 343 (2008) 711–718.
583. I. J. Miller and R. H. Furneaux, The structural determination of the agaroid polysaccharides from four New Zealand algae in the order Ceramiales by means of ^{13}C NMR spectroscopy, *Bot. Mar.*, 40 (1997) 333–339.
584. I. J. Miller, Evaluation of the structures of polysaccharides from two New Zealand members of the Rhodomelaceae by ^{13}C NMR spectroscopy, *Bot. Mar.*, 46 (2003) 386–391.
585. M. E. R. Duarte, M. D. Noseda, M. A. Cardoso, S. Tulio, and A. S. Cerezo, The structure of a galactan sulfate from the red seaweed *Bostrychia montagnei*, *Carbohydr. Res.*, 337 (2002) 1137–1144.
586. D. M. Bowker and J. R. Turvey, Water soluble polysaccharides of the red alga *Laurencia pinnatifida*. Part I. Constituent units, *J. Chem. Soc. C* (1968) 983–988.
587. D. M. Bowker and J. R. Turvey, Water soluble polysaccharides of the red alga *Laurencia pinnatifida*. Part II. Methylation analysis of the galactan sulphate, *J. Chem. Soc. C* (1968) 989–992.
588. A. I. Usov, E. G. Ivanova, and M. Ya Elashvili, Polysaccharides of algae. 41. Characterization of water-soluble polysaccharides from several representatives of the genus *Laurencia* (Ceramiales, Rhodophyta), *Sov. J. Bioorg. Chem.*, 15 (1989) 701–709 (English translation from *Bioorg. Khim.*, 15 (1989) 1259–1267).
589. R. Takano, T. Yokoi, K. Kamei, S. Hara, and S. Hirase, Coexistence of agaroid and carrageenan structures in a polysaccharide from the red seaweed *Rhodomela larix* (Turner) C. Ag., *Bot. Mar.*, 42 (1999) 183–188.
590. I. J. Miller, The chemical structure of galactans from *Sarcodia montagneana* and from *Sarcodia flabellata*, *Bot. Mar.*, 46 (2003) 392–399.

591. A. I. Usov and G. I. Vergunova, Polysaccharides of algae. 39. Study on sulfated galactan from the red seaweed *Palmaria stenogona* Perest, *Sov. J. Bioorg. Chem.*, 15 (1989) 111–118 (English translation from *Bioorg. Khim.*, 15 (1989) 198–207).
592. M. C. Matulewicz, H. H. Haines, and A. S. Cerezo, Sulphated xylogalactans from *Nothogenia fastigiata*, *Phytochemistry*, 36 (1994) 97–103.
593. C. A. Stortz, M. R. Cases, and A. S. Cerezo, Red seaweed galactans. Methodology for the structural determination of corallinan, a different agaroid, in R. R. Townsend and A. T. Hotchkiss, Jr., (Eds.), *Techniques in Glycobiology*, Marcel Dekker, New York, 1997, pp. 567–593.
594. M. I. Bilan and A. I. Usov, Polysaccharides of calcareous algae and their effect on the calcification process, *Russ. J. Bioorg. Chem.*, 27 (2001) 2–16 (English translation from *Bioorg. Khim.*, 27 (2001) 4–20).
595. A. I. Usov, M. I. Bilan, and N. G. Klochkova, Polysaccharides of algae. 48. Polysaccharide composition of several calcareous red algae: Isolation of alginate from *Corallina pilulifera* P. et R. (Rhodophyta, Corallinaceae), *Bot. Mar.*, 38 (1995) 43–51.
596. A. I. Usov and M. I. Bilan, Polysaccharides of algae. 52. The structure of sulfated xylogalactan from the calcareous red alga *Bossiella cretacea* (P. et R.) Johansen (Rhodophyta, Corallinales), *Russ. J. Bioorg. Chem.*, 24 (1998) 123–129 (English translation from *Bioorg. Khim.*, 24 (1998) 139–146).
597. D. A. Navarro and C. A. Stortz, Isolation of xylogalactans from the Corallinales: Influence of theextraction method on yields and composition, *Carbohydr. Polym.*, 49 (2002) 57–62.
598. D. A. Navarro and C. A. Stortz, The system of xylogalactans from the red seaweed *Jania rubens* (Corallinales, Rhodophyta), *Carbohydr. Res.*, 343 (2008) 2613–2622.
599. D. A. Rees, The carrageenan system of polysaccharides. Part I. The relation between the κ- and λ-components, *J. Chem. Soc.* (1963) 1821–1832.
600. L. C.-M. Chen, J. McLachlan, A. C. Neish, and P. F. Shacklock, The ratio of kappa- to lambd-carrageenan in nuclear phases of the rhodophycean algae, *Chondrus crispus* and *Gigartina stellata*, *J. Mar. Biol. Ass. UK*, 53 (1973) 11–16.
601. E. L. McCandless, J. S. Craigie, and J. A. Walter, Carrageenans in the gametophytic and sporophytic stages of *Chondrus crispus*, *Planta*, 112 (1973) 201–212.
602. S. E. Pickmere, M. J. Parsons, and R. W. Bailey, Composition of *Gigartina* carrageenan in relation to sporophyte and gametophyte stages of the life cycle, *Phytochemistry*, 12 (1973) 2441–2444.
603. J. R. Waaland, Experimental studies on propagation of *Iridaea* and *Gigartina*, *J. Phycol.*, 9(Suppl.), (1973) 12.
604. E. L. McCandless, J. A. West, and M. D. Guiry, Carrageenan patterns in the Gigartinaceae, *Biochem. Syst. Ecol.*, 11 (1983) 175–182.
605. E. L. McCandless, J. A. West, and M. D. Guiry, Carrageenan patterns in the Phyllophoraceae, *Biochem. Syst. Ecol.*, 10 (1982) 275–284.
606. R. D. Villanueva and M. N. E. Montaño, Fine chemical structure of carrageenan from the commercially cultivated *Kappaphycus striatum* (sacol varierty) (Solieriaceae, Gigartinales, Rhodophyta), *J. Phycol.*, 39 (2003) 513–518.
607. I. M. Yermak, A. O. Barabanova, V. P. Glazunov, V. V. Isakov, Y. H. Kim, K. S. Shin, T. V. Titlynova, and T. F. Solov'eva, Carrageenans from cystocarpic and sterile plants of *Chondrus pinnulatus* (Gigartinaceae, Rhodophyta) collected from the Russian Pacific coast, *J. Appl. Phycol.*, 18 (2006) 361–368.
608. R. Falshaw, H. J. Bixler, and K. Johndro, Structure and performance of commercial kappa-2 carrageenan extracts. I. Structure analysis, *Food Hydrocolloids*, 15 (2001) 441–452.
609. H. J. Bixler, K. Johndro, and R. Falshaw, Kappa-2 carrageenan: Structure and performance of commercial extracts. II. Performance in two simulated dairy applications, *Food Hydrocolloids*, 15 (2001) 619–630.

610. R. Falshaw, H. J. Bixler, and K. Johndro, Structure and performance of commercial κ-2 carrageenan extracts. Part III. Structure analysis and performance in two dairy applications of extracts from the New Zealand red seaweed, *Gigartina atropurpurea*, *Food Hydrocolloids*, 17 (2003) 129–139.
611. R. D. Villanueva, W. G. Mendoza, M. R. C. Rodrigueza, J. B. Romero, and M. N. E. Montaño, Structure and functional performance of gigartinacean kappa-iota hybrid carrageenan and solieriacean kappa-iota carrageenan blends, *Food Hydrocolloids*, 18 (2004) 283–292.
612. A. O. Barabanova, I. M. Yermak, V. P. Glazunov, V. V. Isakov, E. A. Titlyanov, and T. F. Solov'eva, Comparative srtudy of carrageenans from reproductive and sterile forms of *Tichocarpus crinitus* (Gmel.) Rupr. (Rhodophyta, Tichocarpaceae), *Biochemistry (Moscow)*, 70 (2005) 350–356 (English translation from *Biokhimiya*, 70 (2005) 430–437).
613. R. Tuvikene, K. Truus, M. Robal, O. Volobujeva, E. Mellikov, T. Pehk, A. Kollist, T. Kailas, and M. Vaher, The extraction, structure, and gelling properties of hybrid galactan from the red alga *Furcellaria lumbricalis* (Baltic Sea, Estonia), *J. Appl. Phycol.*, 22 (2010) 51–63.
614. B. Yang, G. Yu, X. Zhao, W. Ren, G. Jiao, L. Fang, Y. Wang, G. Du, C. Tiller, G. Girouard, C. J. Barrow, H. S. Ewart, and J. Zhang, Structural characterisation and bioactivities of hybrid carrageenan-like sulphated galactan from red alga *Furcellaria lumbricalis*, *Food Chem.*, 124 (2011) 50–57.
615. A. O. Barabanova, A. S. Shashkov, V. P. Glazunov, V. V. Isakov, T. B. Nebylovskaya, W. Helbert, T. F. Solov'eva, and I. M. Yermak, Structure and properties of carrageenan-like polysaccharide from the red alga *Tichocarpus crinitus* (Gmel.) Rupr. (Rhodophyta, Tichocarpaceae), *J. Appl. Phycol.*, 20 (2008) 1013–1020.
616. I. J. Miller, The structure of the carrageenan extracted from the tetrasporophytic form of *Stenogramme interrupta* as determined by ^{13}C NMR spectroscopy, *Bot. Mar.*, 44 (2001) 583–587.
617. R. Falshaw, R. H. Furneaux, and D. E. Stevenson, Structural analysis of carrageenans from the red alga, *Callophyllis hombroniana* Mont. Kütz. (Kallymeniaceae, Rhodophyta), *Carbohydr. Res.*, 340 (2005) 1149–1158.
618. J. P. Doyle, P. Giannouli, B. Rudolph, and E. R. Morris, Preparation, authentication, rheology and conformation of theta carrageenan, *Carbohydr. Polym.*, 80 (2010) 648–654.
619. A. Chiovitti, A. Bacic, D. J. Craik, G. T. Kraft, M.-L. Liao, R. Falshaw, and R. H. Furneaux, A pyruvated carrageenan from Australian specimens of the red alga *Sarconema filiforme*, *Carbohydr. Res.*, 310 (1998) 77–83.
620. I. J. Miller, The structure of a pyruvylated carrageenan extracted from *Stenogramme interrupta* as determined by ^{13}C NMR spectroscopy, *Bot. Mar.*, 41 (1998) 305–315.
621. R. Falshaw, R. H. Furneaux, and H. Wong, Analysis of pyruvylated β-carrageenan by 2D NMR spectroscopy and reductive partial hydrolysis, *Carbohydr. Res.*, 338 (2003) 1403–1414.
622. A. Chiovitti, A. Bacic, D. J. Craik, G. T. Kraft, and M.-L. Liao, A nearly idealized 6'-O-methylated ι-carrageenan from the Australian red alga *Claviclonium ovatum* (Acrotylaceae, Gigartinales), *Carbohydr. Res.*, 339 (2004) 1459–1466.
623. R. Falshaw and R. H. Furneaux, Structural analysis of carrageenans from the tetrasporic stages of the red algae, *Gigartina lanceata* and *Gigartina chapmanii* (Gigartinaceae, Rhodophyta), *Carbohydr. Res.*, 307 (1998) 325–331.
624. R. Falshaw and R. H. Furneaux, Carrageenans from the tetrasporic stages of *Gigartina clavifera* and *Gigartina alveata* (Gigartinaceae, Rhodophyta), *Carbohydr. Res.*, 276 (1995) 155–165.
625. A. Amimi, A. Mouradi, T. Givernaud, N. Chadmi, and M. Lahaye, Structural analysis of *Gigartina pistillata* carrageenans (Gigartinaceae, Rhodophyta), *Carbohydr. Res.*, 333 (2001) 271–279.
626. D. B. Smith, A. N. O'Neill, and A. S. Perlin, Studies on the heterogeneity of carrageenin, *Can. J. Chem.*, 33 (1955) 1352–1360.

627. J. M. Estevez, M. Ciancia, and A. S. Cerezo, The system of low-molecular-weight carrageenans and agaroids from the room-temperature-extracted fraction of *Kappaphycus alvarezii*, *Carbohydr. Res.*, 325 (2000) 287–299.
628. J. M. Estevez, M. Ciancia, and A. S. Cerezo, The system of galactans of the red seaweed *Kappaphycus alvarezii* with emphasis on its minor constituent, *Carbohydr. Res.*, 339 (2004) 2575–2592.
629. J. M. Estevez, M. Ciancia, and A. S. Cerezo, DL-Galactan hybrids and agarans from gametophytes of the red seaweed *Gymnogongrus torulosus*, *Carbohydr. Res.*, 331 (2001) 27–41.
630. J. M. Estevez, M. Ciancia, and A. S. Cerezo, The system of sulfated galactans from the red seaweed *Gymnogongrus torulosus* (Phyllophoraceae, Rhodophyta): Location and structural analysis, *Carbohydr. Polym.*, 73 (2008) 594–605.
631. R. Takano, K. Shiomoto, K. Kamei, S. Hara, and S. Hirase, Occurrence of carrageenan structure in an agar from the red seaweed *Digenea simplex* (Wulfen) C. Agardh (Rhodomelaceae, Ceramiales) with a short review of carrageenan-agarocolloid hybrid in the Florideophycidae, *Bot. Mar.*, 46 (2003) 142–150.
632. T. Chopin, M. D. Hanisak, and J. S. Craigie, Carrageenans from *Kallymenia westii* (Rhodophyceae) with a review of the phycocolloids produced by the Cryptonemiales, *Bot. Mar.*, 37 (1994) 433–444.
633. I. J. Miller, R. Falshaw, and R. H. Furneaux, Structural analysis of the polysaccharide from *Pachymenia lusoria* (Cryptonemiales, Rhodophyta), *Carbohydr. Res.*, 268 (1995) 219–232.
634. A. K. Sen, Sr., A. K. Das, K. K. Sarkar, A. K. Siddhanta, R. Takano, K. Kamei, and S. Hara, An agaroid-carrageenan hybrid type backbone structure for the antithrombotic sulfated polysaccharide from *Grateloupia indica* Boergensen (Halymeniales, Rhodophyta), *Bot. Mar.*, 45 (2002) 331–338.
635. I. J. Miller and J. Mollion, Evaluation of the structures of galactans from *Carpopeltis* sp., a member of the Halymeniales in the western Indian Ocean, *Bot. Mar.*, 49 (2006) 79–85.
636. T. A. Fenoradosoa, C. Delattre, C. Laroche, A. Wadouachi, V. Dulong, L. Picton, P. Andriamadio, and P. Michaud, Highly sulphated galactan from *Halymenia durvillei* (Halymeniales, Rhodophyta), a red seaweed of Madagascar marine coast, *Int. J. Biol. Macromol.*, 45 (2009) 140–145.
637. I. J. Miller and R. H. Furneaux, A structural analysis of the polysaccharide from *Kallymenia berggrenii* J. Ag., *Bot. Mar.*, 39 (1996) 141–147.
638. M. C. Rodríguez, E. R. Merino, C. A. Pujol, E. B. Damonte, A. S. Cerezo, and M. C. Matulewicz, Galactans from cystocarpic plants of the red seaweed *Callophyllis variegata* (Kallymeniaceae, Gigartinales), *Carbohydr. Res.*, 340 (2005) 2742–2751.
639. F. R. F. Silva, C. M. P. G. Dore, C. T. Marques, M. S. Nascimento, N. M. B. Benevides, H. A. O. Rocha, S. F. Chavante, and E. L. Leite, Anticoagulant activity, paw edema and pleurisy induced carrageenan: Action of major types of commercial carrageenans, *Carbohydr. Polym.*, 79 (2010) 26–33.
640. D. S. McLellan and K. M. Jurd, Anticoagulants from marine algae, *Blood Coagul. Fibrinolysis*, 3 (1992) 69–77.
641. M. Shanmugam and K. H. Mody, Heparinoid-active sulfated polysaccharides from marine algae as potential blood anticoagulant agents, *Curr. Sci.*, 79 (2000) 1672–1683.
642. V. H. Pomin, An overview about the structure-function relationship of marine sulfated homopolysaccharides with regular chemical structures, *Biopolymers*, 91 (2009) 601–609.
643. G. Opoku, X. Qiu, and V. Doctor, Effect of oversulfation on the chemical and biological properties of kappa carrageenan, *Carbohydr. Polym.*, 65 (2006) 134–138.
644. A. K. Sen, Sr., A. K. Das, N. Banerji, A. K. Siddhanta, K. H. Mody, B. K. Ramavat, V. D. Chauhan, J. R. Vedasiromoni, and D. K. Ganguly, A new sulfated polysaccharide with potent blood anticoagulant activity from the red seaweed *Grateloupia indica*, *Int. J. Biol. Macromol.*, 16 (1994) 279–280.
645. M. J. Carlucci, C. A. Pujol, M. Ciancia, M. D. Noseda, M. C. Matulewicz, E. B. Damonte, and A. S. Cerezo, Antiherpetic and anticoagulant properties of carrageenans from the red seaweed

Gigartina skottsbergii and their cyclized derivatives: Correlation between structure and biological activity, *Int. J. Biol. Macromol.*, 20 (1997) 97–105.
646. W. A. Pushpamali, C. Nikapitiya, M. De Zoysa, I. Whang, S. J. Kim, and J. Lee, Isolation and purification of an anticoagulant from fermented red seaweed *Lomentaria catenata*, *Carbohydr. Polym.*, 73 (2008) 274–279.
647. P. Potin, P. Patier, J.-Y. Floc'h, J.-C. Yvin, C. Rochas, and B. Kloareg, Chemical characterization of cell-wall polysaccharides from tank-cultivated and wild plants of *Delesseria sanguinea* (Hudson) Lamouroux (Ceramiales, Delesseriaceae): Culture patterns and potent anticoagulant activity, *J. Appl. Phycol.*, 4 (1992) 119–128.
648. W. R. L. Farias, A.-P. Valente, M. S. Pereira, and P. A. S. Mourão, Structure and anticoagulant activity of sulfated galactans. Isolation of a unique sulfated galactan from the red algae *Botryocladia accidentalis* and comparison of its anticoagulant action with that of sulfated galactans from invertebrates, *J. Biol. Chem.*, 275 (2000) 29299–29307.
649. M. G. Pereira, N. M. B. Benevides, M. R. S. Melo, A. P. Valente, F. R. Melo, and P. A. S. Mourao, Structure and anticoagulant activity of a sulfated galactan from the red alga, *Gelidium crinale*. Is there a specific structural requirement for the anticoagulant action?*Carbohydr. Res.*, 340 (2005) 2015–2023.
650. M. Witvrouw and E. De Clercq, Sulfated polysaccharides extracted from sea algae as potential antiviral drugs, *Gen. Pharmacol.*, 29 (1997) 497–511.
651. E. B. Damonte, M. C. Matulewicz, and A. S. Cerezo, Sulfated seaweed polysaccharides as antiviral agents, *Curr. Med. Chem.*, 11 (2004) 2399–2419.
652. Y. Chen, T. Maguire, R. E. Hileman, J. R. Fromm, J. D. Esko, R. J. Linhardt, and R. M. Marks, Dengue virus infectivity depends on envelope protein binding to target cell heparan sulfate, *Nat. Med.*, 3 (1997) 866–871.
653. M. J. Carlucci, L. A. Scolaro, M. I. Errea, M. C. Matulewicz, and E. B. Damonte, Antiviral activity of natural sulphated galactans on herpes virus multiplication in cell culture, *Planta Med.*, 63 (1997) 429–432.
654. M. J. Carlucci, M. Ciancia, M. C. Matulewicz, A. S. Cerezo, and E. B. Damonte, Antiherpetic activity and mode of action of natural carrageenans of diverse structural types, *Antiviral Res.*, 43 (1999) 93–102.
655. M. E. R. Duarte, J. P. Cauduro, D. G. Noseda, M. D. Noseda, A. G. Gonçalves, C. A. Pujol, E. B. Damonte, and A. S. Cerezo, The structure of the agaran sulfate from *Acanthophora spicifera* (Rhodomelaceae, Ceramiales) and its antiviral activity. Relation between structure and antiviral activity in agarans, *Carbohydr. Res.*, 339 (2004) 335–347.
656. K. Chattopadhyay, T. Ghosh, C. A. Pujol, M. J. Carlucci, E. B. Damonte, and B. Ray, Polysaccharides from *Gracilaria corticata*: Sulfation, chemical characterization and anti-HSV activities, *Int. J. Biol. Macromol.*, 43 (2008) 346–351.
657. M. E. R. Duarte, D. G. Noseda, M. D. Noseda, S. Tulio, C. A. Pujol, and E. B. Damonte, Inhibitory effect of sulfated galactans from the marine alga *Bostrichia montagnei* on herpes simplex virus replication in vitro, *Phytomedicine*, 8 (2003) 53–58.
658. P. Cáceres, M. J. Carlucci, E. B. Damonte, B. Matsuhiro, and E. A. Zúñiga, Carrageenans from Chilean samples of *Stenogramme interrupta* (Phyllophoraceae): Structural analysis and biological activity, *Phytochemistry*, 53 (2000) 81–86.
659. P. C.de S. F. Tischer, L. B. Talarico, M. D. Noseda, S. M. P. B. Guimarães, E. B. Damonte, and M. E. R. Duarte, Chemical structure and antiviral activity of carrageenans from *Meristiella gelidium* against herpes simplex and dengue virus, *Carbohydr. Polym.*, 63 (2006) 459–465.
660. A. A. Kolender, C. A. Pujol, E. B. Damonte, M. C. Matulewicz, and A. S. Cerezo, The system of sulfated alpha-(1→3)-linked D-mannans from the red seaweed *Nothogenia fastigiata*: Structures, antiherpetic and anticoagulant properties, *Carbohydr. Res.*, 304 (1997) 53–60.

661. H. Nakashima, Y. Kido, N. Kobayashi, Y. Motoki, M. Neushul, and N. Yamamoto, Antiretroviral activity in a marine algae: Reverse transcriptase inhibition by an aqueous extract of *Schizymenia pacifica*, *J. Cancer Res. Clin. Oncol.*, 113 (1987) 413–416.
662. N. Bourgougnon, M. Lahaye, B. Quemener, J.-C. Chermann, M. Rimbert, M. Cormaci, G. Funari, and J.-M. Kornprobst, Annual variation in composition and *in vitro* anti-HIV-1 activity of the sulfated glucuronogalactan from *Schizymenia dubyi* (Rhodophyta, Gigartinales), *J. Appl. Phycol.*, 8 (1996) 155–161.
663. B. Matsuhiro, A. F. Conte, E. B. Damonte, A. A. Kolender, M. C. Matulewicz, E. G. Mejías, C. A. Pujol, and E. A. Zúñiga, Structural analysis and antiviral activity of a sulfated galactan from the red seaweed *Schizymenia binderi* (Gigartinales, Rhodophyta), *Carbohydr. Res.*, 340 (2005) 2392–2402.
664. S. C. Wang, S. W. A. Bligh, S. S. Shi, Z. T. Wang, Z. B. Hu, J. Crowder, C. Branford-White, and C. Vella, Structural features and anti-HIV-1 activity of novel polysaccharides from red algae *Grateloupia longifolia* and *Grateloupia filicina*, *Int. J. Biol. Macromol.*, 41 (2007) 369–375.
665. K. Chattopadhyay, C. G. Mateu, P. Mandal, C. A. Pujol, E. B. Damonte, and B. Ray, Galactan sulfate of *Grateloupia indica*: Isolation, structural features and antiviral activity, *Phytochemistry*, 68 (2007) 1428–1435.
666. T. Yamada, A. Ogamo, T. Saito, J. Watanabe, H. Uchiyama, and Y. Nakagawa, Preparation and anti-HIV activity of low-molecular-weight carrageenans and their sulfated derivatives, *Carbohydr. Polym.*, 32 (1997) 51–55.
667. T. Yamada, A. Ogamo, T. Saito, H. Uchiyama, and Y. Nakagawa, Preparation of O-acylated low-molecular-weight carrageenans with potent anti-HIV activity and low anticoagulant effect, *Carbohydr. Polym.*, 41 (2000) 115–120.
668. I. Groth, N. Grünewald, and S. Alban, Pharmacological profiles of animal- and nonanimal-derived sulfated polysaccharides—Comparison of unfractionated heparin, the semisynthetic glucan sulfate PS3, and the sulfated polysaccharide fraction isolated from *Delesseria sanguinea*, *Glycobiology*, 19 (2009) 408–417.
669. A. Liang, Z. Zhou, Q. Wang, X. Liu, X. Liu, Y. Du, K. Wang, and B. Lin, Structural features in carrageenan that interact with a heparin-binding hematopoietic growth factor and modulate its biological activity, *J. Chromatogr. B*, 843 (2006) 114–119.
670. R. Hoffman, W. W. Burns, and D. H. Paper, Selective inhibition of cell proliferation and DNA synthesis by the polysulfated carbohydrate iota-carrageenan, *Cancer Chem. Pharmacol.*, 36 (1995) 325–334.
671. N. Grünewald, I. Groth, and S. Alban, Evaluation of seasonal variations of the structure and anti-inflammatory activity of sulfated polysaccharides extracted from the red alga *Delesseria sanguinea* (Hudson) Lamouroux (Ceramiales, Delesseriaceae), *Biomacromolecules*, 10 (2009) 1155–1162.
672. N. Grünewald and S. Alban, Optimized and standardized isolation and structural characterization of anti-inflammatory sulfated polysaccharides from the red alga *Delesseria sanguinea* (Hudson) Lamouroux (Ceramiales, Delesseriaceae), *Biomacromolecules*, 10 (2009) 2998–3008.
673. Y. Yoshizawa, J. Tsunehiro, K. Nomura, M. Itoh, F. Fukui, A. Ametani, and S. Kaminogawa, In vivo macrophage-stimulation activity of the enzyme-degraded water-soluble polysaccharide fraction from a marine alga (*Gracilaria verrucosa*), *Biosci. Biotechnol., Biochem.*, 60 (1996) 1667–1671.
674. D. R. Coombe, C. R. Parish, and I. A. Ramshaw, Analysis of the inhibition of tumor metastasis by sulphated polysaccharides, *Int. J. Cancer*, 39 (1987) 82–88.
675. G. Zhou, Y. Sun, H. Xin, Z. Zhang, and Z. Xu, In vivo antitumor and immunomodulation activities of different molecular weight lambda-carrageenans from *Chondrus ocellatus*, *Pharm. Res.*, 50 (2004) 47–53.

676. S. Bondu, E. Deslandes, M. S. Fabre, C. Berthou, and G. Yu, Carrageenan from *Solieria chordalis* (Gigartinales): Structural analysis and immunological activities of the low molecular weight fractions, *Carbohydr. Polym.*, 81 (2010) 448–460.
677. L. V. Abad, H. Kudo, S. Saiki, N. Nagasawa, M. Tamada, Y. Katsumura, C. T. Aranilla, L. S. Relleve, and A. M. De La Rosa, Radiation degradation studies of carrageenans, *Carbohydr. Polym.*, 78 (2009) 100–106.
678. H. Mou, X. Jiang, and H. Guan, A κ-carrageenan derived oligosaccharide prepared by enzymatic degradation containing anti-tumor activity, *J. Appl. Phycol.*, 15 (2003) 297–303.
679. G. Zhou, H. Xin, W. Sheng, S. Yueping, Z. Li, and Z. Xu, In vivo growth-inhibition of S180 tumor by mixture of 5-Fu and low molecular weight λ-carrageenan from *Chondrus ocellatus*, *Pharm. Res.*, 51 (2005) 153–157.
680. Z. Zhang, Q. Zhang, J. Wang, X. Shi, J. Zhang, and H. Song, Synthesis and drug release *in vitro* of porphyran carrying 5-fluorouracil, *Carbohydr. Polym.*, 79 (2010) 628–632.
681. D. H. Paper, H. Vogl, and G. Franz, Defined carrageenan derivatives as angiogenesis inhibitors, *Macromol. Symp.*, 99 (1995) 219–225.
682. H. Chen, X. Yan, J. Lin, F. Wang, and W. Xu, Depolymerized products of λ-carrageenan as a potent angiogenesis inhibitor, *J. Agric. Food Chem.*, 55 (2007) 6910–6917.
683. V. E. Santo, A. M. Frias, M. Carida, R. Cancedda, M. E. Gomes, J. F. Mano, and R. L. Reis, Carrageenan-based hydrogels for the controlled delivery of PDGF-BB in bone tissue engineering applications, *Biomacromolecules*, 10 (2009) 1392–1401.
684. Q. Zhang, P. Yu, Z. Li, H. Zhang, Z. Xu, and P. Li, Antioxidant activities of sulfated polysaccharide fractions from *Porphyra haitanensis*, *J. Appl. Phycol.*, 15 (2003) 305–310.
685. H. Yuan, W. Zhang, X. Li, X. Lü, N. Li, X. Gao, and J. Song, Preparation and in vitro antioxidant activity of κ-carrageenan oligosaccharides and their oversulfated, acetylated, and phosphorylated derivatives, *Carbohydr. Res.*, 340 (2005) 685–692.
686. Z. Zhang, Q. Zhang, J. Wang, H. Zhang, X. Niu, and P. Li, Preparation of the different derivatives of the low-molecular-weight porphyran from *Porphyra haitanensis* and their antioxidant activities *in vitro*, *Int. J. Biol. Macromol.*, 45 (2009) 22–26.
687. Z. Zhang, Q. Zhang, J. Wang, X. Shi, H. Song, and J. Zhang, In vitro antioxidant activities of acetylated, phosphorylated and benzoylated derivatives of porphyran extracted from *Porphyra haitanensis*, *Carbohydr. Polym.*, 78 (2009) 449–453.
688. Z. Zhang, Q. Zhang, J. Wang, H. Song, H. Zhang, and X. Niu, Chemical modification and influence of function groups on the in vitro-antioxidant activities of porphyran from *Porphyra haitanensis*, *Carbohydr. Polym.*, 79 (2010) 290–295.
689. I. M. Ermak, A. O. Barabanova, T. A. Kukarskikh, T. F. Solovyova, R. N. Bogdanovich, A. M. Polyakova, O. P. Astrina, and V. V. Maleev, Natural polysaccharide carrageenan inhibits effect of gram-negative bacterial endotoxins, *Bull. Exp. Biol. Med.*, 141 (2006) 230–232 (in Russian).
690. I. E. Sommsich and K. Halbrock, Pathogen defense in plants: A paradigm of biological complexity, *Trends Plant Sci.*, 3 (1998) 86–90.
691. F. Weinberger, M. Friedlander, and H. G. Hoppe, Oligoagars elicit an oxidative burst in *Gracilaria conferta* (Rhodophyta), *J. Phycol.*, 35 (1999) 747–755.
692. P. Potin, K. Bouarab, F. Küpper, and B. Kloareg, Oligosaccharide recognition signals and defense reactions in marine plant-microbe interactions, *Curr. Opin. Microbiol.*, 2 (1999) 276–283.
693. K. Bouarab, P. Potin, F. Weinberger, J. Correa, and B. Kloareg, The *Chondrus crispus–Acrochaete operculata* host-pathogen association, a novel model in glycobiology and applied phycopathology, *J. Appl. Phycol.*, 13 (2001) 185–193.
694. A. I. Usov, K. S. Adamyants, S. V. Yarotsky, and A. A. Anoshina, Polysaccharides of algae. 16. Study of the structure of the sulfated mannan from the red alga *Nemalion vermiculare* Sur., *Zh. Obshch. Khim.*, 45 (1975) 916–921 (in Russian).

695. A. I. Usov, K. S. Adamyants, and S. V. Yarotsky, Polysaccharides of algae. 18. Acetolysis of sulfated mannan from *Nemalion vermiculare* Sur., *Zh. Obshch. Khim.*, 45 (1975) 1377–1381 (in Russian).
696. A. I. Usov and S. V. Yarotsky, Polysaccharides of algae. 21. Alkaline degradation of sulfated mannan from the red alga *Nemalion vermiculare* Sur, *Sov. J. Bioorg. Chem.*, 1 (1975) 700–703 (English translation from *Bioorg. Khim.*, 1 (1975) 919–922).
697. A. I. Usov, S. V. Yarotsky, and L. K. Vasyanina, Synthesis and carbon-13 NMR spectra of methyl 3-O-methyl-α-D-mannopyranoside sulphates, *Sov. J. Bioorg. Chem.*, 1 (1975) 1137–1141 (English translation from *Bioorg. Khim.*, 1 (1975) 1583–1588).
698. M. C. Matulewicz and A. S. Cerezo, Water-soluble sulfated polysaccharides from the red seaweed *Chaetangium fastigiatum*. Analysis of the system and the structures of the α-D-(1→3)-linked mannans, *Carbohydr. Polym.*, 7 (1987) 121–132.
699. A. A. Kolender, M. C. Matulewicz, and A. S. Cerezo, Structural analysis of antiviral sulfated α-D-(1→3)-linked mannans, *Carbohydr. Res.*, 273 (1995) 179–185.
700. R. Erra-Balsells, A. A. Kolender, M. C. Matulewicz, H. Nonami, and A. S. Cerezo, Matrix-assisted ultraviolet laser-desorption ionization time-of-flight mass spectrometry of sulfated mannans from the red seaweed *Nothogenia fastigiata*, *Carbohydr. Res.*, 329 (2000) 157–167.
701. A. I. Usov and I. M. Dobkina, Polysaccharides of algae. 38. Polysaccharide composition of the red seaweed *Liagora* sp. and the structure of sulfated xylomannan, *Sov. J. Bioorg. Chem.*, 14 (1988) 354–363 (English translation from *Bioorg. Khim.*, 14 (1988) 642–651).
702. P. Mandal, C. A. Pujol, M. J. Carlucci, K. Chattopadhyay, E. B. Damonte, and B. Ray, Anti-herpetic activity of a sulfated xylomannan from *Scinaia hatei*, *Phytochemistry*, 69 (2008) 2193–2199.
703. M. P. Recalde, M. D. Noseda, C. A. Pujol, M. J. Carlucci, and M. C. Matulewiczc, Sulfated mannans from the red seaweed *Nemalion helminthoides* of the South Atlantic, *Phytochemistry*, 70 (2009) 1062–1068.
704. H. H. Haines, M. C. Matulewicz, and A. S. Cerezo, Sulfated galactans from the red seaweed *Nothogenia fastigiata* (Nemaliales, Rhodophyta), *Hydrobiologia*, 204/205 (1990) 637–643.
705. M. A. Cardoso, M. D. Noseda, M. T. Fujii, R. G. M. Zibetti, and M. E. R. Duarte, Sulfated xylomannans isolated from red seaweeds *Chondrophycus papillosus* and *C. flagelliferus* (Ceramiales) from Brazil, *Carbohydr. Res.*, 342 (2007) 2766–2775.
706. C. A. Pujol, M. J. Carlucci, M. C. Matulewicz, and E. B. Damonte, Natural sulfated polysaccharides for the prevention and control of viral infections, *Top. Heterocycl. Chem.*, 11 (2007) 259–281.
707. N. Bourgougnon, M. Lahaye, J.-C. Chermann, and J.-M. Kornprobst, Composition and antiviral activities of sulfated polysaccharide from *Schizymenia dubyi* (Rhodophyta, Gigartinales), *Bioorg. Med. Chem. Lett.*, 3 (1993) 1141–1146.
708. N. Bourgougnon, M. Lahaye, B. Quimener, M. Cormaci, G. Furnari, and J.-M. Kornprobst, Chemical structure analysis of water-soluble sulfated polysaccharide from *Schizymenia dubyi* (Rhodophyta, Gigartinales), *J. Appl. Phycol.*, 8 (1996) 147–153.
709. J. R. Turvey and L. M. Griffiths, Mucilage from a fresh-water red alga of the genus *Batrachospermum*, *Phytochemistry*, 12 (1973) 2901–2907.
710. A. I. Usov, Alginic acids and alginates: Analytical methods used for their estimation and characterization of composition and primary structure, *Russ. Chem. Rev.*, 68 (1999) 957–966 (translated from, *Uspekhi Khimii*, 68 (1999) 1051–1061).
711. M. Okazaki, K. Furuya, K. Tsukayama, and K. Nisizawa, Isolation and identification of alginic acid from a calcareous red alga *Serraticardia maxima*, *Bot. Mar.*, 25 (1982) 123–131.
712. S. Geresh and S. (Malis) Arad, The extracellular polysaccharides of the red microalgae: Chemistry and rheology, *Bioresour. Technol.*, 38 (1991) 195–201.
713. S. Geresh, N. Lupescu, and S. (Malis) Arad, Fractionation and partial characterization of the sulphated polysaccharide of *Porphyridium*, *Phytochemistry*, 31 (1992) 4181–4186.

714. N. Lupescu, S. (Malis) Arad, S. Geresh, M. A. Bernstein, and R. Glaser, Structure of some sulfated sugars isolated after acid hydrolysis of the extracellular polysaccharide of *Porphyridium* sp., a unicellular red alga, *Carbohydr. Res.*, 210 (1991) 349–352.
715. J. Heaney-Kieras and D. J. Chapman, Structural studies on the extracellular polysaccharide of the red alga, *Porphyridium cruentum*, *Carbohydr. Res.*, 52 (1976) 169–177.
716. M. Jaseja, A. S. Perlin, O. Dubinsky, D. Christiaen, S. S. (Malis) Arad, and R. Glaser, N.m.r. structure determination of 3-*O*-(α-D-glucopyranosyluronic acid)-L-galactopyranose, an aldobiuronic acid isolated from the unicellular red alga *Rhodella reticulata*, *Carbohydr. Res.*, 186 (1989) 313–319.
717. S. Geresh, O. Dubinsky, S. (Malis) Arad, D. Christiaen, and R. Glaser, Structure of 3-*O*-(α-D-glucopyranosyluronic acid)-L-galactopyranose, an aldobiouronic acid isolated from the polysaccharides of various unicellular red algae, *Carbohydr. Res.*, 208 (1990) 301–305.
718. N. Lupescu, J. Solo-Kwan, D. Christiaen, H. Morvan, and S. (Malis) Arad, Structural determination by means of gas chromatography—Mass spectrometry of 3-*O*-(α-D-glucopyranosyluronic acid)-galactopyranose, an aldobiuronic acid derived from *Porphyridium* sp. polysaccharide, *Carbohydr. Polym.*, 19 (1992) 131–134.
719. E. Percival and R. A. J. Foyle, The extracellular polysaccharides of *Porphyridium cruentum* and *Porphyridium aerugineum*, *Carbohydr. Res.*, 72 (1979) 165–176.
720. J. Heaney-Kieras, L. Roden, and D. J. Chapman, The covalent linkage of protein to carbohydrate in the extracellular protein polysaccharide from the red alga *Porphyridium cruentum*, *Biochem. J.*, 165 (1977) 1–9.
721. V. Gloaguen, G. Ruiz, H. Morvan, A. Mouradi-Givernaud, E. Maes, P. Krausz, and G. Strecker, The extracellular polysaccharide of *Porphyridium* sp.: An NMR study of lithium-resistant oligosaccharide fragments, *Carbohydr. Res.*, 339 (2004) 97–103.
722. S. Geresh, S. (Malis) Arad, O. Levy-Ontman, W. Zhang, Y. Tekoah, and R. Glaser, Isolation and characterization of poly- and oligosaccharides from the red microalga *Porphyridium* sp, *Carbohydr. Res.*, 344 (2009) 343–349.
723. L. V. Evans, M. E. Callow, E. Percival, and V. Fareed, Studies on the synthesis and composition of extracellular mucilage in the unicellular red alga *Rhodella*, *J. Cell Sci.*, 16 (1974) 1–21.
724. S. V. Fareed and E. Percival, The presence of rhamnose and 3-*O*-methylxylose in the extracellular mucilage from the red alga *Rhodella maculata*, *Carbohydr. Res.*, 53 (1977) 276–277.
725. P. Capek, M. Matulová, and B. Combourieu, The extracellular proteoglycan produced by *Rhodella grisea*, *Int. J. Biol. Macromol.*, 43 (2008) 390–393.
726. S. Geresh, I. Adin, E. Yarmolinsky, and M. Karpasas, Characterization of the extracellular polysaccharide of *Porphyridium* sp.: Molecular weight determination and rheological properties, *Carbohydr. Polym.*, 50 (2002) 183–189.
727. S. (Malis) Arad, G. Keristovesky, B. Simon, Z. Barak, and S. Geresh, Biodegradation of the sulphated polysaccharide of *Porphyridium* by soil bacteria, *Phytochemistry*, 32 (1993) 287–290.
728. E. Cohen and S. (Malis) Arad, A closed system for outdoor cultivation of *Porphyridium*, *Biomass*, 89 (1989) 59–67.
729. S. Geresh, A. Mamontov, and J. Weinstein, Sulfation of extracellular polysaccharides of red microalgae: Preparation, characterization and properties, *J. Biochem. Biophys. Methods*, 50 (2002) 179–187.
730. S. Geresh, R. P. Davadi, and S.(.M.). Arad, Chemical modifications of biopolymers: Quaternization of the extracellular polysaccharide of the red microalga *Porphyridium* sp., *Carbohydr. Polym.*, 63 (2000) 75–80.
731. M. Huleichel, V. Ishanu, J. Tal, and S. (Malis) Arad, Antiviral effect of red microalgal polysaccharides on *Herpes simplex* and *Varicella zoster* viruses, *J. Appl. Phycol.*, 13 (2001) 127–134.
732. L. Sun, C. Wang, Q. Shi, and C. Ma, Preparation of different molecular weight polysaccharides from *Porphyridium cruentum* and their antioxidant activities, *Int. J. Biol. Macromol.*, 45 (2009) 42–47.

733. H. L. Youngs, M. R. Gretz, J. A. West, and M. R. Sommerfeld, The cell wall chemistry of *Bangia atropurpurea* (Bangiales, Rhodophyta) and *Bostrychia moritziana* (Ceramiales, Rhodophyta) from marine and freshwater environments, *Phycol. Res.*, 46 (1998) 63–73.
734. H. Lechat, M. Amat, J. Mazoyer, D. J. Gallant, A. Buléon, and M. Lahaye, Cell wall composition of the carrageenophyte *Kappaphycus alvarezii* (Gigartinales, Rhodophyta) partitioned by wet sieving, *J. Appl. Phycol.*, 9 (1997) 565–572.
735. H. Lechat, M. Amat, J. Mazoyer, A. Buléon, and M. Lahaye, Structure and distribution of glucomannan and sulfated glucan in the cell walls of the red alga *Kappaphycus alvarezii* (Gigartinales, Rhodophyta), *J. Phycol.*, 36 (2000) 891–902.
736. D. K. Watt, S. A. O'Neill, A. E. Percy, and D. J. Brasch, Isolation and characterization of a partially methylated galacto-glucurono-xylo-glycan, a unique polysaccharide from the red seaweed *Apophloea lyallii*, *Carbohydr. Polym.*, 50 (2002) 283–294.
737. B. Matsuhiro and C. C. Urzúa, The acidic polysaccharide from *Palmaria decipiens* (Palmariales, Rhodophyta), *Hydrobiologia*, 326/327 (1996) 491–495.
738. P. A. Sandford and J. Baird, Industrial utilization of polysaccharides, in G. O. Aspinall, (Ed.), *The Polysaccharides*, Vol. 2, Academic Press, New York, 1983, pp. 411–490.
739. W. L. Zemke-White and M. Ohno, World seaweed utilization: An end-of-century summary, *J. Appl. Phycol.*, 11 (1999) 369–376.
740. L. Stoloff and P. Silva, An attempt to determine possible taxonomic significance of the properties of water extractable polysaccharides in red algae, *Econ. Bot.*, 11 (1957) 327–330.
741. L. Stoloff, Algal classification: An aid to improved industrial utilization, *Econ. Bot.*, 16 (1962) 86–94.
742. T. Chopin, B. F. Kerin, and R. Mazerolle, Phycocolloid chemistry as a taxonomic indicator of phylogeny in the Gigartinales, Rhodophyceae: A review and current developments using Fourier transform infrared diffuse reflectance spectroscopy, *Phycol. Res.*, 47 (1999) 167–188.
743. J. A. Hemmingson and W. A. Nelson, Cell wall polysaccharides are informative in *Porphyra* species taxonomy, *J. Appl. Phycol.*, 14 (2002) 357–364.
744. R. Falshaw and R. H. Furneaux, Chemotaxonomy of New Zealand red algae in the family Gigartinaceae (Rhodophyta) based on galactan structures from the tetrasporophyte life-stage, *Carbohydr. Res.*, 344 (2009) 210–216.
745. M. H. Hommersand, S. Fredericq, D. W. Freshwater, and J. Hughey, Recent developments in the systematic of the Gigartinaceae (Gigartinales, Rhodophyta) based on *rbc*L sequence analysis and morphological evidence, *Phycol. Res.*, 47 (1999) 139–151.
746. I. J. Miller, The chemotaxonomic significance of the water-soluble red algal polysaccharides, *Recent Res. Dev. Phytochem.*, 1 (1997) 531–565.
747. A. Chiovitti, G. T. Kraft, G. W. Saunders, M.-L. Liao, and A. Bacic, A revision of the systematic of the Nizymeniaceae (Gigartinales, Rhodophyta) based on polysaccharides, anatomy and nucleotide sequences, *J. Phycol.*, 31 (1995) 153–166.
748. A. Chiovitti, G. T. Kraft, A. Bacic, D. J. Craik, S. L. A. Munro, and M.-L. Liao, Carrageenans from Australian representatives of the family Cystocloniaceae (Gigartinales, Rhodophyta), with description of *Calliblepharis celastospora* sp. nov. and transfer of *Austroclonium* to the family Areschougiaceae, *J. Phycol.*, 34 (1998) 515–535.
749. M.-L. Liao, G. T. Kraft, S. L. A. Munro, D. J. Craik, and A. Bacic, Beta/kappa-carrageenans as evidence for continued separation of the families Dicranemataceae and Sarcodiaceae (Gigartinales, Rhodophyta), *J. Phycol.*, 29 (1993) 833–844.
750. G. W. Saunders and G. T. Kraft, A molecular perspective on red algal evolution: Focus on the Florideophycidae, *Plant Syst. Evol.*, 11(Suppl.), (1997) 115–138.
751. H. S. Yoon, K. M. Müller, R. G. Sheath, F. D. Ott, and D. Bhattacharya, Defining the major lineages of red algae (Rhodophyta), *J. Phycol.*, 42 (2006) 482–492.

TOWARD AUTOMATED GLYCAN ANALYSIS

Shin-Ichiro Nishimura

Field of Drug Discovery Research, Faculty of Advanced Life Science, Hokkaido University, N21, W11, Kita-ku, Sapporo, Japan

I. Introduction	220
1. Background	220
2. Proteomics and Glycomics: Glycan Expression is Not Template-Driven	220
3. Unmet Needs in Glycobiology and Glycotechnology	221
II. Emerging Glycomics Technologies	223
1. Progress in Mass Spectrometry	223
2. The Rate-Determining Stage in Glycomics and Glycoproteomics	234
III. Toward Automated Glycan Analysis	261
References	265

Abbreviations

ABCh, N-(2-aminobenzoyl)cysteine hydrazide; AFP, α-fetoprotein; BOA, benzyloxyamine; CE, collision energy; CHO, Chinese hamster ovary; CID, collision-induced dissociation; DTT, 1,4-dithiothreitol; ECD, electron-capture dissociation; ESI, electrospray ionization; ETD, electron-transfer dissociation; FAB, fast-atom bombardment; GC–MS, gas chromatography MS; GMF, glycofragment mass fingerprinting; GSLs, glycosphingolipids; HCC, hepatocellular carcinoma; HILIC, hydrophilic-interaction chromatography; huEPO, recombinant human erythropoietin; IAA, iodoacetamide; MALDI, matrix-assisted laser-desorption ionization; MRM, multiple-reaction monitoring; MS, mass spectrometry; OPN, human-milk osteopontin; ORF, open reading frame; PC-3, human prostate cancer cells; PCR, polymerase chain reaction; PMF, peptide mass fingerprinting; PNGase F, peptide N-glycosidase F; PrEC, normal prostate epithelial cells; SPOT, solid-phase oligosaccharide tagging; TFA, trifluoroacetic acid; TOFMS, time-of-flight mass spectrometry

I. INTRODUCTION

1. Background

Various glycans that are located at the cellular surfaces, such as glycoproteins and glycosphingolipids (GSLs), or deposited in the extracellular matrices (proteoglycans), or bound to secretory proteins, play a variety of crucial biological roles in the phenotypic expression of cellular genotypes. If satisfactory amounts of glycans, or glycoconjugates themselves, could be isolated by purification from biological samples, such general analytical technologies as spectrometry-based approaches using UV, fluorescence, CD, NMR, MS, and SPR, and X-ray crystallography would greatly contribute to our insight into the structural features and molecular basis of functional roles of many glycoconjugates. In addition, it is clear that synthetic chemistry has provided important key glycans and related derivatives for investigating structure–function relationships. As drastic structural changes in cell-surface glycans of glycoproteins and GSLs, as well as serum glycoproteins, are often observed during cell differentiation and cancer progression, it is considered that glycans may constitute potential candidates for novel diagnostic and therapeutic biomarkers. Although there have been substantial advances in our understanding of the effects of glycosylation on some biological systems, we still do not fully understand the significance and mechanism of glycoform alteration observed widely in many human diseases because of the highly complicated structures of the glycans and their extremely low abundance in common biological samples. As a result, the therapeutic potential of complex glycans has, with a few notable exceptions, not been well exploited.[1,2] This chapter describes the state of the art and current advances in new technologies and efforts toward the development of automated glycan analysis that should greatly accelerate functional glycobiology and their medical/pharmaceutical applications.

2. Proteomics and Glycomics: Glycan Expression is Not Template-Driven

Glycosylation is one of the most important posttranslational modifications of proteins in eukaryotes. This step is essential for modulating a wide range of protein and lipid functions, both on the cellular surfaces and within the cells.[3] The long-term challenge of glycomics is therefore enormous: to define the identities, quantities, relationship between glycoforms and functions of the glycoconjugates, and to characterize how these properties vary in different cellular contexts. Proteomics, the analysis of genomic complements of proteins, has burst onto the broad scientific fields with stunning rapidity over the past few years, perhaps befitting a discipline that can enjoy the virtually instantaneous conversion of the genome sequence into a set of predicted proteins.[4] It is clear that combined use of automated DNA analyzers and synthesizers has allowed for rapid and precise genomics, namely the decoding of the whole human genome. The development of methods for parallel proteomics has relied on the rapid identification of open reading frames (ORFs) and procedures for their facile cloning and expression. Cloning of a genomic set of ORFs, and the chemical synthesis of gene-specific primers, namely synthetic oligonucleotides as fragments of DNA/RNA that are suitable for amplification by the polymerase chain reaction (PCR) [5,6] and for subsequent insertion of the PCR products into appropriate plasmids, permits the widespread use of recombination-based protein engineering and related biotechnologies. The genomic-scale PCR amplification using mouse and human ORFs has also been applied toward large-scale mammalian proteomics. Herein, it may be noted that primers can be made by high-throughput synthetic methods with high fidelity, at reasonable cost and in 96-well format, amenable for robotic manipulation.

Glycomics, a term defining the sequence analysis or profiling of the glycome, differs fundamentally from genomics and proteomics in that the glycan expression mechanism is not template-driven (Fig. 1). In other words, PCR technology cannot be used for preparation of the whole human glycome by amplification from any partial oligosaccharide structures, even if large sets of glycan fragments might be obtained from some biological samples.

3. Unmet Needs in Glycobiology and Glycotechnology

Glycomics may be performed by integrating with proteomics because the glycosylation (affording different glycoforms) greatly influences protein function and structure. However, it should be emphasized that individual glycoforms of the glycoconjugates of interest are impossible to predict based on gene expression patterns as just described, because the biosynthetic process of the glycans is not template-driven and is subject to multiple sequential and competitive enzymatic pathways (Fig. 1).[7–9] At present, there is no PCR-like glycan amplification technology for glycomics. While proteomics as well as genomics can use satisfactory amounts of the genomes and proteins of interest by means of the aforementioned ORF–PCR-based amplification/recombination technology, glycomics requires an enrichment process of glycans from highly complicated mixtures, such as body fluids, cells, tissues, organs, and the like.

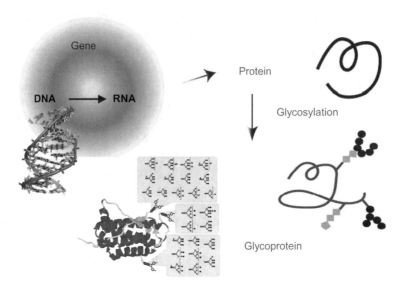

FIG. 1. Biosynthetic pathway of glycoproteins. No one can predict individual glycoforms and the microheterogeneity of glycoproteins, because glycan biosynthesis is not a template-driven posttranslational modification. Recombinant human erythropoietin, the glycoprotein drug produced by mammalian cells, has highly complex N-glycan heterogeneity. (Original figure made by the author). (See Color Insert.)

One of the bottlenecks in structural and functional glycomics is the difficulty of purifying total glycoproteins, or even major glycans, as this requires extremely tedious and time-consuming chromatography-based multistep processes. Glycans, both the N- and O-glycans of glycoproteins, must be prepared and purified from the initial samples by application of further analytical processes. The glycans released through enzymatic digestion from whole glycoproteins (tryptic peptides) in fluids, cells, tissues, and organs are usually obtained as heterogeneous mixtures containing large amounts of such impurities as peptides, nucleotides, lipids, salts, and other components. To date, the technical problems in the purification processes of glycans make it impossible to achieve high-throughput glycomics, namely the study of the structure and function of human whole glycome. Moreover, a lack of general methods for the routine synthesis of glycans remains another obstacle, although extensive efforts have been made toward the development of practical methods for the synthesis of glycans and glycoconjugates.[10] Therefore, there is a pressing need for novel technology that will allow high-throughput glycomics and synthetic protocols for functional glycobiology and glycotechnology that would lead to new disease-relevant glycan biomarkers and the development of carbohydrate-based diagnostic/therapeutic reagents.

II. Emerging Glycomics Technologies

1. Progress in Mass Spectrometry

Mass spectrometry (MS) is an indispensable tool for proteomics, which, in general, deals with the large-scale determination of gene and cellular function directly at the protein level. It seems probable that the ability of MS to identify and quantify precisely many thousands of proteins from complex samples will have broad impact on cell biology and medicine.[4,11] As in proteomics and metabolomics, MS-based analytical methods have come to the fore as a powerful tool for the highly sensitive and definitive analysis of the primary structure of glycans derived from diverse biological sources. Ultra-high sensitivity, coupled with the ability to characterize individual components within a complex mixture of similar glycans, appears to be essential features of MS that are especially advantageous in glycomics.

a. General MS Strategies for Glycomics.—A review article by Dell *et al.*[12] on recent MS techniques widely used in glycomics provides a valuable overview of the repertoire of key MS-based structural analyses best suited for the characterization of complex glycomes. Over the past decade, MS-based glycomics strategies have been devised for screening populations of N- and O-glycans of glycoproteins isolated from a diverse range of biological materials, such as body fluids, secretions, organs, and cultured cell lines. More recently, the protocols have been adapted to include oligosaccharide head groups derived from GSLs.[12]

A typical glycomics strategy is delineated in Fig. 2 as a general protocol when mammalian tissues and cell homogenates are used as the starting biological samples.[12] Glycoproteins and GSLs are first extracted from the biological materials and then separately partitioned. Glycans are liberated from the GSL fraction by treating with ceramide glycanase, and N- and O-glycans are released from the glycoproteins by either peptide: N-glycosidase F (PNGase F) or by reductive β-elimination. A portion of each glycan pool is subjected to permethylation,[13] purification, and then MS analysis. Putative structures are assigned to each molecular-ion peak, based on the known glycan compositions for a given mass and the possible structures in the reported biosynthetic pathways, that is to say, by "mass mapping." This procedure is most conveniently performed by means of matrix-assisted laser-desorption[14,15] ionization (MALDI), because of its very high sensitivity and facile analysis of singly charged precursor ions. MALDI–MS is now firmly established as one of the most convenient technologies and is used widely in glycomics. MALDI-based ionization is a softer and simpler process than other ionization techniques such as fast-atom bombardment (FAB)[16,17] or electrospray ionization (ESI)[18–21] and yields a series of

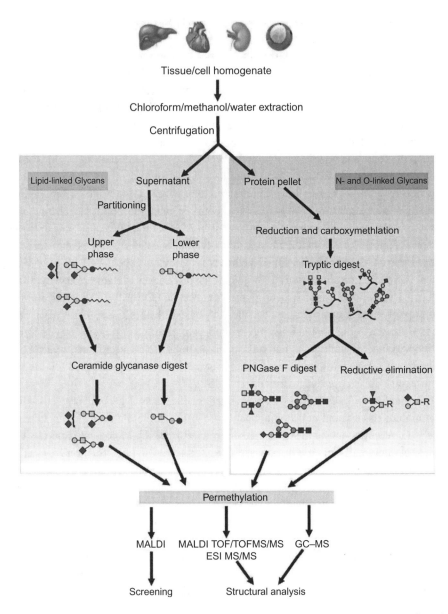

FIG. 2. Schematic representation of MS strategies for glycomics as reported by Dell *et al*. This figure is adapted from ref. 12. (See Color Insert.)

singly charged molecular ions, providing an ideal method for obtaining an overall glycan profile as a fingerprint of glycoforms, even in a complex sample mixture. Moreover, it seems probable that the upper limit of molecular-ion mass detected by MALDI will be considerably higher than those of other ionization methods, and glycans of high-molecular weight, well in excess of 10 kDa, are amenable to this technology.[22] Molecular ions can be profiled as usually unique glycan compositions by means of various tools, such as an algorithm called Cartoonist[23] or other commercially available software such as *GlycoMod*,[24] *GlycoComp (GlycoSuiteDB)*,[25] and *Glyco-Peakfinder*,[26] which are capable of calculating the plausible sugar composition corresponding to a target mass. Elucidation of glycan composition is a combinatorial problem where the number of compositions that must be considered expands exponentially with the number of different monosaccharides that can be considered in the structure. Assignment of the composition for glycans having masses of high-molecular weight is therefore very difficult. In addition, a number of compositions may be indistinguishable within a relatively low mass delta threshold (less than 0.05 Da) when adducts and other spectrometric losses are taken into account. It is clear that some constraints based on taxonomic and biosynthetic information can simplify the compositional analysis by eliminating any sugar residues that are known not to occur in the biological systems, or that require the presence of essential core structures in any solution composition. Glycan composition assignments are confirmed and further refined by tandem MS, both by MALDI–TOF/TOFMS/MS and ESI–MS/MS, in which precursor ions are selected for collision-induced dissociation (CID). To obtain sequence information from the glycan composition and MS/MS fragmentation data, such bioinformatic tools as *GlycosidIQ*[27] have been proposed for automatic evaluation of all the glycans from a structure database on the given MS/MS and MS/MS/MS (MS^n) spectra. The *GlycosidIQ* platform allows for computerized interpretation of glycan MS fragmentation, based on matching experimental data with theoretically fragmented oligosaccharides generated from the structural database *GlycoSuiteDB*. This approach, namely glycofragment mass fingerprinting (GMF), is conceptually similar to the peptide mass fingerprinting (PMF) established for protein sequencing.[28–30] Although it is considered that the versatility of the GMF approach depends significantly on the size and quality of the library of structures stored in the database, the use of such software for GMF analysis should obviate a large portion of the manual, labor intensive, and technically challenging interpretation of glycan fragmentation (Fig. 3).[31] Cross-ring fragments, such as $^{0,2}A_1$, $^{1,5}X_1$, and $^{2,5}A_3$ as represented in Fig. 4, can be extremely useful in identifying the linkage positions of sugars by MS without the need for additional linkage analysis. Permethylation of glycans[13,33] has several advantages, including increased stability of sialic acid

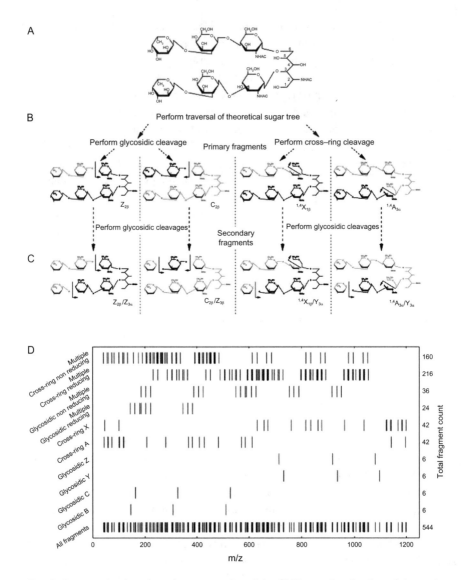

FIG. 3. An example of a schematic representation of the GMF procedure for determining a glycan sequence, based on the theoretical fragments by *GlycosidIQ* platform for the heptasaccharide. This figure is adapted from ref. 27.

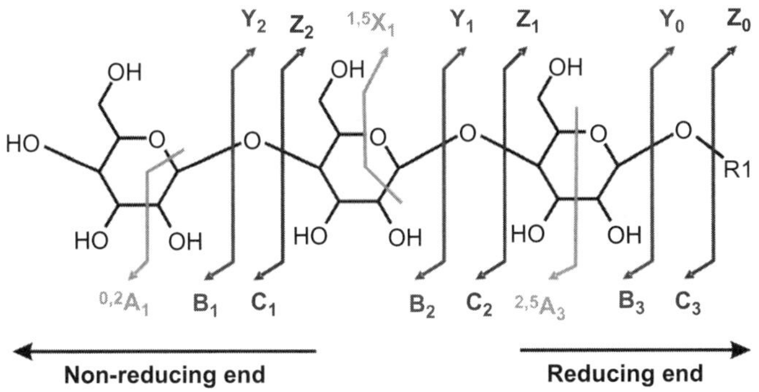

FIG. 4. Nomenclature of fragments of carbohydrates as defined by Domon and Costello.[32] This figure is adapted from ref. 31.

components, improved ionization efficiency, and ease of gas chromatography–MS (GC–MS)-based analysis, as acid hydrolysis gives volatile monosaccharide derivatives. Characterization by means of GC–MS and enzymatic digestion also provides further structural information on the glycosidic linkage and anomeric configurations. It should also be borne in mind that purification and derivatization of whole glycans from biological samples requires tedious and time-consuming processes before subjection to automatable MS-based glycan compositional profiling and further precise structural analyses.

b. Glycoproteomics.—One of the limitations of glycomics after release of the glycan from glycoproteins is the loss of the information regarding N- and O-glycosylation sites, namely the proximal peptide sequence information involving the individual glycosylation site. For this purpose, the structural analysis of the intact glycoproteins and/or digested glycopeptides is required. Glycoproteomics, a term defining the concurrent analyses both of peptide and glycan sequences, is of growing importance, especially in discovery research on biomarkers and in regulatory requirements for therapeutic biopharmaceuticals, including antibody drugs.[34] Alterations of glycan structures and in-site occupancy indicate altered molecular functions and properties of individual glycoproteins, as illustrated in Fig. 1, demonstrating the glycan heterogeneity of recombinant human erythropoietin (huEPO).[35,36]

To acquire information on the glycosylation sites of glycoproteins from biological samples, Zhang *et al.* proposed a convenient method allowing for the large-scale isolation of glycopeptides by periodinate-based nonspecific oxidation at common sugar residues carrying *cis*-diols, such as D-mannopyranose and D-galactopyranose, as well as sialic acids bearing the highly reactive glycerol functional group. The heterogeneous aldehyde derivatives of glycopeptides thus generated are subsequently conjugated to a solid support bearing hydrazide groups.[37–42] Although this technique firmly provides full information on the glycosylation sites (Asn and Ser/Thr residues) of the glycoproteins enriched as a huge glycopeptide pool, it is not possible to achieve any structural information about N- and/or O-glycans because the MS analyses are performed on naked peptides whose glycans are removed preferentially by treating with PNGase F. When glycoproteins can be isolated successfully by conventional separation protocols, glycopeptides should be enriched and separated from complex mixtures of tryptic peptides for further systematic glycoproteomics approach based on LC–MS/MS^2, as illustrated in Fig. 5. It is obvious that glycoproteomics needs an enrichment step of glycopeptides, owing to the loss of MS sensitivity in the presence of large amounts of non-glycosylated tryptic peptides, namely the ion-suppression effect by naked peptides, which show much higher ionization potentials than glycosylated peptides. Therefore, separation methods, in combination with MS analysis, have become one of the most powerful and versatile techniques for glycoproteomics. Hydrophilic-interaction chromatography (HILIC) offers a highly improved separation method for glycopeptides.[44–47] HILIC is one of the simplest technologies for separating and enabling the broad profiling of different oligosaccharides on the basis of the hydrophilic or hydrophobic nature of the glycan side-chain of the individual glycopeptides. Use of activated graphitized carbon is a potential alternative method, when used in conjunction with LC–MS and tandem MS/MS of tryptic glycopeptides.[48] A combination of such HPLC-based separation, in conjunction with MS, could serve for efficient glycopeptide analysis, although several issues remain unresolved, such as the high MS background noise caused by the buffers frequently used in HILIC. Lectin-affinity chromatography also offers several substantial advantages over other common separation methods for glycoproteins/glycopeptides, because this procedure can reduce the sample complexity and increase the detection sensitivity of tryptic glycopeptides for proteins of low abundance, without loss of the glycan moiety.[49–55] However, it should be noted that the method also has serious disadvantages, such as nonspecific binding and the multivalency of lectin binding. As a result, considerable amounts of proteins/peptides enriched by the lectins actually have no designated glycan and linkage that is ascribed to the lectin affinity employed, and further validation of the glycan structure of candidate proteins by means of MS is required. As shown in Fig. 5, only a limited number of mass spectra of glycopeptides can be used for further analyses, both of amino acid sequence and glycan composition, even though automated *in silico* workflows would greatly facilitate the procedures for the structural identification of individual glycopeptides.[43]

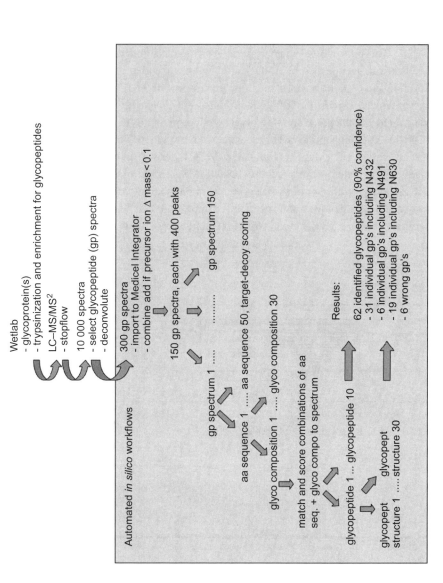

FIG. 5. The glycoprotein analysis workflow proposed by Renkonen et al.[43] As a proof of the principle, the authors showed an example of comprehensive approach for glycoproteomics of purified human plasma serotransferrin.

Electron-induced dissociations, electron-capture dissociation (ECD),[56–66] and electron-transfer dissociation (ETD)[67,68] are unique and promising sequencing techniques for glycoproteomics because they allow specific fragmentation of multiply charged peptide cations into c and $z^·$ ions without any loss in the posttranslational modifications.[69–77] The informative c and $z^·$ series of fragment ions generated by the ECD device is advantageous as it permitted rapid and precise identification of preferential and/or multiple O-glycosylation sites in a mucin peptide MUC4 derivative produced during *in vitro* O-glycosylation by recombinant ppGalNAcT2.[77] For instance, it was demonstrated that the preferential O-glycosylation site in this *in vitro* condition for the MUC4 naked model peptide is Thr10, as indicated in Fig. 6. This approach permitted the identification of a more complicated MUC4 derivative bearing four GalNAc residues. As indicated in Fig. 7, the ECD–MS/MS spectrum was in good agreement (100% coverage) with that of the theoretical fragmentation pattern and was found to bear four GalNAc residues at Thr1, Thr6, Thr10, and Thr15. We demonstrated the versatility of ECD–MS/MS in the structural characterization of a highly complicated

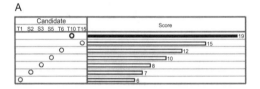

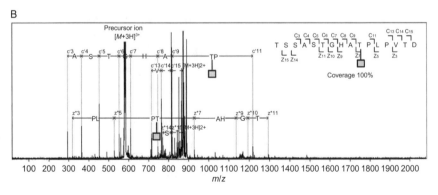

FIG. 6. Identification of the O-glycosylation site of enzymatically glycosylated MUC4 peptide (monoglycosylated MUC4 derivative isolated by HPLC) by ECD–MS/MS. (A) Scoring result between the experimental ECD spectrum and theoretical fragment ions of all candidates. (B) Identified structure and the corresponding fragmentation assignment of the ECD spectrum. This figure is adapted from ref. 77.

glycopeptide analogue related to a cancer-specific MUC1 glycopeptide epitope having both O-glycan and N-glycan chains in a single polypeptide.[75] Structural analysis by ESI–TOFMS (MS/MS) revealed that the ECD device gives highly informative c and $z^{\cdot}$ ions, allowing for the identification of both O- and N-glycosylation sites of the MUC1 glycopeptide without any significant degradation in the glycan chains, whereas typical fragmentation at every glycosidic linkage in the carbohydrate moieties occurred predominantly under conventional CID conditions (Fig. 8). Combining ECD/ETD–MS/MS and NMR studies would provide a comprehensive strategy for the structural characterization of multiply glycosylated glycopeptides/glycoproteins, because it is well documented that peptide conformational alteration induced by multiple and/or site-specific O-glycosylation often influences

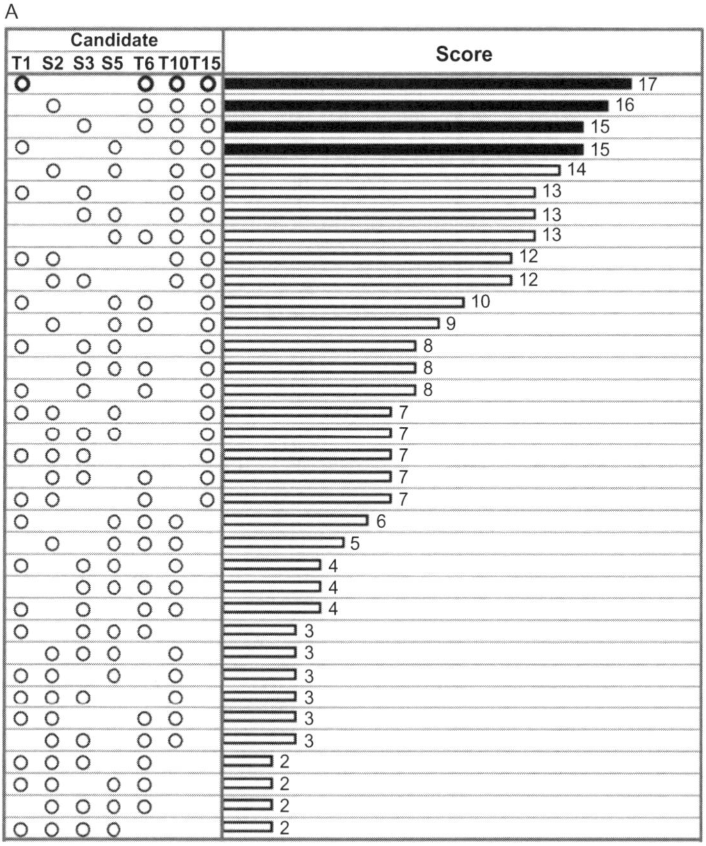

FIG. 7. Identification of four glycosylation sites of the isolated MUC4 derivative by ECD–MS/MS analysis. (A) Results of the spectral matching analysis using 30 possible structures. (B) The top four structures predicted by *in silico* calculation and the spectral coverage (%) when compared with the real experimental spectrum. This figure is adapted from ref. 77. *(Continued)*

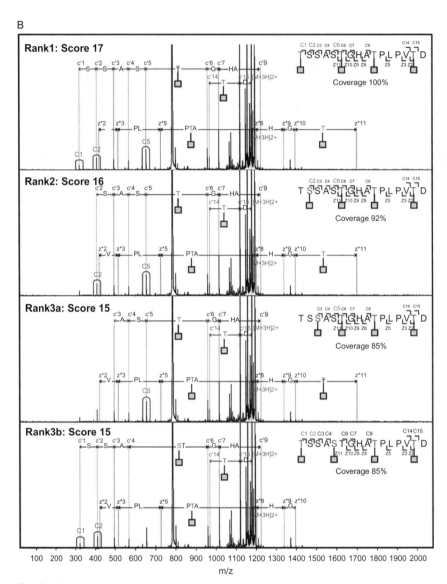

FIG. 7. (B)

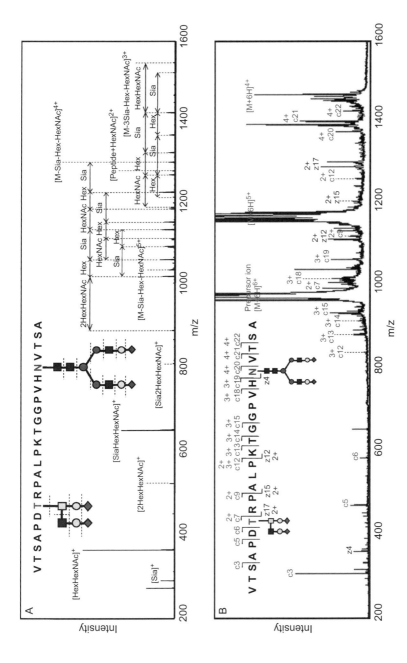

FIG. 8. MS/MS spectra of the 6+ charge state (m/z 982.43) of a synthetic MUC1 glycopeptide carrying N- and O-glycans. (A) CID, (B) ECD spectra. This figure is adapted from ref. 75. (See Color Insert.)

secondary structure and functions of original peptides/proteins.[77–81] It was also reported that ECD–MS/MS under activation by a metal is a potential approach for providing convenient cross-ring cleavages in the sugar moiety, a highly informative fragmentation mode in the context of common N-glycan identification.[82] Zhao et al. also reported the feasibility of ECD–MS/MS to delineate structural analysis of sugar branching and linkage positions of permethylated N-glycans.[83]

2. The Rate-Determining Stage in Glycomics and Glycoproteomics

As stated in the introduction to this chapter, the bottleneck, or rate-limiting stage, in structural and functional glycomics/glycoproteomics relates to the difficulty in the isolation of pure N-/O-glycans and glycopeptides from heterogeneous sample mixtures. The tryptic digests of whole proteins in fluids, cells, tissues, and organs generally contain large amounts of such impurities as naked peptides, nucleotides, lipids, salts, and other products. To separate the glycopeptide component from large amounts of non-glycosylated peptides, HILIC, activated graphitized carbon, and lectin affinity chromatography can be used as already described. When the tryptic digest is subjected directly to optimized conditions for digestion by PNGase F or to conventional alkaline β-elimination, N-/O-glycans can be released successfully from the glycopeptide mixture. However, the free glycans liberated from tryptic glycopeptide mixtures must be chemically modified by permethylation and/or reductive amination with suitable chromophores or fluorophores at the anomeric position, followed by time-consuming HPLC-based purification steps, before conducting the strategic MS analyses that were mentioned in the previous section. It is clear that technical problems in the processes of glycan preparation and purification make high-throughput glycomics and glycoproteomics difficult.

a. Glycoblotting Method.—Carbohydrates are chemically very similar, and this makes purification procedures difficult and derivatization a necessity. Further, those methods that do exist are laborious or inefficient, and the available modifications are application-specific. Recently, we developed *a PCR-like key technology* for glycan-specific enrichment, termed the glycoblotting method[84], which allowed, for the first time, high-throughput/multiple and quantitative glycomics. The original work of Emil Fischer,[85] the reaction of aldoses with phenylhydrazine to give the corresponding phenylhydrazones, motivated us to challenge the chemoselective glycoblotting technique, which has become a key tool in the efficient isolation of glycans in general from complicated mixtures. Fischer found that the reaction of glucose, mannose, and major oligosaccharides with phenylhydrazine proceeds smoothly to give the

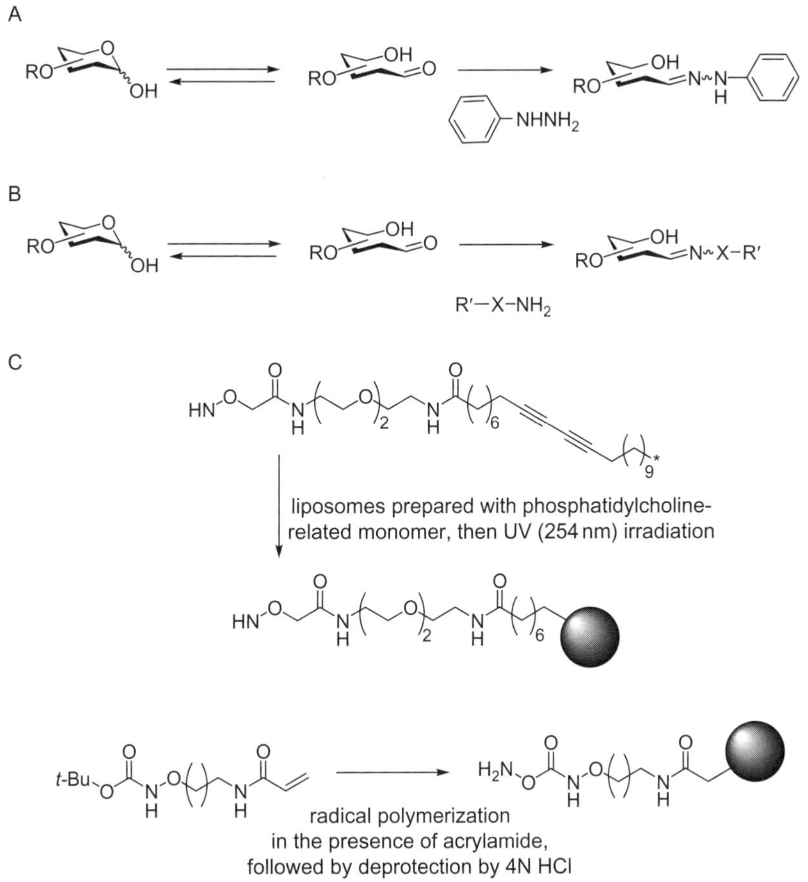

FIG. 9. Fischer's phenylhydrazone reaction[85] and the glycoblotting method first demonstrated by using aminooxy-functionalized synthetic polymers. (A) Reaction of sugars with phenylhydrazine; (B) the general concept of "glycoblotting"; and (C) the aminooxy-functionalized synthetic polymers used in the preliminary study.[84] (Original figure made by the author.)

corresponding stable phenylhydrazone derivatives under mild aqueous conditions (Fig. 9A). Once glycans are released from such glycoconjugates as glycoproteins, GSLs, and proteoglycans, they can be regarded as normal oligosaccharides, namely a class of related compounds having an aldehyde or ketone group at the reducing terminal. As indicated in Fig. 9B, aldehydes react preferentially with reagents bearing hydrazine-like functional groups, such as Fischer's original phenylhydrazine. Such reagents do not need any catalyst or reducing agent to accelerate the coupling reaction with sugars, and the reactions usually proceed spontaneously under mild aqueous conditions. In contrast, the reactions of aldehydes with primary amine groups require activating reagents or conditions for the formation of stable Schiff-base products. In consequence, reducing carbohydrates react preferentially with Fischer-type hydrazine reagents, even in the presence of large amounts of peptides, or amino acids, or nucleotides bearing primary amino groups. We considered that this specific reaction of chemically similar carbohydrates with a hydrazine derivative, or an equivalent aminooxy-functional group, could serve as an essential basis for glycan-selective enrichment technology when these functional groups are present on the polymers or related materials. Taking into account the chemical stability, the handling of reagents, and the feasibility for a generalized synthesis of versatile polymers, we developed some Fischer-type polymer reagents bearing reactive and stable aminooxy-functional groups. These were prepared by polymerization of a diacetylene-containing lipid derivative and an acrylamide-type derivative to serve as first-generation materials for the glycoblotting technique (Fig. 9C).[84]

Liposomes composed of the aminooxy-functionalized diacetylene derivative and a phospholipid derivative [1,2-bis(10,12-tricosadiynoyl)-*sn*-glycero-3-phosphocholine] were polymerized by UV irradiation to afford polymer-based nanoparticles having a diameter of 200–300 nm. A water-soluble polyacrylamide-type reagent was also prepared under a conventional radical copolymerization conditions in the presence of acrylamide. As anticipated, it was observed that N-glycans released by PNGase F from purified human IgG are captured successfully by these aminooxy-functionalized polymers, and after separating the glycan-polymer scaffolds from the mixture, the N-glycans released from the polymers were subjected directly to MALDI–TOF and TOF/TOFMS analysis.[84] The merits of glycoblotting are evident, since the MALDI–TOFMS of major N-glycans captured by the polymer proved to be greatly simplified when compared with the extremely complex spectrum of the crude tryptic digest of the IgG sample (Fig. 10). This glycoblotting method allowed, for the first time, rapid and facile glycan analysis without the need for any derivatization processes and chromatographic purification, although the protocol could not provide quantitative information of the identified N-glycans. It is noteworthy that Hindsgaul *et al.* have independently developed a glycolipid enrichment system, namely the solid-phase oligosaccharide tagging (SPOT) method.[86]

To achieve much higher efficiency of the glycoblotting reaction and to optimize such functions such as feasibility with respect to modification of the carboxyl group of sialic acid residues,[87] and the subsequent labeling of captured glycans for quantitative analysis,[88,89] we have improved and optimized both the Fischer-type polymer reagents and the total protocols for glycoblotting technology. To establish an

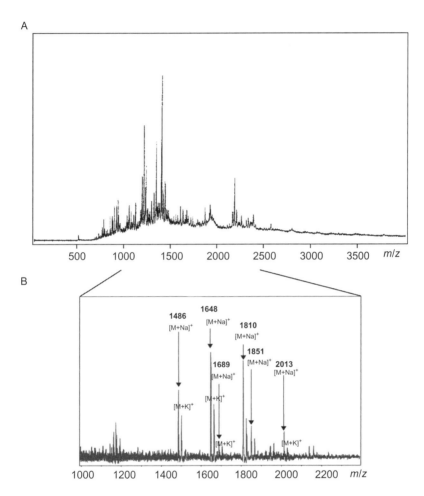

FIG. 10. Glycoblotting of human IgG N-glycans by the aminooxy-functionalized polymer reagent dramatically decreased the complexity of the mass spectra of a tryptic digest. (A) MALDI–TOFMS of crude tryptic digest of human IgG before glycoblotting, and (B) Major precursor-ion peaks attributable to human IgG N-glycans obtained by glycoblotting. This figure is adapted and arranged from ref. 84.

"all-in-one" protocol in the polymer platform, we designed a multifunctional molecular probe, N-(2-aminobenzoyl)cysteine hydrazide (ABCh). This was employed for conjugation with the commercially available Thiopropyl Sepharose 6B, to afford a stable hydrazide-functionalized polymer support, which constitutes a second generation of polymer reagents for glycoblotting, namely the BlotGlycoABC^{TM} bead (Fig. 11).[88] Formation of the hydrazone bond between hydrazide groups of the

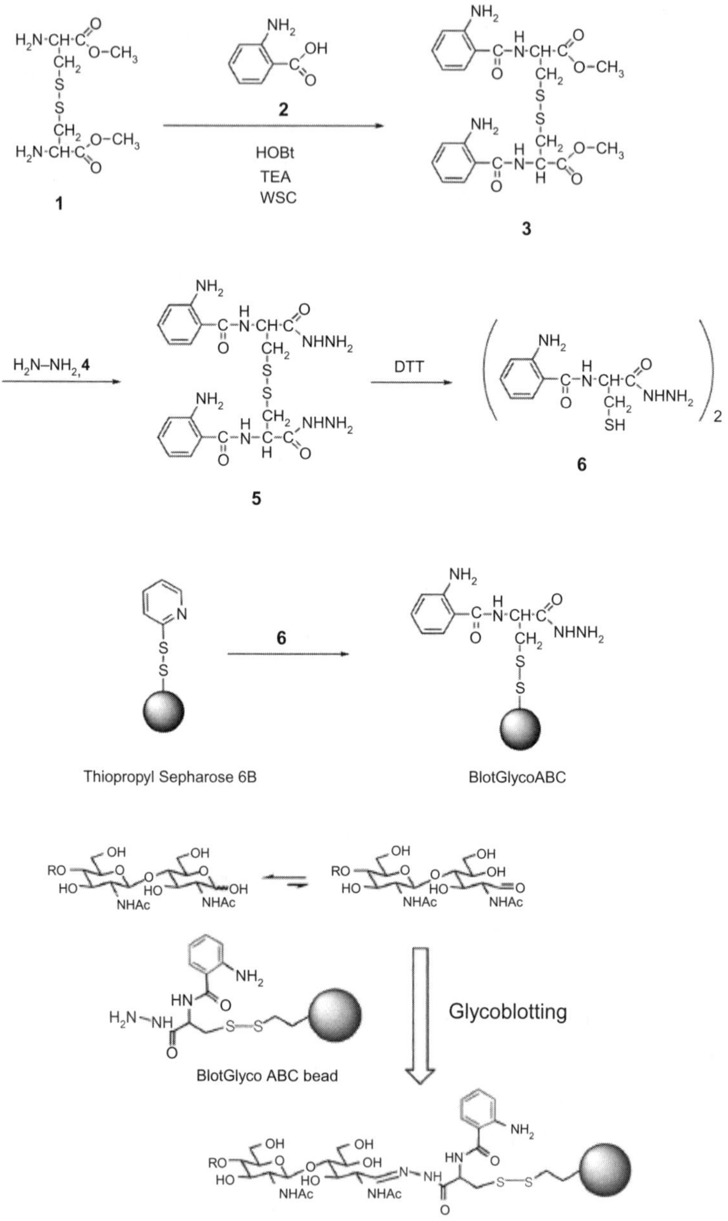

FIG. 11. Synthetic scheme of a multifunctional molecular probe, *N*-(2-aminobenzoyl)cysteine hydrazide (compound **6**; ABCh), and BlotGlycoABC™ bead derived by conjugation with Thiopropyl Sepharose 6B. This figure is adapted and arranged from ref. 88.

BlotGlycoABC™ bead and glycans is reversible, and it allows release of the parent reducing sugars; alternatively, reduction produces the stable C–N bond. This chemistry is suitable for enriching carbohydrates from complex biological materials, even in the presence of various amines and reagents used in preparation of proteomics samples; these include a variety of such reagents as dithiothreitol (DTT), iodoacetamide (IAA), detergents, and others. It was also demonstrated that this improved protocol afforded quantitative ligation of general N-linked-type oligosaccharides without any loss of sialic acids, by on-bead reaction of captured N-glycans with 3-methyl-1-p-tolyltriazene (MTT).[87] The disulfide bond connecting Thiopropyl Sepharose 6B with the ABCh probe allowed quantitative recovery of enriched N-glycans by reductive treatment with DTT in an elution buffer. The recovered N-glycan samples, containing neutral and acidic sugar residues, are ready for subsequent quantitative analysis using conventional HPLC and some of the MS analysis applications. This method can be used not only for analysis of human serum N-glycan analysis but also for profiling human cellular N-glycans, such as human prostate cancer cells (PC-3) and the normal prostate epithelial cells (PrEC). As shown in Fig. 12, the fully synthetic polymer beads prepared from methacrylamide monomer displaying high density-hydrazide groups (BlotGlyco H beads)[89] appear well suited for high-throughput and quantitative glycomics. As illustrated in Fig. 13, BlotGlyco H can serve as an ideal platform for the streamlined "on-bead" chemical manipulations of captured glycans to convert them into tagged derivatives to be subjected to the designated glycan analysis. The advantage of the glycoblotting protocol based on BlotGlyco H is clear, because the enriched glycans can be employed for general solid-phase chemistry with the derivatization required for further structural and functional analyses. As illustrated in Fig. 14, the trans-iminization reactions[89] with the aminooxy derivatives employed for tagging permitted preparation of the desired glycan microarray, as well as for the rapid and facile glycomics of glycoproteins.

The versatility of the glycoblotting method is evident since this technique can be applied for monitoring dynamic N-glycan alteration during mammalian cell differentiation and proliferation. Amano et al.[91] demonstrated the merit of glycoblotting-based quantitative glycomics in their investigation of dynamic glycoform alteration during proliferation and differentiation of mouse P19 cells and ES cells into cardiomyocytes or neural cells. It was demonstrated that the full portrait of N-glycan expression at each cell stage allowed identification of the characteristic glycotypes showing drastic and concerted expression changes during cell differentiation (termed stage-specific embryonic glycotypes). This result indicated for the first time the existence of a "threshold" in expression level of the characteristic glycotypes required for initiating individual cell differentiations, while the functional roles,

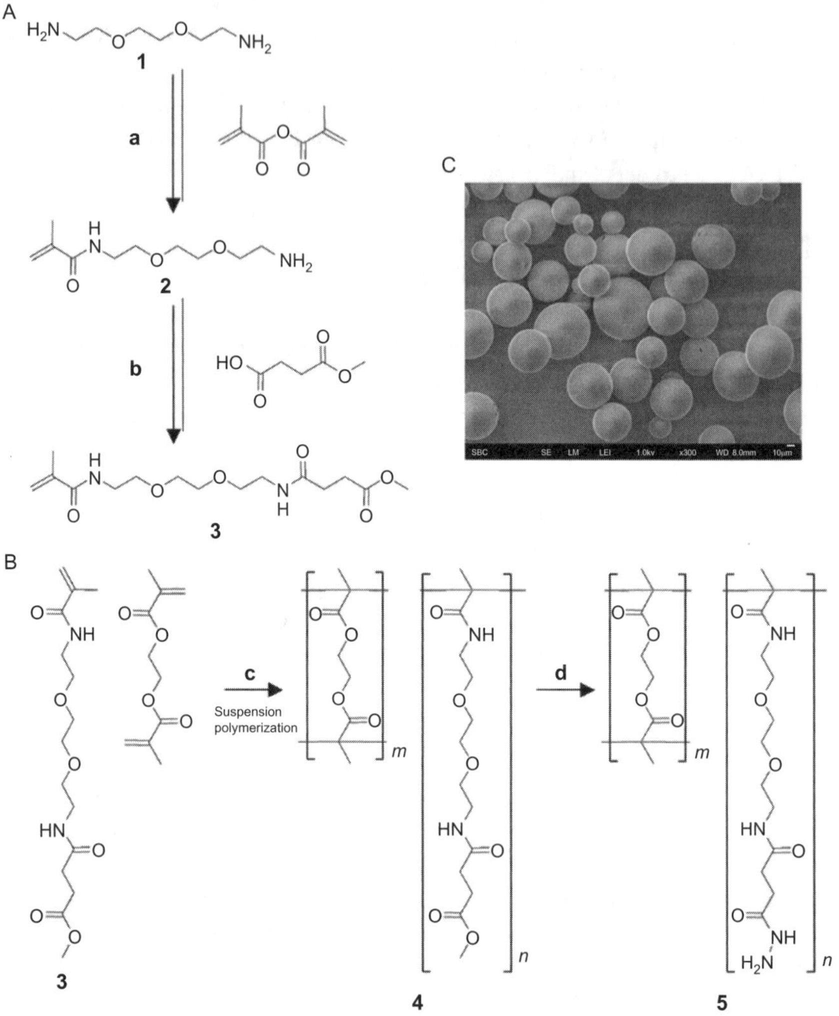

FIG. 12. Synthetic scheme for a hydrazide-functionalized glycoblotting polymer (BlotGlyco H). (A) Preparation of methacrylamide monomer 3. Reagents and conditions: a, CHCl$_3$, 0 °C, 16 h; b, EDC, CHCl$_3$, 0 °C, 16 h. (B) Preparation of hydrazide-functionalized polymer 5. Reagents and conditions: c, AIBN, CHCl$_3$, H$_2$O, 60 °C, 16 h; d, hydrazine monohydrate, room temperature, 2 h. (C) SEM view of BlotGlyco H particles. This figure is adapted from ref. 89.

mechanism, and designated partner molecules remain unknown. Fig. 15 shows an example of drastic glycan alteration during cell differentiation from P19C6 cells to neural cells. It seems likely that drastic enhancement of the expression level in various

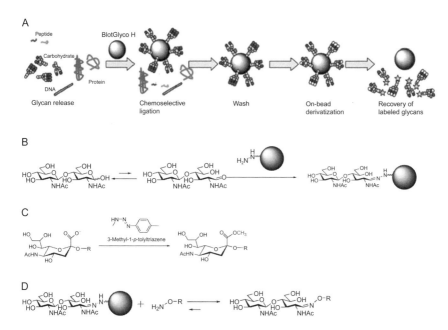

FIG. 13. The glycoblotting protocol, using BlotGlyco H beads. (A) Schematic representation of a practical glycoblotting protocol employing BlotGlyco H beads. The streamlined chemical reactions performed "on-bead" are composed of (B) chemical ligation of glycans by hydrazide groups, (C) methyl esterification of carboxyl group of sialic acid residues,[87] and (D) release of desired glycans as labeled derivatives by trans-iminization reaction[89,90] with designated aminooxy compounds. (This figure was made by the author.) (See Color Insert.)

bisecting-type N-glycans is essential for the differentiation from the progenitor cells into mature neural cells, while the mechanism and key partner molecules remain unclear. However, it is considered that a goal should be set to compare and accumulate a database of whole N-glycan expression levels of feasible human ES cells and iPS cell lines established by different laboratories. These resources should be made readily available to the scientific community as soon as possible, because quantitative cellular glycomics might support quality control of various stem cells of different stages. Microarrays displaying such characteristic cell-surface glycans would become attractive tools for research and discovery of respective receptor molecules/lectins, and for investigating the mechanism and functional roles of cellular N-glycans in cell differentiation and proliferation.

Kimura *et al.* reported the feasibility of the glycoblotting method for rapid and facile N-glycomics of plant glycoproteins, when PNGase A (instead of the PNGase F[94] used for mammalian glycoproteins) was employed for releasing N-glycans from

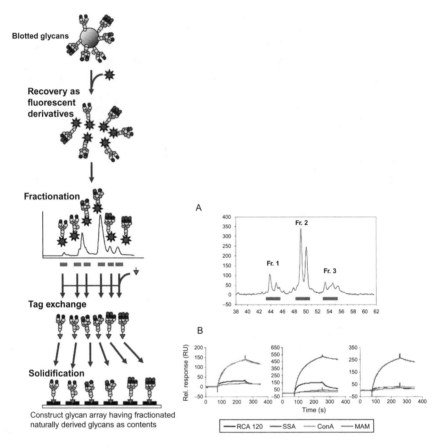

FIG. 14. Schematic diagram showing the sequential multiple-tag conversion, based on trans-iminization of N-glycans enriched by glycoblotting and the construction of a glycan array. (A) Normal-phase chromatography of N-glycans enriched by glycoblotting from human α_1-acid glycoprotein, labeled by transiminization with an anthraniloyl hydrazine derivative. Frs. 1, 2, and 3 correspond to di-, tri-, and tetrasialylated N-glycans. They were then subjected to a tag-exchange reaction with biotin hydrazide for immobilization on the streptavidin-coated surface. The interactions with various lectins were monitored by surface plasmon resonance. (B) Sensorgrams showing interactions of the biotinylated N-glycans of Fr. 1 (left), Fr. 2 (middle), and Fr. 3 (right) with 4 lectins (*Sambucus sieboldiana* lectin (SSA), *Maackia amurensis* lectin (MAM), *Ricinus communis* agglutinin 120 (RCA_{120}), and concanavalin A (Con A)). This figure is adapted and arranged from ref. 89. (See Color Insert.)

the glycoprotein of French-bean 7S globulin, namely phaseolin.[95] Recently, we demonstrated that nonreductive conditions employing a simple ammonium salt, ammonium carbamate, made glycoblotting-based enrichment analysis of O-glycans possible without the significant losses or unfavorable side reactions experienced in

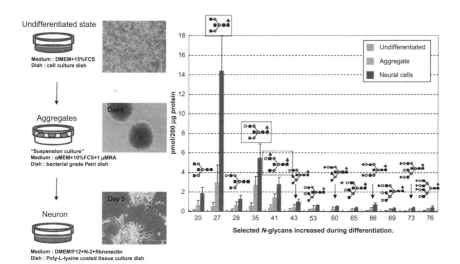

FIG. 15. Differentiation of mouse P19C6 cells to neural cells, and N-glycan changes as monitored by the glycoblotting method. Undifferentiated cells were stimulated by retinoic acid, and the cells were subjected to the glycoblotting protocol optimized for cellular glycomics. It was demonstrated that the expression level of various bisecting-type N-glycans increased drastically during cell differentiation to neural cells.[91] Three glycoforms circled by red lines are known to exist in the mouse brain system.[92,93] (This figure was made by the author.) (See Color Insert.)

conventional conditions (Fig. 16).[96] For example, a general workflow of glycoblotting, using BlotGlyco H beads, on-bead chemical manipulations, and subsequent MS allowed for rapid O-glycomics of human-milk osteopontin (OPN) in a quantitative manner (Fig. 17). It was demonstrated that the structures of O-glycans in human-milk OPN varied with the attachment of fucosyl and N-acetyllactosamine units (Table I). This method proved suitable for the enrichment analysis of common O-glycans from a wide range of biological materials, such as whole serum, cultured cancer cells, and rat-kidney FFPE tissue sections.[96] The protocol for liberating O-glycans safely by simple chemical treatments, in combination with the glycoblotting technique, should contribute significantly to a broad range of studies concerning the biological relevance of O-glycans, especially in relation to the clinical and diagnostic benefits of glycoconjugates containing various O-linked type oligosaccharides. The advantage of this strategy, through combined use of ammonium carbamate and a streamlined chemical manipulation on the glycoblotting platform, is the elimination of both the possible peeling reactions during O-glycan release by nonreductive β-elimination and significant deletion of the acid-labile sialic acid residues.

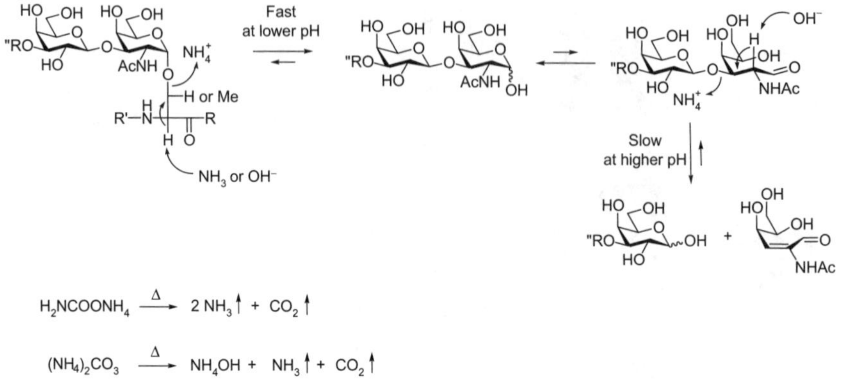

FIG. 16. β-Elimination in common O-glycoside linkages with Ser or Thr residues in alkaline conditions and a plausible mechanism of subsequent peeling reaction. This figure is adapted from ref. 96.

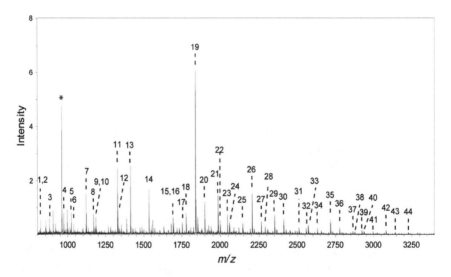

FIG. 17. MALDI–TOFMS of O-glycans enriched by a standard glycoblotting protocol using ammonium carbamate from human-milk OPN. Numbers represent the peaks attributable to O-glycans labeled with benzyloxyamine (BOA), and an asterisk indicates the internal standard. The glycan compositions listed in Table I were estimated from measured masses of the glycans by use of the *GlycoSuite* database. This figure is adapted from ref. 96.

TABLE I
Composition of BOA-Labeled O-Glycans of Human-Milk Osteopontin in Fig. 17

Peak No.	Deduced Composition
1	$(Hex)_1(HexNAc)_1(Neu5Ac)_1$
2	$(Hex)_2(HexNAc)_1(dHex)_1$
3	$(Hex)_2(HexNAc)_1(dHex)_2$
4	$(Hex)_2(HexNAc)_2(dHex)_1$
5	$(Hex)_2(HexNAc)_2(dHex)_1$
6	$(Hex)_3(HexNAc)_2$
7	$(Hex)_1(HexNAc)_1(Neu5Ac)_2$
8	$(Hex)_2(HexNAc)_2(dHex)_2$
9	$(Hex)_2(HexNAc)_2(Neu5Ac)_1$
10	$(Hex)_3(HexNAc)_2(dHex)_1$
11	$(Hex)_2(HexNAc)_2(dHex)_1(Neu5Ac)_1$
12	$(Hex)_3(HexNAc)_2(dHex)_2$
13	$(Hex)_1(HexNAc)_4(dHex)_2$
14	$(Hex)_3(HexNAc)_3(dHex)_2$
15	$(Hex)_3(HexNAc)_3(dHex)_1(Neu5Ac)_1$
16	$(Hex)_4(HexNAc)_3(dHex)_2$
17	$(Hex)_4(HexNAc)_4(dHex)_1$
18	$(Hex)_2(HexNAc)_5(dHex)_2$
19	$(Hex)_3(HexNAc)_3(dHex)_2(Neu5Ac)_1$
20	$(Hex)_4(HexNAc)_4(dHex)_2$
21	$(Hex)_3(HexNAc)_3(dHex)_3(Neu5Ac)_1$
22	$(Hex)_3(HexNAc)_3(dHex)_1(Neu5Ac)_2$
23	$(Hex)_4(HexNAc)_4(dHex)_3$
24	$(Hex)_4(HexNAc)_4(dHex)_1(Neu5Ac)_1$
25	$(Hex)_4(HexNAc)_3(dHex)_3(Neu5Ac)_1$
26	$(Hex)_4(HexNAc)_4(dHex)_2(Neu5Ac)_1$
27	$(Hex)_5(HexNAc)_5(dHex)$
28	$(Hex)_3(HexNAc)_6(dHex)_3$
29	$(Hex)_4(HexNAc)_4(dHex)_3(Neu5Ac)_1$
30	$(Hex)_5(HexNAc)_5(dHex)_3$
31	$(Hex)_4(HexNAc)_4(dHex)_2(Neu5Ac)_2$
32	$(Hex)_5(HexNAc)_5(dHex)_4$
33	$(Hex)_5(HexNAc)_5(dHex)_3(Neu5Ac)_1$
34	$(Hex)_6(HexNAc)_6(dHex)_2$
35	$(Hex)_5(HexNAc)_5(dHex)_3(Neu5Ac)_1$
36	$(Hex)_6(HexNAc)_6(dHex)_3$
37	$(Hex)_5(HexNAc)_5(dHex)_4(Neu5Ac)_1$
38	$(Hex)_5(HexNAc)_5(dHex)_2(Neu5Ac)_2$
39	$(Hex)_6(HexNAc)_6(dHex)_4$
40	$(Hex)_6(HexNAc)_6(dHex)_2(Neu5Ac)_1$
41	$(Hex)_1(HexNAc)_8(dHex)_1(Neu5Ac)_3$
42	$(Hex)_6(HexNAc)_6(dHex)_3(Neu5Ac)_1$
43	$(Hex)_7(HexNAc)_7(dHex)_3$
44	$(Hex)_6(HexNAc)_6(dHex)_4(Neu5Ac)_1$

Hex; hexose, HexNAc; N-acetylhexosamine, Neu5Ac; N-acetylneuraminic acid, dHex; deoxyhexose. This table is adapted from ref. 96.

b. Reverse Glycoblotting Method.—Glycoproteomics, the study of both proteomic and glycomic information of glycoproteins, requires extensive effort for separating the glycopeptides portion from complex tryptic digests containing large amount of non-glycosylated peptides. Chromatography-based enrichment processes, such as HILIC, activated graphitized carbon, and lectins/antibodies have often been used for this purpose, as already described. However, given the fact that the glycoproteins usually have several N- and/or O-glycosylation sites, and with a high level of heterogeneity in the attached glycans, it should be emphasized that the total number of glycopeptides separated makes further structural characterization difficult. We considered that the glycoblotting strategy might become a potential alternative protocol for "glycoform-focused" enrichment analysis of glycopeptides when the terminal sialic acid residues of tryptic glycopeptides can be oxidized selectively to generate reactive aldehyde groups.

In 2007, we reported[97] on the feasibility of this approach by using a simple polyacrylamide polymer reagent having aminooxy-functional groups prepared as for the normal glycoblotting procedure[84] (Fig. 9). As illustrated in Fig. 18, we developed a method for selective enrichment of sialylated glycopeptides from tryptic peptide mixtures by using glycoblotting technology in combination with site-specific oxidation of the terminal sialic acid residues, that is to say, a reverse glycoblotting technique inspired from the original of glycoblotting method concept[84] (Fig. 18A). Considering that periodate oxidation of carbohydrates can be controlled and afford different types of products according to the conditions employed,[98–100] the optimized conditions used in this study (1 mM sodium periodate, 0 °C for 15 min) were chosen. These conditions assure selective and quantitative oxidation for the sialic acid residues, where the terminal linear triols at the C-7, -8, and -9 positions are converted rapidly into reactive derivatives having an aldehyde group at the C-7 position (Fig. 18B). It should be emphasized that *this chemoselective conversion proceeds quantitatively and specifically in all compounds bearing at least one sialic acid residue*. These highly reactive aldehyde groups generated on the sialic acid residues can then be enriched quantitatively at 37 °C for 2 h by chemical ligation with the Fischer-type reagent, the aminooxy-functionalized polyacrylamide, to form stable oxime bonds. After purification to separate non-sialylated glycopeptides and a large excess of non-glycosylated peptides by simple gel filtration, the sialylated glycopeptides captured covalently by the Fischer-type polymer were treated with 3% aqueous trifluoroacetic acid (TFA) at 100 °C for 1 h to cleave selectively the α-glycosidic bonds between sialic acid and adjacent galactose residues. It should be noted that α-L-fucosidic linkages are stable under these conditions, whereas α-sialosides of glycopeptides are labile over the pH 2–3 range, and are selectively hydrolyzed under the conditions optimized in this study. Finally, the recovered asialo-glycopeptides were analyzed by MALDI–TOF and TOF/TOFMS to identify concurrently both the glycan structure and peptide sequence of each glycopeptide. To illustrate this glycoproteomics technique, three glycoproteins, namely, human fetal cord serum α-fetoprotein (AFP), bovine pancreas fibrinogen, and recombinant huEPO expressed in Chinese hamster ovary (CHO) cells, were subjected to protease digestion and subsequent treatment for selective enrichment as already described. Table II summarizes the

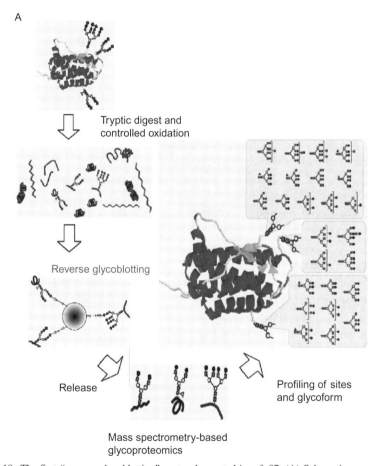

FIG. 18. The first "reverse glycoblotting" protocol reported in ref. 97. (A) Schematic representation, showing the procedure for glycoproteomics, profiling both the glycan heterogeneity and sites of glycosylation. (B) General conditions for selective oxidation and reverse glycoblotting of the terminal sialic acids. This figure is adapted from ref. 97. (See Color Insert.) *(Continued)*

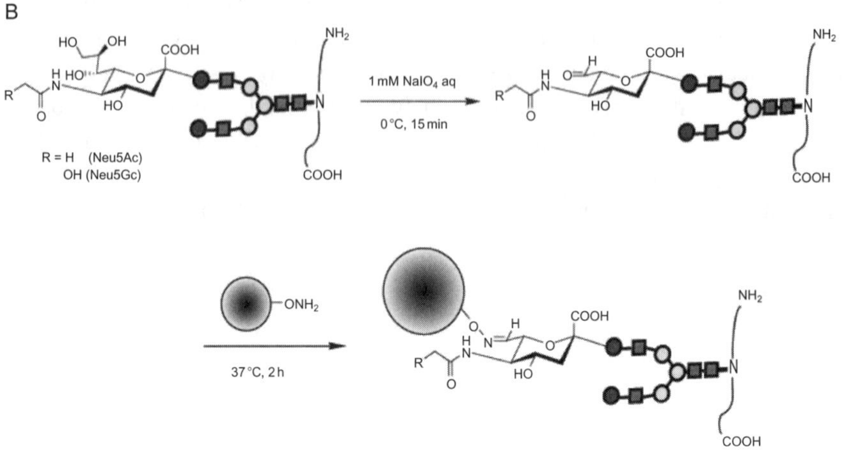

FIG. 18. (B)

results of N-glycosylation sites and asialoglycan structures for AFP, fibrinogen, and huEPO, as revealed by the reverse glycoblotting method.

Recently, Nilsson *et al.* have reported that glycopeptides from human cerebrospinal fluid can be enriched on the basis of the same principle as the reverse glycoblotting protocol. They used commercially available hydrazide beads (Bio-Rad), and the captured glycopeptides were released by treating the beads with 0.1 M formic acid at 80 °C for 1 h and the products analyzed by reversed-phase LC coupled with ESI FT-ICR MS.[101] This protocol also afforded "asialo" glycopeptides on account of the already-described acid-based hydrolysis at the sialoside residues, indicating that both protocols lost the information on the intact sialic acid residues in the parent glycopeptides. As it is well known that sialic acids play important roles in various biological processes, including cell differentiation, the immune response, and oncogenesis,[3,102–105] our attention was directed toward the feasibility of using the reverse glycoblotting technique in quantitative analysis of the original glycopeptides carrying sialic acid(s), notably "sialic acid-focused" quantitative glycoproteomics.

Targeted proteomics employing multiple-reaction monitoring (MRM) technology, complementing the discovery capabilities of shotgun strategies as well as an alternative powerful MS-based approach to measure a series of candidate biomarkers,[106–112] motivated us to integrate this emerging technique with the reverse glycoblotting method. MRM exploits the unique functions of triple-quadrupole (QQQ) MS for quantitative analysis. In MRM, the first and third quadrupoles act as filters to specifically select predefined m/z values corresponding to the peptide precursor ion

Glycoproteomics of AFP, Fibrinogen, and Recombinant Human EPO, Based On the Reverse Glycoblotting Protocol[a]

Protein	Peptide Sequence peptide mol. weight (theor.)	Glycoform glycan mol. weight (theor.)	Glycopeptide found mass (theor. mass [M+H]⁺)
alpha-fetoprotein	232VNFTEIQK239 977.52	1640.59 1786.57	2747.45 (2747.16) 2601.40 (2601.10)
Fibrinogen	73QVENK77 616.32	1640.59 1478.55	2222.80 (2239.91) 2060.90 (2077.87)
	368VGENR372 573.29	1640.59 1478.55 1275.48	2196.80 (2196.87) 2034.78 (2034.83) 1831.30 (1831.81)
huEPO	43LLEAKEAENI52 1128.60	2151.68 2005.65 1843.61 2370.71 1640.59 1478.55 1989.64 2208.67	3483.52 (3482.30) 3116.76 (3117.24) 3320.34 (3319.26) 2751.42 (2751.18) 3262.65 (3262.27) 2955.13 (2954.2) 3160.04 (3158.22) 3101.03 (3100.23)
	63NENI66 488.22	2046.63 2735.77 1884.59 2370.71	3206.56 (3205.98) 2841.24 (2840.92) 2516.86 (2516.84) 2354.89 (2354.80)
	108LVNSSQPW115 929.46	2573.73 2735.77 2776.75 2354.70 2370.71 2719.76 2516.74 2208.67 1843.61 2046.63 1681.57 2005.65	3689.51 3429.14 2958.31 (3688.20) (3428.19) (2958.08) 3648.68 3281.53 2917.24 (3647.22) (3282.16) (2917.10) 3631.87 3266.65 2754.32 (3631.21) (3266.15) (2755.06) 3485.47 3120.53 2593.51 (3485.18) (3120.12) (2593.02)

[a] In EPO, Asn51, Asn65, and Asn110 proved to be glycosylated. All glycoforms were identified as "asialo" N-glycans, because captured glycopeptides were released by selective acid hydrolysis of the linkage between sialic acid and adjacent galactose residues in this preliminary study. This table is adapted from ref. 97.

and specific fragment ion of the peptide, whereas the second quadrupole serves as a collision cell. Several such transitions (precursor/fragment-ion pairs) are monitored over time, yielding a set of chromatographic traces with retention time and signal intensity for a specific transition as coordinates. These measurements have been multiplexed to provide 30 or more specific assays in one run. The basic strategy for multiplexed quantitative glycoproteomics of mouse serum samples (50 μL), based on the reverse glycoblotting method and LC–MRM assays, is outlined in Fig. 19A.[113] The objective of this approach is to generate reproducible glycopeptide patterns representing the serum glycoproteome, allowing quantitative analysis of specific glycopeptides that discriminate between related groups of glycoproteomes and the subsequent identification of these discriminatory glycopeptides. We thereby established a novel method for the enrichment analysis of serum glycopeptides bearing sialic acids, in which nonreducing sugar residues are labeled selectively by reductive amination with 2-aminopyridine of the aldehyde group generated at the C-7 position of sialic acid (Fig. 19B). This improved process greatly facilitated further LC–MS and MS/MS characterization involving MRM-based measurements of mouse serum

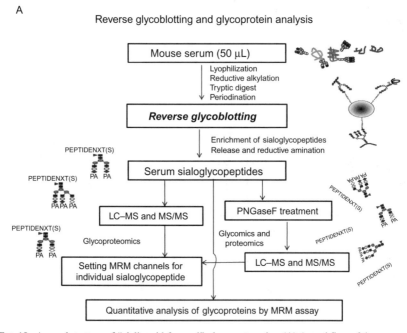

FIG. 19. A novel strategy of "sialic acid-focused" glycoproteomics. (A) A workflow of the comprehensive approach of the reverse glycoblotting-assisted MRM assays. (B) An improved protocol for reverse glycoblotting, release, and fluorescence tagging of enriched glycopeptides. These figures are adapted and arranged from ref. 113. (See Color Insert.) *(Continued)*

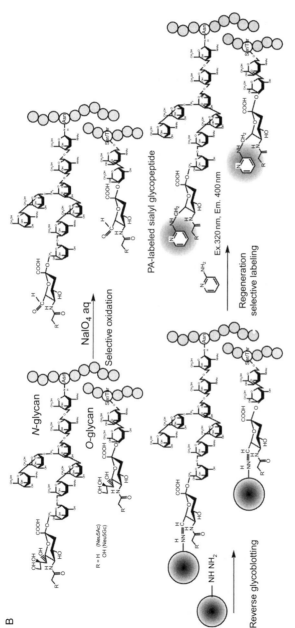

FIG. 19. (B)

glycopeptides, because pyridyl aminated glycopeptides are amenable to general fluorescence detection. Our attention was thus directed toward the chemical flexibility of the hydrazone linkage generated from hydrazide-functionalized polymer reagents. We considered that the hydrazone linkage between glycopeptides and the polymer supports would be much more sensitive to general conditions of acid hydrolysis than the α-(2→3)/(2→6) glycoside bonds between sialic acid and the adjacent galactose residue. As anticipated, an improved protocol allowed for the reliable enrichment of glycopeptides bearing sialic acids. Upon rapid, selective, and quantitative oxidation of the tryptic digests of whole glycoproteins derived from 50 μL of mouse serum, by treating them with 30 mM $NaIO_4$ (final concentration) at 4 °C for 60 min, the aldehydes generated at terminal sialic acid residues were selectively captured at 37 °C during 2 h by chemical ligation with commercially available hydrazide modified polymer (Affi-Gel Hz, Bio-Rad). After thorough washing to remove nonspecifically bound molecules, the covalently enriched glycopeptides were released as free aldehydes by treatment with ice-cold aqueous 1 M HCl. One-pot reductive amination of the regenerated aldehydes with 2-aminopyridine was performed in the presence of 2-picoline-borane (pH 7.0) at room temperature. The advantage of PA-labeling with sialyl glycopeptides is clear, because the glycopeptides can be monitored and quantified on HPLC by common fluorescence detection, independent of the peptide sequence. It is also expected that the enhanced ionization potency introduced by PA-derivatization should significantly facilitate the structural characterization of sialyl glycopeptides through general ESI–MS and MS/MS analysis of ideal fragment ions.

Mouse-serum PA-labeled sialyl glycopeptides enriched by this novel reverse glycoblotting protocol were subjected to LC–MS analysis according to the systematic strategy shown in Fig. 19A. Structural assignment of the enriched glycopeptides having sialic acid-containing glycans was performed by direct LC–MS/MS analysis of PA-labeled sialyl glycopeptides, with or without PNGase treatment. In the typical MS/MS spectra of PA-labeled sialyl glycopeptides, we observed the generation of specific carbohydrate fragment ions (namely, m/z 324, PA-NeuGc; m/z 486, PA-NeuGc-Gal; m/z 689, PA-NeuGc-Gal-GlcNAc; and others), as well as peptide fragments with and without an innermost GlcNAc residue. It was suggested that these signature signals in MS/MS spectra can be used for rapid identification of PA-labeled sialyl glycopeptides. Eventually, we could identify, by the improved reverse glycoblotting protocol, 270 unique sialylated glycopeptides mapping to 95 sialylated glycoproteins.[113] The MS/MS spectra of 67 glycopeptides from 26 mouse serum glycoproteins were actually obtained, and the information on peptide sequence, glycoform, observed parent mass, and retention times on reversed-phase chromatography are also summarized in Table III. In MRM experiments, two mass analyzers

TABLE III
The 67 Glycopeptides Identified in Mouse Serum[a]

No.		Protein Name	Peptide	Glycoform	Q1	Q3	time (min)	Optimal CE
1	1	Thrombospondin-1	K.VV STTGPGEHLR.N	(PA-NeuGc-Gal-GlcNAc)2-Man3GlcNAc2 + Fuc2	983.0 (+4)	1569.8	34.8	
2	2	Serotransferrin	NSTLCDLCIGPLK	(PA-NeuGc-Gal-GlcNAc)2-Man3GlcNAc2	941.0 (+4)	1693.8	47	39
	3			(PA-NeuGc-Gal-GlcNAc)2-Man3GlcNAc2 + Fuc	977.5 (+4)	1693.8	47	44
	4			(PA-NeuGc)3-(Gal-GlcNAc)2-Man3GlcNAc2	1021.8 (+4)	1693.8	46.7	
3	5	Serine protease inhibitor A3K	K.YTGNASALLILPDQGR.M	(PA-NeuGc)2-(Gal-GlcNAc)3-Man3GlcNAc2 + Fuc2	1154.75 (+4)	1892.2	54.5	
	6			(PA-NeuGc-Gal-GlcNAc)2-Man3GlcNAc2	990.50 (+4)	1892.8	48.7	50
	7		K.FNLTETPEADIHQG.F	(PA-NeuGc-Gal-GlcNAc)2-Man3GlcNAc2	961.25 (+4)	1775	46.3	
	8		K.FNLTETPEADIH.Q	(PA-NeuGc)2-(Gal-GlcNAc)3-Man3GlcNAc2 + Fuc2	863.8 (+5)	1590	45.9	
4	9	Putative uncharacterized protein	DLLFSDDTECLSNLQNK	DLLFSDDTECLSNLQNK	857.4 (+5)	1108.3 (+2)	51.7	35
5	10	Protein TANC2	H.GMLANGSRGDLLER.V	(PA-NeuGc-Gal-GlcNAc)2-Man3GlcNAc2	945.0 (+4)	1709.9	45.9	
6	11	Protein AMBP	EDSCQLNYSEGPCLGMQER	(PA-NeuGc-Gal-GlcNAc)2-Man3GlcNAc2	1136.75 (+4)	1238.3 (+2)		
7	12	Phosphatidylinositol-glycan-specific phospholipase D	LSSSPNVTISCK	(PA-NeuGc-Gal-GlcNAc)2-Man3GlcNAc2	891.5 (+4)	1495.9		
8	13	Murinoglobulin 1	K.YLNETQQLTQK.I	(PA-NeuGc-Gal-GlcNAc)2-Man3GlcNAc2	909.75 (+4)	1568.8	33.1	40

Continued

TABLE III (Continued)

No.	Protein Name	Peptide	Glycoform	Q1	Q3	time (min)	Optimal CE
14			PA-NeuGc-Gal-(Gal-GlcNAc)2-Man3GlcNAc2 + Fuc	1207.33 (+3)	1570	36.2	60
15			(PA-NeuGc)2-Gal-(Gal-GlcNAc)3-Man3GlcNAc2 + Fuc	1078.0 (+4)	1568.9	34.8	
16		F.HVNATVTEEGTGLEFSR.S	(PA-NeuGc-Gal-GlcNAc)2-Man3GlcNAc2	1030 (+4)	2050	37.5	50
16a					1025.5		39
17	Liver carboxylesterase N	R.FHSELNISES.M	(PA-NeuGc-Gal-GlcNAc)2-Man3GlcNAc2	859.0 (+4)	1365.6	36.2	31
18	Inter-alpha trypsin inhibitor, heavy chain 3	K.ENITAEALDLSLK.Y	(PA-NeuGc-Gal-GlcNAc)2-Man3GlcNAc2	1229.1 (+4)	1619.6	43.5	63
18a				922.3 (+4)			
19	Inter alpha-trypsin inhibitor, heavy chain 4	H.MQNITFQTEASVAQQEK.E	(PA-NeuGc-Gal-GlcNAc)2-Man3GlcNAc2	1060.5 (+4)	1086.5 (+2)	37.2	50
20	Inner centromere protein	Q.LEDEELQPCQNK.T	(PA-NeuGc)2-(Gal-GlcNAc)3GlcNAc-Man3GlcNAc2 + Fuc	886.8 (+5)	1647.6	37.4	
21	Immunoglobulin J chain	R.ENISDPTSPLR.R	(PA-NeuGc-Gal-GlcNAc)2-Man3GlcNAc2 + Fuc	911.7 (+4)	1431.8	36.5	40
22			(PA-NeuGc-Gal-GlcNAc)2-Man3GlcNAc2	875.5 (+4)	1431.8	34.5	43

14	23	Ig gamma-2B chain C region	R.EDYNSTIR.V	(PA-NeuGc-Gal-GlcNAc)GlcNAc-Man3GlcNAc2 + Fuc	976.67 (+3)	1200.5	29.2	46
	23a					1346.7		41
	24			(PA-NeuGc-Gal-GlcNAc)2-Man3GlcNAc2 + Fuc	854.25 (+4)	1200.5	30.6	40
	25			PA-NeuGc-(Gal-GlcNAc)2-Man3GlcNAc2 + Fuc	1030.67 (+3)	1200.5	29.7	52
	25a					1346.5		47
15	26	Ig gamma-2 chain C region	R.EEQFNSTFR.V	PA-NeuGc-(Gal-GlcNAc)2-Man3GlcNAc2 + Fuc	1084.0 (+3)	1360.4	34.3	
	27			(PA-NeuGc-Gal-GlcNAc)2-Man3GlcNAc2 + Fuc	894.25 (+4)	1360.4	34.6	43
	28			(PA-NeuGc-Gal-GlcNAc)GlcNAc-Man3GlcNAc2 + Fuc	1030.0 (+3)	1360.6	34.6	
16	29	Haptoglobin	K.NLFLNHSETASAK.D	(PA-NeuGc-Gal-GlcNAc)2-Man3GlcNAc2	926.25 (+4)	1635.8	31.6	48
	30		Y.ENSTVPEK.K	(PA-NeuGc-Gal-GlcNAc)2-Man3GlcNAc2	794.25 (+4)	1106.5	24.9	32
17	31	Complement factor H	WDPEPNCTSK	(PA-NeuGc-Gal-GlcNAc)2-Man3GlcNAc2	876.75 (+4)	1437.6	31.4	37
	32		K.DNSCVDPPHVPNAT.I	(PA-NeuGc-Gal-GlcNAc)2-Man3GlcNAc2	949.0 (+4)	1725.8	37.5	45
	33			(PA-NeuGc-Gal-GlcNAc)2-Man3GlcNAc2 + Fuc	985.5 (+4)	1725.8	33.4	
18	34	Complement C4-B	K.NTTCQDLQIEVK.V	(PA-NeuGc-Gal-GlcNAc)2-Man3GlcNAc2	930.5 (+4)	1652	35.8	
	35			(PA-NeuGc-Gal-GlcNAc)2-Man3GlcNAc2 + Fuc	967.0 (+4)	1652.3	36.5	
19	36	Clusterin	R.QELNDSLQVAER.L	(PA-NeuGc-Gal-GlcNAc)2-Man3GlcNAc2	1219.33 (+3)	1588.9	43.9	

Continued

TABLE III (Continued)

No.	Protein Name	Peptide	Glycoform	Q1	Q3	time (min)	Optimal CE
37			PA-NeuGc-Gal-(Gal-GlcNAc)2-Man3GlcNAc2 + Fuc	910.75 (+4)	1588	47.1	
20	38	Ceruloplasmin	A.VYPDNTTDFQR.A	(PA-NeuGc-Gal-GlcNAc)2-Man3GlcNAc2	907.5 (+4)	1560.7	41.3	40
	39			(PA-NeuGc-Gal-GlcNAc)2-Man3GlcNAc2 + Fuc	943.75 (+4)	1558.9	41.3	
	40		V.YPDNTTDFQR.A	(PA-NeuGc-Gal-GlcNAc)2-Man3GlcNAc2	882.5 (+4)	1459.8	39.5	40
21	41	C4b-binding protein	N.ALPASDVNR.T	(PA-NeuGc-Gal-GlcNAc)2-Man3GlcNAc2	804.0 (+4)	1145.5	28.5	32
22	42	Alpha-2-macroglobulin	R.PLNETFPVVYIETPK.R	(PA-NeuGc-Gal-GlcNAc)2-Man3GlcNAc2	1005.0 (+4)	1949.9	53.8	50
	42a					976.8 (+2)		45
	43		K.VNLSFPSAQSLPASDTHLK.V	(PA-NeuGc-Gal-GlcNAc)2-Man3GlcNAc2	1071.25 (+4)	1108.7 (+2)	46.7	50
	44		N.YLNETQQLTEAIK.S	(PA-NeuGc-Gal-GlcNAc)2-Man3GlcNAc2	956.0 (+4)	1754	42.8	45
	45		L.NYLNETQQLTEAIK.S	PA-NeuGc-PA-NeuAc-(Gal-GlcNAc)2-Man3GlcNAc2	980.75 (+4)	1868	52.5	43
	46			(PA-NeuGc-Gal-GlcNAc)2-Man3GlcNAc2	984.75 (+4)	1869	47.2	45
	47		N.IYVLNYLNETQQLTEAIK.S	(PA-NeuGc-Gal-GlcNAc)2-Man3GlcNAc2	1106.75 (+4)	1177.8 (+2)	61.7	
	48		R.INVSYTGER.P	(PA-NeuGc-Gal-GlcNAc)2-Man3GlcNAc2	828.0 (+4)	1241.6	33.4	
	49			(PA-NeuGc)3-(Gal-GlcNAc)4-Man3GlcNAc2	873.4 (+5)	1241.9	30.5	43

				Glycan	m/z	Mass		
	50			(PA-NeuGc-Gal-GlcNAc)2GlcNAc-Man3GlcNAc2+Fuc	915.25 (+4)	1241.9	31.5	
	51		R.I VSYTGERPSSN.M	(PA-NeuGc)2-(Gal-GlcNAc)3-Man3GlcNAc2	919.25 (+4)	1241.9	31.5	45
	52			(PA-NeuGc-Gal-GlcNAc)2-Man3GlcNAc2	924.25 (+4)	1626.1	33.4	
	53		R.PLNETFPVVYIE.T	(PA-NeuGc-Gal-GlcNAc)2-Man3GlcNAc2	1231.0 (+3)	1623.8	60.3	57
	54		Y.LNETQQLTEAIK.S	(PA-NeuGc-Gal-GlcNAc)2-Man3GlcNAc2+Fuc	951.75 (+4)	1591.9	38	
	55			(PA-NeuGc-Gal-GlcNAc)2-Man3GlcNAc2	915.25 (+4)	1591	38	
23	56	Alpha-2-HS-glycoprotein	H.ALDPTPLANCSVR.Q	PA-NeuGc-Gal-(Gal-GlcNAc)2-Man3GlcNAc2+Fuc	917.75 (+4)	1616.9	38.9	
	57			(PA-NeuGc-Gal-GlcNAc)2-Man3GlcNAc2+Fuc	958.5 (+4)	1617.1	38.1	
24	58	Alpha-1-antitrypsin 1-2	R.ELVHQS TSNIF.F	(PA-NeuGc-Gal-GlcNAc)2-Man3GlcNAc2+Fuc	911.25 (+4)	1573.9	42.2	50
	59			(PA-NeuGc-Gal-GlcNAc)2-Man3GlcNAc2+Fuc	947.75 (+4)	1573.9	43.2	
	60		L.QFNLTQTSEADIHK.S	(PA-NeuGc-Gal-GlcNAc)2-Man3GlcNAc2+Fuc	1008.75 (+4)	1819.1	43.7	
	61		L.I DTIELR.E	(PA-NeuGc-Gal-GlcNAc)2-Man3GlcNAc2+Fuc	811.75 (+4)	1176.8	39.2	35
25	62	Alpha-1-acid glycoprotein 1	N.LINDTIELR.E	(PA-NeuGc-Gal-GlcNAc)2-Man3GlcNAc2	810.0 (+5)	1289.8	39.1	
	63		T.NLINDTIELR.E	(PA-NeuGc-Gal-GlcNAc)3-Man3GlcNAc2		1404.8	45.2	45

Continued

TABLE III (*Continued*)

No.	Protein Name	Peptide	Glycoform	Q1	Q3	time (min)	Optimal CE
64			(PA-NeuGc)3-(Gal-GlcNAc)2-Man3GlcNAc2	949.75 (+4)			
			(PA-NeuGc-Gal-GlcNAc)2-Man3GlcNAc2	868.75 (+4)	1404.8	45.5	
65			(PA-NeuGc-Gal-GlcNAc)2-Man3GlcNAc2+Fuc	905.25 (+4)	1404.8	45.2	
66		Y.LTTNLINDTIELR.E	(PA-NeuGc-Gal-GlcNAc)2-Man3GlcNAc2	947.25 (+4)	1718	52.8	
26 67	Beta-2-glycoprotein 1	K.DYRPSAG NNSLY.Q	(PA-NeuGc-Gal-GlcNAc)2-Man3GlcNAc2+Fuc	944.1 (+4)	1558.9	34.5	45

[a] No., protein name, peptide sequence, glycoform, MRM setting [Q1 filter: parent ion, Q3: specific fragment ion (peptide+GlcNAc residue or peptide+GlcNAc+Fuc residues), LC retention time, Optimal CE]. This table is adapted from ref. 113.

(Q1 and Q3) are used as static mass filters to monitor a particular fragment ion of a selected precursor ion (Fig. 20A). Unlike conventional shotgun proteomic studies, MRM measurements strictly target a predetermined set of peptides and depend strongly on specific MRM transitions (a specific pair of m/z values associated with the precursor and fragment ions selected) for each targeted peptide. Since the intensities of individual fragments derived from one precursor ion differ substantially, to obtain a high-sensitivity assay, it is essential to select transitions specific for the most intense fragments. The MS/MS data of 67 glycopeptides, and their N-deglycosylated species, allowed us to generate MRM parameters in a straightforward manner.

Regardless of the peptide sequences, the MS/MS cleavage occurs preferentially between the two GlcNAc residues of the core chitobiose, resulting in a peptide fragment ion of high intensity that retains one GlcNAc residue on the asparagine residue, such as the KVAN(GlcNAc)KT ion indicated in Fig. 20B. Therefore, we selected a fragment ion containing the innermost GlcNAc residue as a key transition (Q3). Basically, the observed choice of transitions is the parent mass as the Q1 filter, the observed fragment ions as the Q3 filter, and the collision energy (CE) is selected from an optimal CE guided by several measurements with different CEs on the basis of observed MS/MS analysis with a rolling collision-energy program (Fig. 20C and D). As anticipated, the MRM setting obtained, named as the "MRM channel" for PA-labeled yolk glycopeptide as an internal standard, showed high sensitivity and reproducibility, as shown in Fig. 20D and E. It was also demonstrated that relationship between fragment-ion counts and the amount of PA-labeled glycopeptide tested exhibited good linearity over the 1–6 pmol range. We constructed for the first time 25 MRM channels for PA-labeled sialyl glycopeptides enriched from mouse serum, by setting this highly specific fragment ion containing the innermost GlcNAc residue as the Q3 filter, the precursor glycopeptide ion as the Q1 filter, and the CE was optimized. The versatility of MRM channels for these 25 targeted glycopeptides was demonstrated by applying the assay to multiplex quantitative measurements of targeted glycoproteins having sialylated N-glycans that were of interest in diabetic model mice in comparison with normal mice. Reverse glycoblotting-assisted MRM assays uncovered changes in the expression level of some sialyl glycopeptides in mouse serum derived in serotransferrin, murinoglobulin 1, and serine protease inhibitor A3K, during disease development in diabetic model male mice. Surprisingly, no significant alteration was observed between diabetic and normal female mice in these 25 targeted glycoproteins. It should be noted that MRM assays focusing on non-sialylated glycopeptides, Gal, and/or GlcNAc-terminated glycopeptides remain unaccessible by the technique, although glycopeptides bearing high-mannose type

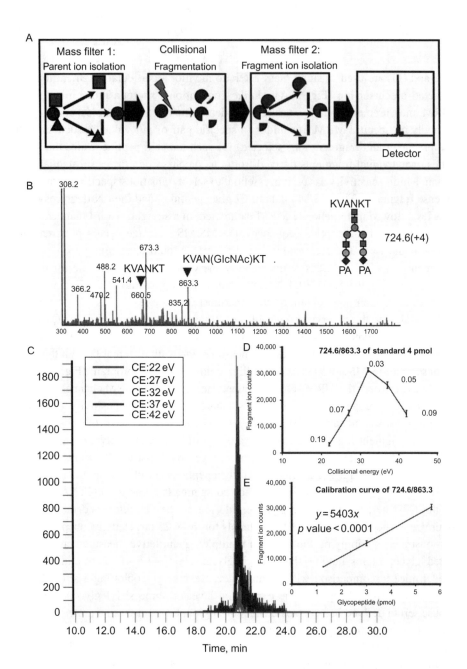

FIG. 20. Schematic diagram of MRM assays and a typical procedure for MRM setting. (A) The general concept of MRM. (B) The MS/MS spectrum of PA-labeled sialyl glycopeptide from egg yolk. (C) The MRM chromatogram under different collision energies (CE) at 22, 27, 32, 37, and 42 eV for 724.6/863.3 as Q1/Q3. (D) Relationship between fragment-ion counts versus CE for MRM assay [the numerical value means each coefficient of variation (CV)]. (E) Relationship between fragment-ion counts versus the amount of PA-labeled glycopeptide. This figure is adapted from ref. 113. (See Color Insert.)

N-glycans can be isolated by means of conventional lectin-based affinity chromatography, as described previously.

Focused glycoproteomics is based on chemical ligation with reactive aldehyde groups specifically generated at sialic acid residues of glycopeptides. We also demonstrated that GSLs can be selectively captured when lipid mixtures or membrane fractions are subjected to selective ozonolysis of the C–C double bond in the ceramide moiety, with subsequent enrichment of the GSL-aldehydes thus generated, employing chemical ligation with aminooxy-functionalized gold nanoparticles. This protocol should be of widespread utility for identifying and characterizing whole GSLs present in the surfaces of living cells, notably as new approach for glycosphingolipidomics.[114]

III. Toward Automated Glycan Analysis

It seems probable that advancements in structural glycobiology made in the past 30 years have been driven in large part by advances in MS technology and application techniques. As described in this chapter, as modern mass spectrometers are fully automated and are suitably equipped with LC systems, MS plays an integral and key role in the structural and functional characterization of glycoproteins, GSLs, and the glycosaminoglycans of proteoglycans. In addition to the emerging MS instrumentation, progress in software tools and computer platforms contributes greatly to the analysis of glycan structure from MS data, hitherto a major bottleneck in high-throughput MS-based glycomics. In fact, some software tools may permit (semi) automated/computer-assisted interpretation of MS data to estimate glycan structures, including composition analysis and structure sequencing. For example, *GlycoWorkbench*, as developed by EUROCarbDB, provides an integrated environment with an easy-to-use graphical interface, a comprehensive and increasing set of structural databases, an exhaustive collection of fragmentation data, and a broad list of annotation options.[31] This platform provides an assignment tool that is completely automatic to support the routine interpretation of MS data, although considerably more intelligent features based on a valuable source of expert knowledge on glycan biosynthesis are needed to increase the level of automation in the annotation process. In addition, *SysBioWare* is a platform that allows importation of the raw MS data into the spectrum browser and to perform isotopic grouping of detected ion peaks, after denoising and wavelet analysis.[115] Thus, monoisotopic m/z values enable peak-list association with the raw MS spectrum and make compositional assignment based on the tuned building-block library possible. This platform was applied for human urine glycomics as a potent tool for rapid assignment of known and/or unknown glycomes.

In contrast, the enrichment process for glycans and glycopeptides remains a critical obstacle in high-throughput glycomics and glycoproteomics. The glycoblotting technology seems to be only method currently available that allows large-scale clinical glycomics of human whole-serum glycoproteins,[88] because it requires very little material (such as 5–10 μL of serum) and, when combined with the automated system SweetBlot, takes only 14 h to complete (Fig. 21). SweetBlot is a sample-processing machine containing a standard multi-well filter-plate format suited for an "all-in-one" glycoblotting protocol in a single workflow capable of automation.[88] The researcher needs only to transfer the MALDI plate carrying spotted test samples, enriched N-glycans labeled, to the mass spectrometer after setting human serum in the 96-well plate. The validity of the automated protocol was assessed by a simultaneous run of the same human serum digests, and its reliability was confirmed by the observed good reproducibility. To demonstrate the versatility of the present protocol in large-scale clinical glycomics, a preliminary investigation of this technique was conducted with multiple sera of patients suffering from hepatocellular carcinoma (HCC).

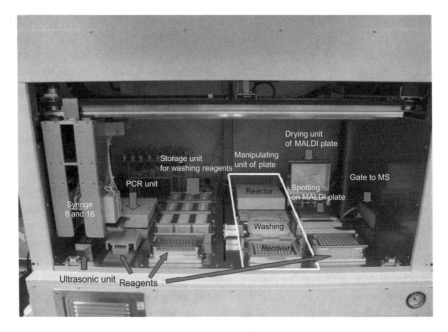

FIG. 21. Photograph of the SweetBlot machine, an automated glycan-processing system based on the glycoblotting method. This prototype machine is the latest version, designed for eventual combination with MALDI–TOFMS.

As anticipated, use of the automated glycoblotting protocol based on BlotGlycoABC™ and MALDI–TOF analysis permitted rapid and quantitative N-glycan profiling of 103 human serum samples, as shown in the raw MS data for 83 patients and 20 normal donors (Fig. 22A). To identify the essential features for optimal classification of the sera between the two relevant classes (disease and normal), we applied a sequential forward-selection algorithm that sequentially selected a better combination of N-glycan peaks, based on leave-one-out (LOO) error rates of a k-nearest neighbor classifier ($k=3$). When we selected the ratio of every two peaks among those acquired that show a significant difference (two-sided t-test, $P < 0.001$) between disease and control, the algorithm finally selected three combinations of N-glycan ratio features that distinguished, with 99% accuracy, the HCC samples from normal controls (Fig. 22B). Thus, the glycoblotting-based whole-serum N-glycan profiling demonstrably provides a convenient, noninvasive diagnostic tool for

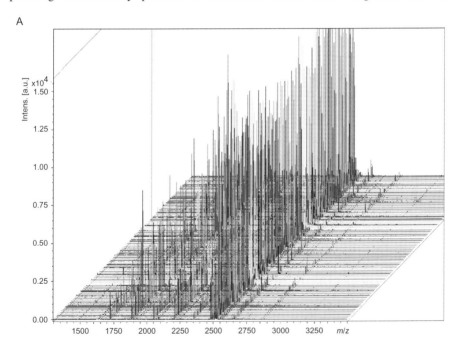

FIG. 22. Large-scale glycomics performed by means of the automated glycan-processing machine "SweetBlot" in a test with HCC patients (80 samples) and healthy donors (28 samples). (A) The raw mass spectra were subjected to compositional analysis and structural elucidation. (B) upper panel; process of the feature-subset selection in which sequential addition of the most significant N-glycans ratios at each step, testing a total of 276 models, resulted in a minimal error rate of 0%; lower panel; heat-map view of the selected three N-glycan ratios as features for the classification (brighter color indicates higher ratio of glycan abundance), and box-plot expression of the selected features (ratios of N-glycan abundance and oligosaccharide structures are shown in the figure). Parts of these figures are adapted from ref. 88. (See Color Insert.) *(Continued)*

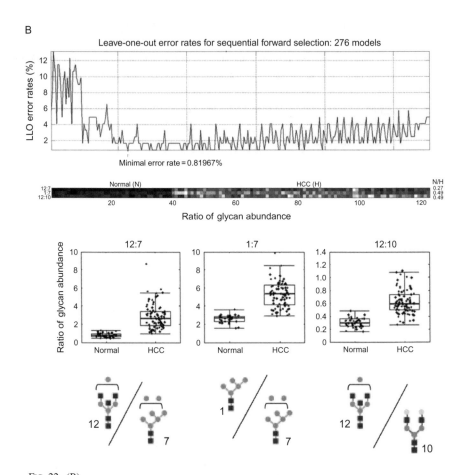

FIG. 22. (B)

diseases whose early diagnosis was previously difficult, and differentiation from situations when further large-scale and extensive evaluation steps may be required.

Our ongoing target is the development of fully automated glycan analysis and structural profiling systems, by combining SweetBlot and such versatile mass spectrometers as MALDI–TOFMS and other analytical methods used in general biochemistry/medical laboratories, via development of a designated interface, user-friendly software, and computer-assisted platforms, for structural characterization based on the raw mass spectra obtained.[116]

REFERENCES

1. D. H. Dube and C. R. Bertozzi, Glycans in cancer and inflammation-potential for therapeutics and diagnostics, *Nat. Rev. Drug Discov.*, 4 (2005) 477–488.
2. M. Petitou and C. A. van Boeckel, A synthetic antithrombin III binding pentasaccharide is now a drug! What comes next?*Angew. Chem. Int. Ed.*, 43 (2004) 3118–3133.
3. R. A. Dwek, Glycobiology: Toward understanding the function of sugars, *Chem. Rev.*, 96 (1996) 683–720.
4. R. Aebersold and M. Mann, Mass spectrometry-based proteomics, *Nature*, 422 (2003) 198–207.
5. R. K. Saiki, S. Scharf, F. Faloona, K. B. Mullis, G. T. Horn, H. A. Erlich, and N. Arnheim, Enzymatic amplification of β-globin genomic sequences and restriction site analysis for diagnosis of sickle cell anemia, *Science*, 230 (1985) 1350–1354.
6. R. K. Saiki, D. H. Gelfand, S. Stoffel, S. J. Scharf, R. Higuchi, G. T. Horn, K. B. Mullis, and H. A. Erlich, Primer-directed enzymatic amplification of DNA with a thermostable DNA polymerase, *Science*, 239 (1988) 487–491.
7. N. Taniguchi, K. Honke, and M. Fukuda, Handbook of Glycosyltransferase and Related Genes, (2002) Springer, Tokyo.
8. C. Breton, L. Snajdrova, C. Jeanneau, J. Koca, and A. Imberty, Structures and mechanisms of glycosyltransferases, *Glycobiology*, 16 (2006) 29R–37R.
9. M. R. Pratt, H. C. Hang, H. K. G. Ten Hagen, J. Rarick, T. A. Gerken, L. A. Tabak, and C. R. Bertozzi, Deconvoluting the functions of polypeptide N-α-acetylgalactosaminyltransferase family members by glycopeptide substrate profiling, *Chem. Biol.*, 11 (2004) 1009–1016.
10. T. J. Boltje, T. Buskas, and G.-J. Boons, Oppotunities and challenges in synthetic oligosaccharide and glycoconjugate research, *Nat. Chem.*, 1 (2009) 611–622.
11. N. L. Anderson and N. G. Anderson, The human plasma proteome: History, character, and diagnostic prospects, *Mol. Cell. Proteomics*, 1 (2002) 845–867.
12. A. Dell, J. J. Lee, P.-C. Pang, S. Parry, M. S. Smith, B. Tissot, H. R. Morris, M. Panico, and S. M. Haslam, Glycomics and mass spectrometry, in B. O. Fraser-Reid, K. Tatsuta, and J. Thiem, (Eds.), *Glycoscience, Chemistry and Chemical Biology, Vol. 3*, Springer-Verlag, Berlin, Heidelberg, New York, 2008, pp. 2191–2217.
13. I. Ciucanu and F. Kerek, A simple and rapid method for the permethylation of carbohydrates, *Carbohydr. Res.*, 131 (1984) 209–217.
14. K. Tanaka, H. Waki, Y. Ido, S. Akita, Y. Yoshida, T. Yoshida, and T. Matsuo, Protein and polymer analyses up to m/z 100,000 by laser ionization time-of-flight mass spectrometry, *Rapid Commun. Mass Spectrom.*, 2 (1988) 151–153.
15. M. Karas and F. Hillenkamp, Laser desorption ionization of proteins with molecular masses exceeding 10,000 daltons, *Anal. Chem.*, 60 (1988) 2299–2301.
16. M. Barber, R. S. Bordoli, G. V. Garner, D. B. Gordon, R. D. Sedgwick, L. W. Tetler, and A. N. Tyler, Fast-atom-bombardment mass spectra of enkephalins, *Biochem. J.*, 197 (1981) 401–404.

17. H. Egge and J. P. Katalini, *Mass Spectrom. Rev.*, 6 (1987) 331–393.
18. M. Dole, L. L. Mack, R. L. Hines, R. C. Mobley, L. D. Ferguson, and M. B. Alice, Molecular beams of macroions, *J. Chem. Phys.*, 49 (1968) 2240–2249.
19. M. Yamashita and J. B. Fenn, Electrospray ion source. Another varlation on the free-jet theme, *J. Phys. Chem.*, 88 (1984) 4451–4459.
20. M. Yamashita and J. B. Fenn, Negative ion production with the electrospray ion source, *J. Phys. Chem.*, 88 (1984) 4671–4675.
21. J. B. Fenn, M. Mann, C. K. Meng, S. F. Wong, and C. M. Whitehouse, Electrospray ionization for mass spectrometry of large biomolecules, *Science*, 246 (1989) 64–71.
22. M. Terada, K.-H. Khoo, R. Inoue, C.-I. Chen, K. Yamada, H. Sakaguchi, N. Kadowaki, B. Y. Ma, S. Oka, T. Kawasaki, and N. Kawasaki, Characterization of oligosaccharide ligands expressed on SW1116 cells recognized by mannan-binding protein, *J. Biol. Chem.*, 280 (2005) 10897–10913.
23. D. Goldberg, M. Sutton-Smith, J. C. Paulson, and A. Dell, Automatic annotation of matrix-assisted laser desorption/ionization *N*-glycan spectra, *Proteomics*, 5 (2005) 865–875.
24. C. A. Cooper, E. Gasteiger, and N. H. Packer, GlycoMod: A software tool for determining glycosylation compositions from mass spectrometric data, *Proteomics*, 1 (2001) 340–349.
25. C. A. Cooper, H. J. Joshi, M. J. Harrison, M. R. Wilkins, and N. H. Packer, GlycoSuiteDB: A curated relational database of glycoprotein glycan structures and their biological sources. 2003 update, *Nucleic Acids Res.*, 31 (2003) 511–513.
26. K. Maass, R. Ranzinger, H. Geyer, C.-W. Von der Lieth, and R. Geyer, "Glyco-peakfinder"-de novo composition analysis of glycoconjugates, *Proteomics*, 7 (2007) 4435–4444.
27. H. J. Joshi, M. J. Harrison, B. L. Schulz, C. A. Cooper, N. H. Packer, and N. G. Karlsson, Development of a mass fingerprinting tool for automated interpretation of oligosaccharide fragmentation data, *Proteomics*, 4 (2004) 1650–1664.
28. W. J. Henzel, T. M. Billeci, J. T. Stults, S. C. Wong, C. Grimley, and C. Watanabe, Identifying proteins from two-dimensional gels by molecular mass searching of peptide fragments in protein sequence databases, *Proc. Natl. Acad. Sci. USA*, 90 (1993) 5011–5015.
29. M. Mann, P. Højrup, and P. Roepstorff, Use of mass spectrometric molecular weight information to identify proteins in sequence databases, *Biol. Mass Spectrom.*, 6 (1993) 338–345.
30. J. K. Eng, A. L. McCormack, and J. R. Yates, III,, An approach to correlate tandem mass spectral data of peptides with amino acid sequences in a protein database, *J. Am. Soc. Mass Spectrom.*, 5 (1994) 976–989.
31. A. Ceroni, K. Maass, H. Geyer, R. Geyer, A. Dell, and S. M. Haslam, GlycoWorkbench: A tool for the computer-assisted annotation of mass spectra of glycans, *J. Proteome Res.*, 7 (2008) 1650–1659.
32. B. Domon and C. E. Costello, A systematic nomenclature for carbohydrate fragmentations in FAB-MS/MS spectra of glycoconjugates, *Glycoconj. J.*, 5 (1988) 397.
33. M. Guillard, J. Gloerich, H. J. C. T. Wessels, E. Morava, R. A. Wevers, and D. J. Lefeber, Automated measurement of permethylated serum *N*-glycans by MALDI-linear ion trap mass spectrometry, *Carbohydr. Res.*, 344 (2009) 1550–1557.
34. G. Walsh, Biopharmaceutical benchmarks, *Nat. Biotechnol.*, 21 (2003) 865–870.
35. N. Casadevall, J. Nataf, B. Viron, A. Kolta, J.-J. Kiladjian, P. Martin-Dupont, P. Michaud, T. Papo, V. Ugo, I. Teyssandier, B. Varet, and P. Mayeux, Pure red-cell aplasia and antierythropoietin antibodies in patients treated with recombinant erythropoietin, *N. Engl. J. Med.*, 346 (2002) 469–475.
36. J. Vansteenkiste, G. Rossi, and M. Foote, Darbepoetin alpha: A new approach to the treatment of chemotherapy-induced anaemia, *Expert Opin. Biol. Ther.*, 3 (2003) 501–508.
37. H. Zhang, X.-J. Li, D. B. Martin, and R. Aerbersold, Identification and quantification of *N*-linked glycoproteins using hydrazide chemistry, stable isotope labeling and mass spectrometry, *Nat. Biotechnol.*, 21 (2003) 660–666.

38. H. Zhang, E. C. Yi, X.-J. Li, P. Mallick, K. S. Kelly-Spratt, C. D. Masselon, D. G. Camp, R. D. Smith, C. J. Kemp, and R. Aebersold, High throughput quantitative analysis of serum proteins using glycopeptide capture and liquid chromatography mass spectrometry, *Mol. Cell. Proteomics*, 4 (2005) 144–155.
39. T. Liu, W.-J. Qian, M. A. Gritsenko, D. G. Camp, M. E. Monroe, R. J. Moore, and R. D. Smith, Human plasma *N*-glycoproteome analysis by immunoaffinity subtraction, hydrazide chemistry, and mass spectrometry, *J. Proteome Res.*, 4 (2005) 2070–2080.
40. Y. Zhou, R. Aebersold, and H. Zhang, Isolation of *N*-linked glycopeptides from plasma, *Anal. Chem.*, 79 (2007) 5826–5837.
41. R. Chen, X. Jiang, D. Sun, G. Han, F. Wang, M. Ye, L. Wang, and H. Zou, Glycoproteomics analysis of human liver tissue by combination of multiple enzyme digestion and hydrazide chemistry, *J. Proteome Res.*, 8 (2009) 651–661.
42. L. Zhou, R. Beuerman, A. P. Chew, S. K. Koh, T. Cafaro, E. Urrets-Zavalia, J. Urrets-Zavalia, S. Li, and H. Serra, Quantitative analysis of *N*-linked glycoproteins in tear fluid of climatic droplet keratopathy by glycopeptide capture and iTRAQ, *J. Proteome Res.*, 8 (2009) 1992–2003.
43. S. Joenväärä, I. Ritamo, H. Peltoniemi, and R. Renkonen, *N*-Glycoproteomics: An automated workflow approach, *Glycobiology*, 18 (2008) 339–349.
44. Y. Wada, M. Tajiri, and S. Yoshida, Hydrophilic affinity isolation and MALDI multiple-stage tandem mass spectrometry of glycopeptides for glycoproteomics, *Anal. Chem.*, 76 (2004) 6560–6565.
45. J. Wohlgemuth, M. Karas, W. Jiang, R. Hendriks, and S. Andrecht, Enhanced glyco-profiling by specific glycopeptide enrichment and complementary monolithic nano-LC (ZIC-HILIC/RP18e)/ESI-MS analysis, *J. Sep. Sci.*, 33 (2010) 880–890.
46. Y. Takegawa, H. Ito, T. Keira, K. Deguchi, H. Nakagawa, and S.-I. Nishimura, Profiling of *N*- and *O*-glycopeptides of erythropoietin by capillary zwitterionic type of hydrophilic interaction chromatography, *J. Sep. Sci.*, 31 (2008) 1585–1593.
47. Y. Takegawa, M. Hato, K. Deguchi, H. Nakagawa, and S.-I. Nishimura, Chromatographic deuterium isotope effects of derivatized *N*-glycans and *N*-glycopeptides in a zwitterionic type of hydrophilic interaction chromatography, *J. Sep. Sci.*, 31 (2008) 1594–1597.
48. W. R. Alley, Y. Mechref, and M. V. Novotny, Use of activated graphitized carbon chips for liquid chromatography/mass spectrometric and tandem mass spectrometric analysis of tryptic peptides, *Rapid Commun. Mass Spectrom.*, 23 (2009) 495–505.
49. H. Kaji, H. Saito, Y. Yamauchi, T. Shinkawa, M. Taoka, J. Hirabayashi, K. Kasai, N. Takahashi, and T. Isobe, Lectin affinity capture, isotope-coaded tagging and mass spectrometry to identify *N*-linked glycoproteins, *Nat. Biotechnol.*, 21 (2003) 667–672.
50. D. Ghosh, O. Krokhin, M. Antonovici, W. Ens, K. G. Standing, R. C. Beavis, and J. A. Wilkins, Lectin affinity as an approach to the proteomic analysis of membrane glycoproteins, *J. Proteome Res.*, 3 (2004) 841–850.
51. R. Uematsu, Y. Shinohara, H. Nakagawa, M. Kurogochi, J.-i. Furukawa, Y. Miura, M. Akiyama, H. Shimizu, and S.-I. Nishimura, Glycosylation specific for adhesion molecules in epidermis and its receptor revealed by glycoform-focused reverse genomics, *Mol. Cell. Proteomics*, 8 (2009) 232–244.
52. R. Uematsu, Y. Shinohara, J. Furukawa, H. Nakagawa, H. Shimaoka, K. Deguchi, K. Monde, and S.-I. Nishimura, High throughput quantitative glycomics and glycoform-focused proteomics of murine dermis and epidermis, *Mol. Cell. Proteomics*, 4 (2005) 1977–1989.
53. Z. Dai, J. Zhou, S.-J. Qiu, Y.-K. Liu, and J. Fan, Lectin-based glycoproteomics to explore and analyze hepatocellular carcinoma-related glycoprotein biomarkers, *Electrophoresis*, 30 (2009) 2957–2966.
54. Z. Darula and K. F. Medzihradszky, Affinity enrichment and characterization of mucin core-1 type glycopeptides from bovine serum, *Mol. Cell. Proteomics*, 8 (2009) 2515–2526.
55. M. Kullolli, W. S. Hancock, and M. Hincapie, Automated platform for fractionation of human plasma glycoproteome in clinical proteomics, *Anal. Chem.*, 82 (2010) 115–120.

56. R. A. Zubarev, N. L. Kelleher, and F. W. McLafferty, Electron capture dissociation of multiply charged protein cations. A nonergodic process, *J. Am. Chem. Soc.*, 120 (1998) 3265–3266.
57. R. A. Zubarev, D. M. Horn, E. K. Fridriksson, N. L. Kelleher, N. A. Kruger, M. A. Lewis, B. K. Carpenter, and F. W. McLafferty, Electron capture dissociation for structural characterization of multiply charged protein cations, *Anal. Chem.*, 72 (2000) 563–573.
58. K. F. Haselmann, B. A. Budnik, J. V. Olsen, M. L. Nielsen, C. Reis, H. Clausen, A. H. Johnsen, and R. A. Zubarev, Advantages of external accumulation for electron capture dissociation in Fourier transform mass spectrometry, *Anal. Chem.*, 73 (2001) 2998–3005.
59. F. Kjeldsen, K. F. Haselmann, B. A. Budnik, F. Jensen, and R. A. Zubarev, Dissociative capture of hot (3-13 eV) electrons by polypeptide polycations: An efficient process accompanied by secondary fragmentation, *Chem. Phys. Lett.*, 356 (2002) 201–206.
60. F. Kjeldsen and R. A. Zubarev, Secondary losses via γ-lactam formation in hot electron capture dissociation: A missing link to complete *de Novo* sequencing of proteins?*J. Am. Chem. Soc.*, 125 (2003) 6628–6629.
61. F. Kjeldsen, K. F. Sørensen, and R. A. Zubarev, Distribution of Iie/Leu amino acid residues in the PP3 protein by (hot) electron capture dissociation in Fourier transform ion cyclotron resonance mass spectrometry, *Anal. Chem.*, 75 (2003) 1267–1274.
62. R. A. Zubarev, Reactions of polypeptide ions with electrons in the gas phase, *Mass Spectrom. Rev.*, 22 (2003) 57–77.
63. R. A. Zubarev, Electron-capture dissociation tandem mass spectrometry, *Curr. Opin. Biotechnol.*, 15 (2004) 12–16.
64. T. Baba, Y. Hashimoto, H. Hasegawa, A. Hirabayashi, and I. Waki, Electron capture dissociation in a radio frequency ion trap, *Anal. Chem.*, 76 (2004) 4263–4266.
65. H. Satake, H. Hasegawa, A. Hirabayashi, Y. Hashimoto, T. Baba, and K. Masuda, Fast multiple electron capture dissociation in a linear radio frequency quadrupole ion trap, *Anal. Chem.*, 79 (2007) 8755–8761.
66. M. Mann and O. N. Jensen, Proteomic analysis of post-translational modifications, *Nat. Biotechnol.*, 21 (2003) 225–261.
67. J. E. P. Syka, J. J. Coon, M. J. Schroender, J. Shabanowitz, and D. F. Hunt, Peptide and protein sequence analysis by electron transfer dissociation mass spectrometry, *Proc. Natl. Acad. Sci. USA*, 101 (2004) 9528–9533.
68. J. J. Coon, B. Ueberheide, J. Syka, D. D. Dryhust, J. Ausio, J. Shabanowitz, and D. F. Hunt, Protein identification using sequential ion/ion reactions and tandem mass spectrometry, *Proc. Natl. Acad. Sci. USA*, 102 (2005) 9463–9468.
69. P. Mirgorodskaya, P. Roepstorff, and R. A. Zubarev, Localization of *O*-glycosylation sites in peptides by electron capture dissociation in a Fourier transform mass spectrometer, *Anal. Chem.*, 71 (1999) 4431–4436.
70. F. Kjeldsen, K. F. Haselmann, B. A. Budnik, E. S. Sørensen, and R. A. Zubarev, Complete characterization of posttranslational modification sites in the bovine milk protein PP3 by tandem mass spectrometry with electron capture dissociation as the last stage, *Anal. Chem.*, 75 (2003) 2355–2361.
71. M. Mormann, B. Maček, A. G. De Peredo, J. Hofsteenge, and J. P. Katalinić, Structural studies on protein *O*-fucosylation by electron capture dissociation, *Int. J. Mass Spectrom.*, 234 (2004) 11–21.
72. J. M. Hogan, S. J. Pitteri, P. A. Chrisman, and S. A. McLuckey, Complementary structural information from a tryptic *N*-linked glycopeptides via electron transfer ion/ion reactions and collision-induced dissociation, *J. Proteome Res.*, 4 (2005) 628–632.
73. S.-I. Wu, A. F. R. Huhmer, Z. Hao, and B. L. Karger, On-line LCMS approach combining collision-induced dissociation (CID), electron transfer dissociation (ETD), and CID of an isolated charge-reduced species for the trace-level characterization of proteins with post-translational modifications, *J. Proteome Res.*, 6 (2007) 4230–4244.

74. C. Sihlbom, I. D. Härd, M. E. Lidell, T. Noll, G. C. Hansson, and M. Bäckström, Localization of O-glycans in MUC1 glycoproteins using electron-capture dissociation fragmentation mass spectrometry, *Glycobiology*, 19 (2009) 375–381.
75. T. Matsushita, R. Sadamoto, N. Ohyabu, H. Nakata, M. Fumoto, N. Fujitani, Y. Takegawa, T. Sakamoto, M. Kurogochi, H. Hinou, H. Shimizu, T. Ito, K. Naruchi, H. Togame, H. Takemoto, H. Kondo, and S.-I. Nishimura, Functional neoglycopeptides: Synthesis and characterization of new class MUC1 glycoprotein models having core 2-based O-glycan and complex-type N-glycan chains, *Biochemistry*, 48 (2009) 11117–11133.
76. Y. Yoshimura, T. Matsushita, N. Fujitani, Y. Takegawa, H. Fujihira, K. Naruchi, X.-G. Gao, N. Manri, T. Sakamoto, K. Kato, H. Hinou, and S.-I. Nishimura, *Biochemistry*, 49 (2010) 5229–5941.
77. R. Hashimoto, N. Fujitani, Y. Takegawa, M. Kurogochi, T. Matsushita, K. Naruchi, N. Ohyabu, H. Hinou, X. D. Gao, N. Manri, H. Satake, A. Kaneko, T. Sakamoto, and S.-I. Nishimura, An efficient approach for the characterization of mucin type glycopeptides: The effect of O-glycosylation on the conformation of synthetic mucin peptides, *Chem. Eur. J.*, 17 (2011) 2393–2404.
78. M. R. Wormald, A. J. Petrescu, Y.-L. Pao, A. Glithero, T. Elliott, and R. A. Dwek, Conformational studies of oligosaccharides and glycopeptides: Complementarity of NMR, X-ray crystallography, and molecular modeling, *Chem. Rev.*, 102 (2002) 371–386.
79. D. M. Coltart, A. K. Royyuru, L. J. Williams, P. W. Glunz, D. Sames, S. D. Kuduk, J. B. Schwarz, X.-T. Chen, S. J. Danishefsky, and D. H. Live, Principles of mucin architecture: Structural studies on synthetic glycopeptides bearing clustered mono-, di-, tri-, and hexasaccharide glycodomains, *J. Am. Chem. Soc.*, 124 (2002) 9833–9844.
80. L. Kinarsky, G. Suryanarayanan, O. Prakash, H. Paulsen, H. Clausen, F.-G. Hanisch, M. A. Hollingsworth, and S. Sherman, Conformational studies on the MUC1 tandem repeat glycopeptides: Implication for the enzymatic O-glycosylation of the mucin protein core, *Glycobiology*, 13 (2003) 929–939.
81. Y. Tachibana, G. L. Fletcher, N. Fujitani, S. Tsuda, K. Monde, and S.-I. Nishimura, Antifreeze glycoproteins: Elucidation of the structural motifs that are essential for antifreeze activity, *Angew. Chem. Int. Ed.*, 43 (2004) 856–862.
82. J. T. Adamson and K. Håkansson, Electron capture dissociation of oligosaccharides ionized with alkali, alkaline earth, and transition metals, *Anal. Chem.*, 79 (2007) 2901–2910.
83. C. Zhao, B. Xie, S.-Y. Chan, C. E. Costello, and P. B. O'Connor, Collisionally activated dissociation and electron capture dissociation provide complementary structural information for branched permethylated oligosaccharides, *J. Am. Soc. Mass Spectrom.*, 19 (2008) 138–150.
84. S.-I. Nishimura, K. Niikura, M. Kurogochi, T. Matsushita, M. Fumoto, H. Hinou, R. Kamitani, H. Nakagawa, K. Deguchi, N. Miura, K. Monde, and H. Kondo, High-throughput protein glycomics: Combined use of chemoselective glycoblotting and MALDI-TOF/TOF mass spectrometry, *Angew. Chem. Int. Ed.*, 44 (2005) 91–96.
85. E. Fischer, Verbindungen des phenylhydrazines mit den zuckerarten, *Ber. Dtsch. Chem. Ges.*, 17 (1884) 579–584.
86. A. Lohse, R. Martins, M. R. Jorgensen, and O. Hindsgaul, Solid-phase oligosaccharide tagging (SPOT): Validation on glycolipid-derived structures, *Angew. Chem. Int. Ed.*, 45 (2006) 4167–4172.
87. Y. Miura, Y. Shinohara, J.-i. Furukawa, N. Nagahori, and S.-I. Nishimura, Rapid and simple solid-phase esterification of sialic acid residues for quantitative glycomics by mass spectrometry, *Chem. Eur. J.*, 13 (2007) 4797–4804.
88. Y. Miura, M. Hato, Y. Shinohara, H. Kuramoto, J. Furukawa, M. Kurogochi, H. Shimaoka, M. Tada, K. Nakanishi, M. Ozaki, S. Todo, and S.-I. Nishimura, BlotGlycoABC^{TM}, an integrated glycoblotting technique for rapid and large scale clinical glycomics, *Mol. Cell. Proteomics*, 7 (2008) 370–377.

89. J.-i. Furukawa, Y. Shinohara, H. Kuramoto, Y. Miura, H. Shimaoka, M. Kurogochi, M. Nakano, and S.-I. Nishimura, Comprehensive approach to structural and functional glycomics based on chemoselective glycoblotting and sequential tag conversion, *Anal. Chem.*, 80 (2008) 1094–1101.
90. H. Shimaoka, H. Kuramoto, J.-i. Furukawa, Y. Miura, M. Kurogochi, Y. Kita, H. Hinou, Y. Shinohara, and S.-I. Nishimura, One-pot solid-phase glycoblotting and probing by transoximization for highthroughput glycomics and glycoproteomics, *Chem. Eur. J.*, 13 (2007) 1664–1673.
91. M. Amano, M. Yamaguchi, M. Y. Takegawa, T. Yamashita, M. Terashima, J.-i. Furukawa, Y. Miura, Y. Shinohara, N. Iwasaki, A. Minami, and S.-I. Nishimura, Threshold in stage-specific embryonic glycotypes uncovered by a full portrait of dynamic N-glycan expression during cell differentiation, *Mol Cell Proteomics*, 9 (2010) 523–537.
92. H. Shimizu, K. Ochiai, K. Ikenaka, K. Mikoshiba, and S. Hase, Structures of N-linked sugar chains expressed mainly in mouse brain, *J. Biochem.*, 114 (1993) 334–338.
93. S. Zamze, D. J. Harvey, P. Pesheva, T. S. Mattu, M. Schachner, R. A. Dwek, and D. R. Wing, Glycosylation of a CNS-specific extracellular matrix glycoprotein, tenascin-R, is dominated by O-linked sialylated glycans and "brain-type" neutral N-glycans, *Glycobiology*, 9 (1999) 823–831.
94. Y. Kita, Y. Miura, J.-i. Furukawa, M. Nakano, Y. Shinohara, M. Ohno, A. Takimoto, and S.-I. Nishimura, Quantitative glycomics of human whole serum glycoproteins based on the standardized protocol for liberating N-glycans, *Mol. Cell. Proteomics*, 6 (2007) 1437–1445.
95. A. Kimura, M.-R. Tandang, T. Fukuda, C. Cabanos, Y. Takegawa, M. Amano, S.-I. Nishimura, Y. Matsumura, N. Maruyama, and S. Ustumi, Carbohydrate moieties contribute significantly to the excellent physicochemical properties of French bean 7S globulin phaseolin, *J. Agric. Food Chem.*, 58 (2010) 2923–2930.
96. Y. Miura, K. Kato, Y. Takegawa, M. Kurogochi, J.-i. Furukawa, Y. Shinohara, N. Nagahori, M. Amano, H. Hinou, and S.-I. Nishimura, Glycoblotting-assisted O-glycomics: Ammonium carbamate allows for highly efficient O-glycan release from glycoproteins, *Anal. Chem.*, 82 (2010) 10021–10029.
97. M. Kurogochi, M. Amano, M. Fumoto, A. Takimoto, H. Kondo, and S.-I. Nishimura, Reverse glycoblotting allows rapid enrichment glycoproteomics of biopharmaceuticals and disease-related biomarkers, *Angew. Chem. Int. Ed.*, 46 (2007) 8808–8813.
98. L. V. Lenten and G. Ashwell, Studies on the chemical and enzymatic modification of glycoproteins, *J. Biol. Chem.*, 246 (1971) 1889–1894.
99. M. S. Quesenberry and Y. C. Lee, A rapid formaldehyde assay using purpald reagent: Application under periodation conditions, *Anal. Biochem.*, 234 (1996) 50–55.
100. K. B. Lee and Y. C. Lee, Conformational studies of glycopeptides by energy transfer, *J. Biol. Chem.*, 271 (1996) 1462–1469.
101. J. Nilsson, U. Rüetschi, A. Halim, C. Hesse, E. Carlsohn, G. Brinkmalm, and G. Larson, Enrichment of glycopeptides for glycan structure and attachment site identification, *Nat. Methods*, 6 (2009) 809–811.
102. H. Lis and N. Sharon, Lectins: Carbohydrate-specific proteins that mediate cellular recognition, *Chem. Rev.*, 98 (1998) 637–674.
103. R. Schauer, Sialic acids as regulators of molecular and cellular interactions, *Curr. Opin. Struct. Biol.*, 19 (2009) 507–514.
104. A. Varki, Sialic acids in human health and disease, *Trends Mol. Med.*, 14 (2008) 351–360.
105. A. Varki, Multiple changes in sialic acid biology during human evolution, *Glycoconj. J.*, 26 (2009) 231–245.
106. D. B. Martin, T. Holzman, D. May, A. Peterson, A. Eastham, J. Eng, and M. McIntosh, MRMer, an interactive open source and cross-platform system for data extraction and visualization of multiple reaction monitoring experiments, *Mol. Cell. Proteomics*, 7 (2008) 2270–2278.

107. M. A. Kuzyk, D. Smith, J. Yang, T. J. Cross, A. M. Jackson, D. B. Hardie, N. L. Anderson, and C. H. Borchers, Multiple reaction monitoring-based, multiplexed, absolute quantitation of 45 proteins in human plasma, *Mol. Cell. Proteomics*, 8 (2009) 1860–1877.
108. L. Leigh Anderson and C. L. Hunter, Quantitative mass spectrometric multiple reaction monitoring assays for major plasma proteins, *Mol. Cell. Proteomics*, 5 (2006) 573–588.
109. R. G. Kay, B. Gregory, P. B. Grace, and S. Pleasance, The application of ultra-performance liquid chromatography/tandem mass spectrometry to the detection and quantitation of apolipoproteins in human serum, *Rapid Commun. Mass Spectrom.*, 21 (2007) 2585–2593.
110. J. Stahl-Zeng, V. Lange, R. Ossola, K. Eckhardt, W. Krek, R. Aebersold, and B. Domon, High sensitivity detection of plasma proteins by multiple reaction monitoring of N-glycosites, *Mol. Cell. Proteomics*, 6 (2007) 1809–1817.
111. H. Keshishian, T. Addona, M. Burgess, E. Kuhn, and S. A. Carr, Quantitative, multiplexed assays for low abundance proteins in plasma by targeted mass spectrometry and stable isotope dilution, *Mol. Cell. Proteomics*, 6 (2007) 2212–2229.
112. T. A. Addona, et al., Multi-site assessment of the precision and reproducibility of multiple reaction monitoring-based measurements of proteins in plasma, *Nat. Biotechnol.*, 27 (2009) 660–666.
113. M. Kurogochi, T. Matsushista, M. Amano, J.-i. Furukawa, Y. Shinohara, M. Aoshima, and S.-I. Nishimura, Sialic acid-focused quantitative mouse serum glycoproteomics by multiple reaction monitoring assay, *Mol. Cell. Proteomics*, 9 (2010) 2354–2368.
114. N. Nagahori, M. Abe, and S.-I. Nishimura, Structural and functional glycosphingolipidomics by glycoblotting with aminooxy-functionalized gold nanoparticle, *Biochemistry*, 48 (2009) 583–594.
115. S. Y. Vakhrushev, D. Dadimov, and J. Peter-Katalinic, Software platform for high-throughput glycomics, *Anal. Chem.*, 81 (2009) 3252–3260.
116. This project has been supported partly by a grant for "Development of Systems and Technology for Advanced Measurement and Analysis (SENTAN)" from the Japan Science and Technology Agency (JST) from the Ministry of Education, Culture, Science, and Technology, Japan (http://www.jst.go.jp/sentan/en/)

AUTHOR INDEX

Page numbers in roman type indicate that the listed author is cited on that page; page numbers in italic denote the page where the literature citation is given.

A

Aasen, I. M., 160, *205*
Abad, L. V., 171, *214*
Abe, M., 261, *271*
Acquistapace, S., 146, *196*
Adams, N. M., 121, 179, *183*
Adamson, J. T., 234, *269*
Adamyants, K. S., 121, 138–139, 171–172, *183*, *192*, *214–215*
Addona, T. A., 248, *271*
Aebersold, R., 221, 223, 228, 248, *265–267*, *271*
Aerbersold, R., 228, *266*
Aguilan, J. T., 140, 166, *193*
Ahluwalia, R., 70–72, *113*
Akanuma, H., 119, *182*
Akita, S., 223, *265*
Akiyama, M., 82, *114*, 228, *267*
Al-Abed, Y., 65, *112–113*
Alban, S., 170–171, *213*
Alice, M. B., 223, *265*
Aliganga, A. K. A., 163, *207*
Alley, W. R., 228, *267*
Allouch, J., 157, *203*
Allsobrook, A. J. R., 135, 138, *192*
Amachi, T., 48, *110*
Amado, A. M., 146, *196*
Amano, H., 156, *202*
Amano, M., 239, 242–252, 258–260, *270–271*
Amat, M., 176, *217*
Ametani, A., 159, 171, *204*, *213*

Amimi, A., 167, *210*
Amin, M. H. G., *42–43*
Amster, J., 148, *197*
Anderson, N. G., 223, *265*
Anderson, N. L., 223, 248, *265*, *271*
Anderson, N. S., 116, 124, 126, 131, 133, 143, 145, 162, 165, 167, *180*, *184–185*, *189*, *191*, *195*
Anderson, R. J., 155, *202*
Anderson, W., 128, 130, *186*
Andolfo, P., 162, *206*
Ando, O., 47, 51, 58, 87, 100–101, 106–107, *109–112*, *114*
Andrade, J. M., 146, *196*
Andrecht, S., 228, *267*
Andriamadio, P., 168, *211*
Andriamanantoanina, H., 163, *207*
Andries, M., 118–119, *181*
Angyal, S. J., 68, 70–72, *113*
Anoshina, A. A., 121, 171–172, *183*, *214*
Antipova, A. S., 150, 166, *199*
Antonopoulos, A., 148, 150, 155, 160, 166, *197*, *199*
Antonovici, M., 228, *267*
Aoki, T., 155, *202*
Aoki, Y., 158, *204*
Aono, R., 47, *109*
Aoshima, M., 250–252, 258–260, *271*
Apt, A. S., 48, *110*
Aqeel, A., 58, *112*
Arad, S. (.M.)., 174–177, *215–216*

Arai, K., 129–130, 155–156, *187, 202–203*
Arai, M., 66, 74–76, 96, *113*
Arakawa, S., 161, *206*
Araki, C., 116, 124, 127, 129–131, 133, 142, 155–156, *180, 184–185, 187–191, 194, 202–203*
Araki, T., 155, 158, *202, 204*
Aranilla, T., 171, *214*
Araño, K. G., 163, *207*
Arce Vera, F., 146, *196*
Ar Gall, E., 146, *196*
Arison, B. H., 66–67, *113*
Arkhipova, V. S., 143, 166, *194*
Armisen, R., 117, 163, *180, 207*
Arnheim, N., 221, *265*
Aronson, J. M., 120, 162, *182, 207*
Arsenault, G. P., 128, 130, *186*
Asano, H., 51–52, 58, *111*
Asano, N., 47, *109*
Ascêncio, S. D., 117, 163, *181, 207*
Ashwell, G., 246, *270*
Aspinall, G. O., 144, *195*
Astrina, O. P., 171, *214*
Attard, J. J., *41–42*
Ausio, J., 230, *268*
Austin, S., 128, *186*
Avonce, N., 48, *110*
Awamura, T., 82, *114*

B

Baba, T., 230, *268*
Bache, B. W., *42*
Bacic, A., 129–130, 153, 155, 162–163, 166–167, 179, *187–189, 200–201, 206–207, 210, 217*
Bäckström, M., 230, *269*
Bacon, B. E., 153, 167, *200*
Bailey, R. W., 165, *209*
Baird, J., 177, *217*
Baldan, B., 162, *206*
Baldwin, J. J., 66–67, *113*
Ball, S. G., 119, *181*
Banerji, N., 169, *211*

Barabanova, A. O., 166, 171, *209–210, 214*
Barak, Z., 176, *216*
Barbakadze, V. V., 135, 137, 144, 147, 152, 168, *192, 195, 199*
Barber, M., 223, *265*
Barbeyron, T., 150, 155, 157, 160, *199, 202–203, 205*
Bar-Guilloux, E., 47, *109*
Barnes, J. A., *42*
Barros, A. S., 146, *196*
Barrow, C. J., 166, *210*
Barry, V. C., 118, *181*
Batey, J. F., 138, 163, *192*
Bauknecht, N., 128, *187*
Baute, M. A., 119, *182*
Baute, R., 119, *182*
Baxter, B., 159, *204*
Beavis, R. C., 228, *267*
Bellanger, F., 128–129, *186, 188*
Bellion, C., 130, 152, 159–160, *188, 204–205*
Bell, R. A., 126, 130, 147, 167, *185, 197*
BeMiller, J. N., 127, 130, 133, *185*
Benevides, N. M. B., 168–170, *211–212*
Benitez, L. V., 159, *204*
Berecibar, A., 53, *111*
Bernabe, M., 55, 61, 63–64, *111–112*
Bernanbe, M., 55, *111*
Bernet, B., 57, *112*
Bernstein, M. A., 174, *216*
Berry, J. M., *41*
Berthou, C., 171, *214*
Bertocchi, C., 120, *182*
Bertozzi, C. R., 220–221, *265*
Bessières, M.-A., 117, *181*
Best, M. D., 161, *205*
Beuerman, R., 228, *267*
Bhattacharjee, S. S., 116, 150–152, 157, *180, 198*
Bhattacharya, D., 179, *217*
Biemann, K., 147, *197*
Bilan, M. I., 129–130, 143–144, 150, 154–155, 164–165, 173–175, *187, 195, 209*
Billeci, T. M., 225, *266*

Bird, K. T., 129, *188*
Bixler, H. J., 117, 166, *180*, *209–210*
Bjoerk, M., 159, *204*
Björndal, H., 120, 129, 143, 146, *182*, *187*
Blattner, R., 59, *112*
Bleehen, N. M., *42*
Blennow, A., 118, *181*
Bligh, S. W. A., 170, *213*
Bloys van Treslong, C. J., 128, *186*
Blunt, J. W., 138, 144, 155, 164, 168, *192*, *195*, *201–202*
Bobo, S., 54–55, 60–61, 66, 89–90, 104, *111–112*
Bociek, S., 130, 152, *188*
Bock, K., 149, *198*
Bogdanovich, R. N., 171, *214*
Böhm, M., 46, *108*
Boiron, A., 48–49, 54, *110*
Bojko, M., 118, *181*
Boltje, T. J., 222, *265*
Bondu, S., 117, 171, *181*, *214*
Boons, G.-J., 222, *265*
Borchers, C. H., 248, *271*
Bordoli, R. S., 223, *265*
Borgström, J., 128, *186*
Bosco, M., 150, *199*
Bouarab, K., 171, *214*
Boulenguer, P., 141, 150, 155, 159–160, 166, *194*, *199*, *205*
Bourgougnon, N., 117, 170, 173, *181*, *213*, *215*
Bowker, D. M., 163, *208*
Bowtle, W., 128, 130, *186*
Branford-White, C., 170, *213*
Brasch, D. J., 152, 162, 177, *200*, *217*
Breitmaier, E., 152, *200*
Breton, C., 221, *265*
Brigand, G., 130, 152, *188*
Brinkmalm, G., 248, *270*
Broberg, A., 121, *184*
Broom, J. E., 140, 166, *193*
Brown, I. N., 48, *110*
Brown, R. G., 120, 122, *182*
Buchanan, J. G., 63, *112*

Budnik, B. A., 230, *268*
Buléon, A., 119, 121, 176, *181*, *183*, *217*
Burgess, M., 248, *271*
Burns, W. W., 170, *213*
Burova, T. V., 150, 166, *199*
Buskas, T., 222, *265*

C

Cabanos, C., 242, *270*
Cáceres, P., 170, *212*
Cafaro, T., 228, *267*
Callow, M. E., 176, *216*
Calumpong, H. P., 163, *207*
Camarasa, M. J., 51, 89–90, 92, 99–100, *111*
Camp, D. G., 228, *266–267*
Campo, V. L., 117, *180*
Cancedda, R., 171, *214*
Capek, P., 176, *216*
Caram-Lelham, N., 128, *186*
Cardoso, M. A., 140, 163, 173, *194*, *208*, *215*
Carida, M., 173, *214*
Carlsohn, E., 248, *270*
Carlucci, M. A., 163, 170, *207*
Carlucci, M. J., 169–170, 172–173, *211–212*, *215*
Carpenter, B. K., 230, *268*
Carpenter, T. A., *41–42*
Carpita, R., 143, *195*
Carreira, E. M., 50, 66, 89, *110*
Carroll, D., 47, *109*
Carroll, J. D., 48, *110*
Carr, S. A., 248, *271*
Carvalho, I., 117, *180*
Casadevall, N., 227, *266*
Cases, M. R., 124, 128–130, 143, 164, 168, *184*, *186–188*, *195*, *209*
Cauduro, J. P., 170, *212*
Cerezo, A. S., 121, 123–124, 128–133, 140, 143, 148–150, 153, 155, 163–164, 166–170, 172–173, *183–184*, *186–190*, *193–195*, *197–198*, *200–201*, *208–209*, *211–212*, *215*
Cerná, M., 146, *196*

Ceroni, A., 225, 227, 261, *266*
Cesàro, A., 150, *199*
Chadmi, N., 167, *210*
Chai, W., 148, *197*
Chambat, G., 163, *207*
Chandrkrachang, S., 155, 166, *201*
Chang, H. M., 152, 162, *200*
Chan, S.-Y., 234, *269*
Chantalat, L., 160, *205*
Chaplin, M. F., 129, *187*
Chapman, D. J., 174, *216*
Chattopadhyay, K., 170, 172–173, *212–213, 215*
Chauhan, V. D., 169, *211*
Chavante, S. F., 168–169, *211*
Chaves, L. S., 163, *208*
Chen, C.-I., 225, *266*
Chen, H., 133, 171, *190, 214*
Chen, L. C.-M., 165, *209*
Chen, R., 228, *267*
Chen, X.-T., 234, *269*
Chen, Y., 170, *212*
Chermann, J. C., 170, 173, *213, 215*
Cherniak, R., 153, 167, *200*
Chevolot, Y., 150, 160, 167, *199*
Chew, A. P., 228, *267*
Chiara, J. L., 48, 54–55, 60–61, 63–64, 66, 89–90, 104, *110–112*
Chida, N., 51, 53, 59, *111–112*
Chi, L., 148, *197*
Chiovitti, A., 130, 153, 155, 162, 166–167, 179, *188–189, 200, 206, 210, 217*
Chisholm, J. D., 82, 84, *114*
Chizhov, O. S., 138, 146–147, *192, 196*
Chopin, T., 146, 168, 178, *196, 211, 217*
Chorley, R. J., *42*
Chrisman, P. A., 230, *268*
Christensen, B. E., 128, *186*
Christiaen, D., 126, 129, 166, 174, *185, 188, 216*, 717
Christison, J., 155–156, *202–203*
Chuah, C. T., 152, 162, *200*
Chu, Y., 148, *197*

Ciancia, M., 130–132, 140, 143, 148, 150, 163, 166, 168–170, *188–190, 193–194, 197–198, 208, 211–212*
Ciucanu, I., 143, *195*, 223, 225, *265*
Claeyssens, M., 121, *184*
Clancy, M. J., 144, *195*
Claude, A., 117, *181*
Clausen, H., 230, 234, *268–269*
Clayden, N. J., *42*
Clegg, J. S., 47, *109*
Cleland, R. L., 128, *186*
Clements, K. D., 156, *203*
Clifford, K. H., 47, *109*
Clingman, A. L., 131, 133, *189*
Coelho, H., 146, *196*
Cohen, E., 176, *216*
Coimbra, M. A., 146, *196*
Colleoni, C., 119, *181*
Colodi, F. G., 140, *194*
Coltart, D. M., 234, *269*
Combourieu, B., 176, *216*
Conte, A. F., 170, *213*
Convay, E., 162, *206*
Cooke, D., 127, 166, *185*
Cook, P. A., 156, *203*
Cook, W. H., 116, 128, 133, 165, *180*
Coombe, D. R., 171, *213*
Coon, J. J., 230, *268*
Cooper, C. A., 225–226, *266*
Copiková, J., 146, *196*
Corey, E. J., 77, *114*
Cormaci, M., 170, 173, *213, 215*
Correa, J., 171, *214*
Correc, G., 157, 168, *203*
Cosson, J., 162, *206*
Costello, C. E., 227, 234, *266, 269*
Coxon, B., 149, *198*
Coyne, V. E., 155–156, *202–203*
Craigie, J. S., 117, 120, 122, 126–130, 132, 140, 159, 161, 165, 168, *180, 182, 185, 188, 190, 193, 209, 211*
Craik, D. J., 130, 153, 155, 166–167, 179, *188–189, 200, 210, 217*
Crimmins, M. T., 81–82, *114*

Critchley, A. T., 146, 162, *196*, *207*
Cross, T. J., 248, *271*
Crowder, J., 170, *213*
Culoso, F., 162, *206*
Czjzek, M., 155, 157, 168, *202–203*

D

Dadimov, D., 261, *271*
Dairi, T., 48, *110*
Dai, Z., 228, *267*
D'Amato, M., 127, *185*
Damonte, E. B., 122, 163, 168–170, 172–173, *184*, *207*, *211–213*, *215*
Dancel, M. C. A., 140, 166, *193*
Daniel, C., 117, *181*
Daniellou, R., 160, *205*
Danishefsky, S. J., 234, *269*
Darula, Z., 228, *267*
Das, A. K., 168–169, *211*
da Silva, D. B. Jr., 117, *180*
Davadi, R. P., 176, *216*
Davidson, T. C., 120, *182*
Davies, G. J., 46, 48, *108*, *110*
Day, D. F., 157, *203*
Dayrit, F. M., 140, 166, *193*
De Clercq, E., 170, *212*
Decoster, E., 54, 57, *111*
Defaye, J., 47, *109*
Deffieux, G., 119, *182*
de Flores, E. A., 155, 159, *202*
de Gracia, I. S., 48, 54–55, 60–61, 66, 89–90, 104, *110–112*
Deguchi, K., 228, 234–237, 246, *267*, *269*
De Jongh, D. C., 147, *197*
De La Rosa, A. M., 171, *214*
de las Heras, F. G., 51, 89–90, 92, 99–100, *111*
Delattre, C., 168, *211*
Delgadillo, I., 146, *196*
Dell, A., 146–147, *196*, 223–225, 227, 261, *265–266*
de los Reyes, F. N., 162, *206*
den Hollander, J. A., 47, *109*
Deniaud, E., 121, *183*

de Paula, R. C. M., 163, *207–208*
de, P. C., 170, *212*
De Peredo, A. G., 230, *268*
Derbyshire, J. A., *42*
De Ruiter, G. A., 117, 131, *180*, *190*
Deslandes, E., 117, 140, 146, 161–162, 171, *181*, *193*, *196*, *207*, *214*
De Smet, K. A. L., 48, *110*
Destabel, C., 55, 61, 63–64, *111–112*
Dewar, E. T., 142, *194*
De Zoysa, M., 169, *212*
Dhar, T. G. M., 51, 101, *111*
Dideberg, O., 160, *205*
Dietrich, H., 54–55, *111*
Dijk, J. A. P. P., 128, *186*
DiNinno, V. L., 126, 130, 147, 161, 167, *185*, *197*, *205–206*
Diouris, M., 140, 161, *193*
Dixon, D. W. J., 54, *111*
Dmitriev, B. A., 152, *199*
Dobkina, I. M., 121, 133, 172, *184*, *191*, *215*
Doctor, V., 169, *211*
Dodgson, K. S., 132, 145, *190*, *195*
Dolan, T. C. S., 116, 126, 131, 138, 143, 145, 165, 167, *180*, *184–185*, *189*, *195*
Dole, M., 223, *265*
Domon, B., 227, 248, *266*, *271*
Dore, C. M. P. G., 168–169, *211*
Doyle, J. P., 167, *210*
Driguez, H., 47, *109*
Driscoll, P. C., 48, *110*
Dryhust, D. D., 230, *268*
Duarte, H. S., 152, *200*
Duarte, J. H., 152, *200*
Duarte, M. E. R., 117, 133, 135, 140, 148–149, 155, 163, 170, 173, *181*, *190–191*, *194*, *207–208*, *212*, *215*
Dube, D. H., 220, *265*
Dubinsky, O., 174, *216*, 717
Ducatti, D. R. B., 133, 135, 140, 148–149, 155, *190–191*, *194*
Duce, S. L., *41–42*
Duckworth, M., 123, 128, 130, 155–157, 162, *184–185*, *189*, *202–203*

Duee, E., 160, *205*
Duesler, E. N., 85–86, *114*
Du, G., 166, *210*
Dulong, V., 168, *211*
Dumont, L. E., 130, *189*
Dunn, A. D., 63, *112*
Duus, J. O., 121, *184*
Du, Y., 170, *213*
Dwek, R. A., 46, *108*, 220, 234, 243, 248, *265*, *269–270*
Dyrby, M., 146, *196*

Esko, J. D., 170, *212*
Esteves, M. L., 120, 152, *182*
Estevez, J. M., 168–169, *211*
Estevez, M. L., 121, 173, *183*
Eswaran, K., 163, *208*
Evans, D. R., 47, *109*
Evans, J. M., 118, *181*
Evans, L. V., 176, *216*
Evans, S., *43*
Evelegh, M. J., 161, *206*
Ewart, H. S., 166, *210*

E

Eastham, A., 248, *270*
Eckhardt, K., 248, *271*
Edavana, V. K., 48, *110*
Edgar, A. R., 63, *112*
Egge, H., 223, *265*
Ekeberg, D., 148, *197*
Ek, M., 51–52, *111*
El Ashry, E. S. H., 45–108, *108*
Elashvili, M. Ya., 130–133, 135–136, 148, 154, 163–164, 178, *188*, *190–191*, *197*
Elbein, A. D., 46–48, *109–110*
Elliot, R. P., 66, 70, *113*
Elliott, T., 234, *269*
El Nemr, A., 45–108, *108*
Elsley, D. A., 62, *112*
Engelsen, S. B., 118, 146, *181*, *196*
Eng, J. K., 225, 248, *266*, *270*
Englar, J. R., 120, 133, 145, *182*, *191*
Enokita, R., 47, *109–110*
Enriquez, E. P., 163, *207*
Ens, W., 228, *267*
Enzymic, N. M. R., 150, *198*
Erasmus, J. H., 156, *203*
Eriksson, K. E., 120, *182*
Erlich, H. A., 221, *265*
Ermak, I. M., 171, *214*
Erra-Balsells, R., 121, 132–133, 148–149, 172, *184*, *190*, *197*, *215*
Errea, M. I., 127–128, 131, 140, 162, 170, *185–186*, *189*, *193*, *212*

F

Faber, D., 48–49, 54, *110*
Fabre, M. S., 171, *214*
Fairbanks, A. J., 66, *113*
Faloona, F., 221, *265*
Falshaw, A., 130, *189*
Falshaw, R., 130, 134, 143, 147, 153, 155, 163–164, 166–168, 178–179, *188–189*, *191*, *200–201*, *207–211*, *217*
Fang, H., 47, *109*
Fang, L., 166, *210*
Fan, J., 228, *267*
Fareed, S. V., 176, *216*
Farias, W. R. L., 169–170, *212*
Farrant, A. J., 129, 135, 138, *187*, *192*
Fatema, M. K., 133, 149, *190*
Favetta, P., 148, 160, *197*
Feitosa, J. P. A., 163, *207–208*
Felix, A. S., 51, 89–90, 92, 99–100, *111*
Feng, X., 85–86, *114*
Fenn, J. B., 223, *266*
Fenoradosoa, T. A., 168, *211*
Ferguson, L. D., 223, *265*
Fernández-Resa, P., 51, 89–90, 92, 99–100, *111*
Ferrari, B., 54, 57, *111*
Ferreira, L. G., 140, *194*
Ferrier, R. J., 59, *112*
Finch, P., 150, 166, *198*
Fischer, E., 234–235, *269*
Flament, D., 157, *203*

Flatt, P. M., 48, *110*
Fleet, G. W. J., 46, 66, 70, *109*, *113*
Fletcher, G. L., 234, *269*
Fleurence, J., 121, *183*
Floc'h, J. Y., 140, 146, 161, 169, *193*, *196*, *212*
Flores, M. L., 138, *193*
Foote, M., 227, *266*
Fordham, E. J., 42
Fotedar, R., 163, *208*
Fournet, I., 146, *196*
Foyle, R. A. J., 174, *216*
França, R. A., 117, *181*
Francisco, P. B. Jr., 119, *181*
Frangou, S. A., 144, *195*
Franz, G., 171, *214*
Fredericq, S., 178–179, *217*
Freitas, A. L. P., 163, *207*–*208*
Freshwater, D. W., 178–179, *217*
Frias, A. M., 171, *214*
Fridriksson, E. K., 230, *268*
Friendlander, M., 171, *214*
Frith, W. J., 127, 166, *185*
Fromm, J. R., 170, *212*
Fujihira, H., 230, *269*
Fujii, M. T., 173, *215*
Fuji, K., 58, *112*
Fujita, E., 58, *112*
Fujita, N., 119, *181*
Fujitani, N., 230–234, *269*
Fujiwara, S., 119, *181*
Fukuda, M., 221, *265*
Fukuda, T., 242, *270*
Fukui, F., 159, 171, *204*, *213*
Fukuyama, Y., 121, 148, *184*, *197*
Fumoto, M., 230–231, 233–237, 246–249, *269*–*270*
Funari, G., 170, *213*
Funatsu, M., 155, *202*
Furfine, E., 54, *111*
Furnari, G., 173, *215*
Furneaux, R. H., 121, 130–131, 134, 140, 143, 147, 153, 155, 161–164, 166–168, 178–179, *183*, *188*–*191*, *193*, *200*–*201*, *205*–*208*, *210*–*211*, *217*

Furukawa, J.-i., 228, 236–245, 250–252, 258–260, 262, 264, *267*, *269*–*271*
Furukawa, Y., 66, 74–76, 96, *113*
Furuya, K., 173, *215*
Fu, X., 148, *197*

G

Galatas, F., 117, *180*
Gallant, D. J., 176, *217*
Gallego, P., 55, 61, 63–64, *111*–*112*
Ganem, B., 77, 79, 99, *114*
Ganesan, M., 162–163, *206*, *208*
Ganguly, D. K., 169, *211*
Ganzon-Fortes, E. T., 162–163, *206*, *208*
Gao, X., 171, *214*
Gao, X. D., 230, 232, 234, *269*
García, A., 48, 55, *110*–*111*
García-Lopez, M. T., 51, 89–90, 92, 99–100, *111*
Garcia Reina, G., 156, 159, *203*–*204*
Garegg, P. J., 51–52, *111*, 120, 131, 152, *182*, *190*, *200*
Garner, G. V., 223, *265*
Gasteiger, E., 225, *266*
Gelas J., 61, *112*
Gelfand, D. H., 221, *265*
Genicot-Joncour, S., 161–162, *205*
Génicot, S., 150, 160, 167, *199*
Geresh, S., 174–176, *215*–*216*, 717
Gerken, T. A., 221, *265*
Gero, S. D., 68, 70–72, *113*
Gerwig, G. J., 129, *187*
Geyer, H., 225, 227, 261, *266*
Geyer, R., 225, 227, 261, *266*
Ghosal, P. K., 163, 170, *207*
Ghosh, D., 228, *267*
Ghosh, P. K., 162, *206*
Ghosh, T., 122, 170, *184*, *212*
Giannouli, P., 167, *210*
Gibbs, S. J., 42
Gibo, H., 48, *110*
Gibor, A., 156, *202*
Gibson, R. P., 48, *110*

Gidley, M. J., 127, 166, *185*
Giese, B., 48–49, 54, *110*
Gillhouley, J. G., 62, *112*
Gilsenan, P. M., 150, 166, *199*
Girard, P., 54, *111*
Girouard, G., 166, *210*
Givernaud, T., 162, 167, *206*, *210*
Glaser, R., 174–175, *216*, 717
Glazunov, V. P., 166, *209–210*
Glithero, A., 234, *269*
Gloaguen, V., 174–175, *216*
Gloerich, J., 225, *266*
Gloster, T. M., 48, *110*
Glunz, P. W., 234, *269*
Goldberg, D., 225, *266*
Goldstein, M. E., 160, *205*
Gomersall, M., 157, *203*
Gomes, M. E., 171, *214*
Gomez-Pinchetti, J. L., 156, 159, *203–204*
Gonçalves, A. G., 133, 135, 148–149, 155, 170, *190–191*, *212*
Gonçalves, M. P., 163, *208*
Gong, Q., 159, *204*
Goodall, D. M., 128, *186*
Gordon, D. B., 223, *265*
Gordon-Mills, E. M., 161, *206*
Gorin, P. A. J., 149, 152, *198*, *200*
Grace, P. B., 248, *271*
Granjean, C., 53, *111*
Grant, D., 144, *195*
Grasdalen, H., 127, 150, 159–160, 166, *185*, *198–199*
Greer, C. W., 126, 153, 159–160, 166, *184*, *200*, *204–205*
Gregory, B., 248, *271*
Gretz, M. R., 120, 161–162, 176, *182*, *206–207*, *217*
Griffiths, L. M., 173, 176, *215*
Grimley, C., 225, *266*
Grinberg, N. V., 150, 166, *199*
Grinberg, V. Ya., 150, 166, *199*
Gritsenko, M. A., 228, *267*
Groleau, D., 157, *203*
Groth, I., 170–171, *213*

Grünewald, N., 170–171, *213*
Guan, H., 148, 150, 155, 171, *197*, *199*, *214*
Guéant, J. L., 121, *183*
Guégan, J.-P., 160, *205*
Guibet, M., 141, 150, 155, 160, 166–167, *194*, *199*
Guillard, M., 225, *266*
Guimarães, M., 163, *207*
Guimarães, S. M. P. B., 170, *212*
Guiry, M. D., 165, *209*
Guiseley, K. B., 130, 162, *189*, *206*
Guist, G., 124, *184*

H

Hagman, A., 146, *196*
Haines, H. H., 164, 172, *209*, *215*
Haines, S. R., 59, *112*
Håkansson, K., 234, *269*
Hakomori, S., 143, *195*
Halbrock, K., 171, *214*
Halim, A., 248, *270*
Hall, L. D., *41–43*, 149, *198*
Halsall, T. G., 118, *181*
Hamano, K., 47, *110*
Hama, Y., 131, *189*
Hamer, G. K., 116, 151–152, 159–160, *180*, *204–205*
Hancock, W. S., 228, *267*
Han, F., 148, *197*
Han, G., 228, *267*
Hang, H. C., 221, *265*
Hanisak, M. D., 168, *211*
Hanisch, F.-G., 234, *269*
Hanson, S. R., 161, *205*
Hansson, G. C., 230, *269*
Hanzawa, H., 58, 100, *112*
Hao, Z., 230, *268*
Hara, S., 129–130, 139, 162, 164, 168–169, *187–188*, *193*, *206*, *208*, *211*
Hara, Y., 119, *181*
Härd, I. D., 230, *269*
Hardie, D. B., 248, *271*

Harris, M. J., 149, *198*
Harrison, M. J., 225–226, *266*
Harris, R., 48, *110*
Hart, R. J., 128, *186*
Haruyama, H., 47, 57–58, 66, 74–76, 87, 90, 96, 100, 106–107, *109–110*, *112–114*
Harvey, D. J., 148, *197*, 243, *270*
Hasegawa, H., 230, *268*
Haselmann, K. F., 230, *268*
Hase, S., 243, *270*
Hashimoto, H., 57–58, *111–112*
Hashimoto, R., 230, 232, 234, *269*
Hashimoto, Y., 230, *268*
Haslam, S. M., 223–225, 227, 261, *265–266*
Hato, M., 228, 236–238, 262, 264, *267*, *269*
Haug, A., 128, *186*
Hausleden, E., 54, *111*
Hawkes, R. C., *41*
Hayashi, J., 129–130, 139, 164, *187*
Hayashi, K., 129–130, 139, 162, 164, 168–169, *187–188*, *193*, *206*
Heaney-Kieras, J., 174, *216*
Hecker, L. I., 47, *109*
Hehemann, J.-H., 157, 168, *203*
Helbert, W., 141, 148, 150, 155, 157, 159–162, 166–168, *194*, *197–199*, *202–203*, *205*, *210*
Hellerqvist, C. G., 129, 143, 146, *187*
Hemmingson, J. A., 140, 161–162, 166, 178, *193*, *205*, *207*, *217*
Hendriks, R., 228, *267*
Henrissat, B., 47, *109*, 160, *205*
Henzel, W. J., 225, *266*
Herrod, J. J., *42*
Herrod, N., *41*
Herth, W., 120, *182*
Hesse, C., 248, *270*
Heyraud, A., 160, *204*
Higa, M., 162, *206*
Higuchi, R., 221, *265*
Hileman, R. E., 170, *212*
Hillenkamp, F., 223, *265*
Hilliou, L., 140, 163, *193*, *208*
Hincapie, M., 228, *267*

Hindsgaul, O., 236, *269*
Hines, R. L., 223, *265*
Hinou, H., 230–237, 241, 243–246, *269–270*
Hirabayashi, A., 230, *268*
Hirabayashi, J., 228, *267*
Hirase, S., 124, 127, 129–130, 133, 139, 142, 164, 167–169, *184*, *187–189*, *191*, *193*, *208*, *211*
Hirsch, J., 121, *183*
Hirst, E. L., 118, *181*
Hjerde, T., 128, *186*
Hjerten, S., 127, *185*
Hodgson, R. J., *42*
Hoffmann, E. M., 128, *187*
Hoffmann, R. A., 127, 166, *185*
Hoffman, R., 170, *213*
Hofsteenge, J., 230, *268*
Hogan, J. M., 230, *268*
Hogan, P. G., *41*
Højrup, P., 225, *266*
Hollingsworth, M. A., 234, *269*
Holt, A., 144, *195*
Holzman, T., 248, *270*
Hommersand, M. H., 178–179, *217*
Honda, T., 60, *112*
Hong, K. C., 128, *185*
Honke, K., 221, *265*
Hoppe, H. G., 171, *214*
Horn, D. M., 230, *268*
Horn, G. T., 221, *265*
Horton, D., 61, *112*
Hosford, S. P. C., 133, 145, 161, *191*, *205*
Hosomi, Y., 82, *114*
Hosoyama, S., 139, *193*
Howard, B. H., 120, *182*
Hu, B., 159, *204*
Hughey, J., 178–179, *217*
Huhmer, A. F. R., 230, *268*
Hui, A., 66, *113*
Huleichel, M., 176, *216*
Hultberg, H., 51–52, *111*
Hunt, D. F., 230, *268*
Hunter, C. L., 248, *271*
Hurd, C. L., 163, 179, *208*

Hurtado, A. Q., 163, *207*
Huvenne, J. P., 146, *196*
Hu, Z. B., 170, *213*

I

Iacomini, M., 152, *200*
Ibanez, C. M., 130, *188*
Ido, Y., 223, *265*
Ikenaka, K., 243, *270*
Imberty, A., 221, *265*
Imeson, A., 117, *180*
Inoue, H., 47, *109*
Inoue, R., 225, *266*
Inoue, Y., 138, *192*
Ioanoviciu, A. S., 150, 155, *199*
Ireland, R. E., 54, *111*
Isakov, V. V., 166, *209–210*
Ishanu, V., 176, *216*
Ishihara, H., 161, *206*
Ishikawa, T., 149, *198*
Ismaeli, N. M., 162, *207*
Isobe, T., 228, *267*
Isomura, S., 161, *206*
Ito, H., 228, *267*
Itoh, M., 159, 171, *204*, *213*
Itoh, T., 120, *182*
Itoi, K., 47, *109–110*
Ito, T., 48, *110*, 133, *191*, 230–231, 233, *269*
Iturriaga, G., 48, *110*
Ivanova, E. G., 128, 130, 133, 135, 147, 152–153, 157–158, 162–164, *185*, *188*, *191*, *197*, *203*, *207–208*
Iwaisaki, Y., 73, 93–94, 102–103, *113–114*
Iwane-Sakata, H., 130, 168–169, *188*
Iwasaki, N., 239, 243, *270*
Iwashita, T., 48, *110*
Izumi, K., 127, 149, 162, *185*, *198*, *206*
Izumo, A., 119, *181*

J

Jackson, A. M., 248, *271*
Jackson, P., *42*
Jacobsson, S. P., 146, *196*
Jacquinet, J.-C., 148, *198*
Jaffray, A. E., 155, *202*
Jam, M., 157, *203*
Jansson, P.-E., 154, *201*
Jarret, R. M., 121, *183*
Jaseja, M., 142, 174, *194*, *216*
Jeanneau, C., 221, *265*
Jensen, F., 230, *268*
Jensen, O. N., 230, *268*
Jerez, J. R., 121, *183*
Jezzard, P., *42*
Jha, B., 162, *206*
Jiang, W., 228, *267*
Jiang, X., 171, *214*, 228, *267*
Jiao, G., 166, *210*
Ji, M., 153, *200*
Joenväärä, S., 228–229, *267*
Johndro, K. D., 117, 166, *180*, *209–210*
Johnsen, A. H., 230, *268*
Johnston, K. H., 159, 161, *204–205*
Jol, C. N., 131, *190*
Jones, J. K. N., 118, 120, *181–182*
Jones, K. A., 47, *109*
Jorgensen, M. R., 236, *269*
Joshi, H. J., 225–226, *266*
Jouanneau, D., 141, 148, 150, 155, 159–160, *194*, *198–199*, *205*
Joubert, Y., 121, *183*
Jr, 163, *207*
Jung, G. L., 103–104, *114*
Jung, M., 82, 84, *114*
Jurd, K. M., 168–169, *211*
Jurgens, A., 129, *188*

K

Kadowaki, N., 225, *266*
Kagan, H. B., 54, *111*
Kahn, R., 160, *205*
Kailas, T., 162, 166, *206*, *210*
Kaji, H., 228, *267*
Kakimoto, D., 159, *204*
Kamata, T., 138, *192*
Kamei-Hayashi, K., 129, 139, 164, 168, *187*, *193*, *208*, *211*

Kamei, Y., 158, *204*
Kamerling, J. P., 129, *187*
Kametani, S., 119, *182*
Kametani, T., 60, *112*
Kaminogawa, S., 159, 171, *204*, *213*
Kamitani, R., 234–237, 246, *269*
Kanda, T., 139, *193*
Kaneko, A., 230, 232, 234, *269*
Kaneko, M., 119, *181*
Kantor, T. G., 138, *192*
Karamanos, Y., 129, *188*
Karas, M., 223, 228, *265*, *267*
Karger, B. L., 230, *268*
Kariya, Y., 139, *193*
Karlsson, A., 138, *192*
Karlsson, N. G., 225–226, *266*
Karsten, U., 117, *181*
Kasai, K., 228, *267*
Kasanah, N., 48, *110*
Kaskar, B., 73, 75, *113*
Kasuya, A., 66, 74–76, 96, *113*
Katalinić, J. P., 223, 230, *265*, *268*
Kato, A., 47, *109*
Kato, K., 230, 243–245, *269–270*
Katsumura, Y., 171, *214*
Kawamura, K., 60, *112*
Kawano, D. F., 117, *180*
Kawasaki, N., 225, *266*
Kawasaki, T., 225, *266*
Kay, R. G., 248, *271*
Kazlowski, B., 133, *190*
Keeling, P. J., 118, *181*
Keira, T., 228, *267*
Kelleher, N. L., 230, *267–268*
Kelly-Spratt, K. S., 228, *266*
Kemp, C. J., 228, *266*
Kenne, L., 119, 154, *182*, *201*
Kenny, P. T. M., 48, *110*
Kerek, F., 143, *195*, 223, 225, *265*
Kerin, B. F., 178, *217*
Keristovesky, G., 176, *216*
Kervarec, N., 117, 150, 155, 160, 166–167, *181*, *199*
Keshishian, H., 248, *271*
Khan, M. S., 73, 75, *113*

Khiar, N., 55, *111*
Khira, N., 55, *111*
Khoo, K.-H., 225, *266*
Kido, Y., 170, *213*
Kifune, M., 47, *109*
Kiladjian, J.-J., 227, *266*
Kim, S. J., 169, *212*
Kimura, A., 242, *270*
Kimura, H., 48, *110*
Kimura, K., 82, *114*
Kim, Y. H., 166, *209*
Kinarsky, L., 234, *269*
King, C. H. R., 47, *109*
Kinoshita, T., 47, *109–110*
Kirimura, K., 159, *204*
Kirst, G. O., 117, *180–181*
Kitahashi, H., 47, 50, 73, 93–94, 102–103, *109*, *113–114*
Kitamikado, M., 155, *202*
Kita, Y., 241, *270*
Kjeldsen, F., 230, *268*
Kloareg, B., 117, 150, 156–157, 160–162, 169, 171, *180*, *199*, *203–205*, *212*, *214*
Klochkova, N. G., 131, 133, 164, 175, 178, *190*, *209*
Klunder, J. M., 66–67, *113*
Knapp, S., 51, 73, 101, *111*, *113*
Knight, G. A., 130, *189*
Knirel, Yu. A., 152, 154, *199–201*
Knutsen, S. H., 120, 124, 126–127, 148–150, 159–160, 166, *182*, *184–185*, *197–199*, *205*
Kobayashi, A., 82, *114*
Kobayashi, H., 47, *109*
Kobayashi, M., 47, *109*
Kobayashi, N., 170, *213*
Kobayashi, R., 159, *204*
Kobayashi, Y., 48, 51, 57–58, 66, 74–76, 89–90, 96, 99–101, *110–114*
Koca, J., 221, *265*
Kochetkov, N. K., 121, 131, 135, 138–139, 143–144, 146–147, 152, 154, 156, 171–172, *183*, *189*, *192–197*, *199–202*
Koelliker, U., 77, *114*
Koenig, J. L., 145, *195*

Koh, S. K., 228, *267*
Kolender, A. A., 139–140, 168, 170, 172–173, *193*, *212–213*, *215*
Kolender, A. D., 121, *184*
Kollist, A., 166, *210*
Kolta, A., 227, *266*
Komura, H., 48, *110*
Kondo, H., 230–231, 233–237, 246–249, *269–270*
Konishi, M., 119, *181*
Konradsson, P., 131, *190*
Kornprobst, J. M., 170, 173, *213*, *215*
Koushal, G. P., 46, *109*
Ková, P., 103–104, *114*, 121, *183*
Koyama, M., 120, *182*
Ko, Y. T., 133, *190*
Kozlova, E. G., 124, 131, 147, 163, *184*
Kraft, G. T., 130, 153, 155, 162, 166–167, 179, *188–189*, *200*, *206*, *210*, *217*, 719
Krausz, P., 174–175, *216*
Krek, W., 248, *271*
Kristensen, T., 150, 160, *199*
Krokhin, O., 228, *267*
Kruger, N. A., 230, *268*
Kuboki, A., 82, *114*
Kubo, N., 58, *112*
Kudo, H., 171, *214*
Kuduk, S. D., 234, *269*
Kuhn, E., 248, *271*
Kukarskikh, T. A., 171, *214*
Kullolli, M., 228, *267*
Kumar, V., 163, *208*
Küpper, F., 171, *214*
Kuramoto, H., 236–242, 262, 264, *269–270*
Kurogochi, M., 228, 230–252, 258–260, 262, 264, *267*, *269–271*
Kuzyk, M. A., 248, *271*
Kvarnström, I., 131, 152, *190*, *200*
Kyosseva, S. V., 47, *109*
Kyossev, Z. N., 47, *109*

L

Labriola, R., 121, *183*
Lacombe, J.-M., 54, 57, *111*
Laetsch, W. M., 161, *205*
Lafosse, M., 148, 150, 155, 160, 166, *197*, *199*
Lahaye, M., 117, 121, 127–129, 131, 133, 146, 153, 162, 167, 173, 176, *180*, *183*, *185–186*, *188–189*, *195*, *200*, *210*, *213*, *215*, *217*
Lamont, R. B., 66, *113*
Lange, V., 248, *271*
Lang, F., 77, 79, 99, *114*
Lapshina, A. A., 163, *207*
Laroche, C., 168, *211*
Larsen, B., 124, 126, 128, *184*, *186*
Larsen, J., 146, *196*
Larson, G., 248, *270*
Lau, E., 129, *187*
Lawsom, A. M., 148, *197*
Lawson, C. J., 122, 126, 133, 135, 138, 140, 161, 167, *184–185*, *191–192*
Lechat, H., 176, *217*
Ledford, B. E., 50, 66, 89, *110*
Lee, B., 61, *112*
Lee, G. A., 57, 59, *112*
Lee, J. J., 169, *212*, 223–224, *265*
Lee, K. B., 246, *270*
Lee, Y. C., 246, *270*
Lefeber, D. J., 225, *266*
Léger, J.-M., 117, *181*
Legrand, P., 146, *196*
Leigh Anderson, L., 248, *271*
Leigh, C., 127–128, *185*
Leite, E. L., 168–169, *211*
Lemoine, M., 160, *205*
Lenten, L. V., 246, *270*
Leontein, K., 129, *187*
Levy-Ontman, O., 175, *216*
Lewis, G., 124, *184*
Lewis, M. A., 230, *268*
Lezerovich, A., 121, *183*
Liang, A., 170, *213*
Liao, M.-L., 130, 153, 155, 162–163, 166–167, 179, *188–189*, *200–201*, *206–207*, *210*, *217*
Lichtenthaler, F. W., 70–72, *113*

Lidell, M. E., 230, *269*
Li, J., 77, 79, 99, *114*, 148, 159, *197*, *204*
Lim, M., 73, 75, *113*
Li, N., 162, 171, *207*, *214*
Lin, B., *170*, *213*
Lindberg, B., 120, 129, 131, 143, 146, 152, *182*, *187*, *190*, *200*
Linhardt, R. J., 139, 148, 150, 155, 170, *193*, *197*, *199*, *212*
Lin, J., 133, 171, *190*, *214*
Li, P., 171, *214*
Lipkind, G. M., 152, 154, *199–201*
Li, S., 228, *267*
Lis, H., 248, *270*
Liu, L., 54, *111*
Liu, P. S., 47, *109*
Liu, T., 228, *267*
Liu, X., 162, 170, *207*, *213*
Liu, Y-K., 228, *267*
Live, D. H., 234, *269*
Li, X.-J., 171, *214*, 228, *266*
Li, Z., 162, 171, *207*, *214*
Lloyd, A. G., 145, *195*
Lohse, A., 236, *269*
Long, W. F., 155–156, 162, *202–203*
Lönngren, J., 129, 146, *187*, *196*
Lotov, R. A., 131, *189*
Lucas, A. J., *42*
Lundt, I., 119, *182*
Lupescu, N., 174, *215–216*
Luttrell, B. M., 70–72, *113*
Lu, W., 48, *110*
Lü, X., 148, 171, *197*, *214*
Lu, Z., 158, *204*
Lynds, S. M., 130, *189*

M

Maass, K., 225, 227, 261, *266*
Ma, B. Y., 225, *266*
Ma, C., 148, 176, *197*, *216*
Macaranas, J. M., 159, *204*
Macaskie, L. E., *42*
Maciel, J. S., 163, *208*
Mack, L. L., 223, *265*
Madden, J. K., 117, 145, 162, *180*
Madsen, F., 118, *181*
Maeda, M., 118, *181*
Maček, B., 230, *268*
Maes, E., 174–175, *216*
Magbanua, M., 163, *207*
Maguire, T., 170, *212*
Mahmud, T., 48, *110*
Makienko, V. F., 128, 163, *185*
Maleev, V. V., 171, *214*
Malfait, T., 146, *196*
Mallick, P., 228, *266*
Malmqvist, M., 155–156, *202–203*
Mamontov, A., 176, *216*
Mandal, P., 122, 170, 172–173, *184*, *213*, *215*
Mann, M., 221, 223, 225, 230, *265–266*, *268*
Mano, J. F., 171, *214*
Manri, N., 230, 232, 234, *269*
Mantri, V. A., 163, *208*
Manville, J. F., *41*
Marchessault, R. H., 119, *182*
Marco-Contelles, J., 55, 61, 63–64, *111–112*
Mariani, P., 162, *206*
Mariano, P. S., 85–86, *114*
Marinez-Grau, A., 55, *111*
Marinho-Soriano, E., 163, *207*
Marks, R. M., 170, *212*
Marques, C. T., 168–169, *211*
Marquez, V. E., 73, 75, *113*
Marrion, O., 121, *183*
Marrs, W. M., 128, *186*
Martin, D. B., 228, 248, *266*, *270*
Martin-Dupont, P., 227, *266*
Martin-Ortega, M. D., 60, 66, 89–90, 104, *112*
Martins, R., 236, *269*
Martynova, M. D., 156, *202*
Maruyama, N., 242, *270*
Masselon, C. D., 228, *266*
Masuda, K., 230, *268*
Mateu, C. G., 170, *213*
Mathlouthi, M., 145, *195*
Matilewicz, M. C., 140, 143, 166, *193*

Matsuhiro, B., 121, 130, 146, 170, 177, *183, 188–189, 196, 212–213, 217*
Matsui, K., 47, *109*
Matsumoto, T., 155, *202*
Matsumura, Y., 242, *270*
Matsuo, M., 139, *193*
Matsuo, T., 223, *265*
Matsushita, T., 230–237, 246, 250–252, 258–260, *269, 271*
Mattu, T. S., 243, *270*
Matulewicz, M. C., 121, 127–133, 139–140, 143, 148–150, 162–164, 166, 168–170, 172–173, *183–190, 193–194, 197–198, 208–209, 211–213, 215*
Matulová, M., 176, *216*
Ma, Y., 159, *204*
May, D., 248, *270*
Mayeux, P., 227, *266*
Maypa, A., 163, *207*
Mazerolle, R., 178, *217*
Mazoyer, J., 141, 150, 155, 159–160, 166, 176, *194, 199, 205, 217*
Mazumder, S., 163, 170, *207*
Mazurek, M., 152, *200*
McAlister, M. S. B., 48, *110*
McAllister, G. D., 57, *111*
McCandless, E. L., 126, 130, 147, 159, 161–162, 165, 167, *185, 197, 204–207, 209*
McClure, D. E., 66–67, *113*
McCormack, A. L., 225, *266*
McDowell, R. H., 116, *180*
McIntosh, M., 248, *270*
McKenna, P. J., *42*
McLachlan, J., 129, 165, *188, 209*
McLafferty, F. W., 230, *267–268*
McLean, M. W., 155–156, 159, 161–162, *202–204*
McLellan, D. S., 168–169, *211*
McLuckey, S. A., 230, *268*
McManus, L. J., 162, *206*
Mechref, Y., 228, *267*
Medoruma, K., 162, *206*
Medzihradszky, K. F., 228, *267*

Meena, R., 163, *208*
Meeuse, B. J. D., 118–119, *181*
Mehta, G., 85–86, *114*
Mejías, E. G., 170, *213*
Mellado, R. P., 155, *202*
Mellikov, E., 166, *210*
Melo, F. R., 169–170, *212*
Melo, M. R. S., 163, 169–170, *207, 212*
Melton, L. D., 152, 162, *200*
Méndez-Castrillón, P. P., 51, 89–90, 92, 99–100, *111*
Mendoza-Vargas, A., 48, *110*
Mendoza, W. G., 166, *210*
Meng, C. K., 223, *266*
Mensler, K., 66–67, *113*
Merca, F. E., 163, *207*
Merino, E. R., 168, *211*
Mesquita, J. F., 146, *196*
Metro, F., 131, 133, *189*
Michaud, F., 117, *181*
Michaud, P., 168, *211*, 227, *266*
Michel, G., 155, 157, 160, *202–203, 205*
Miertus, S., 150, *199*
Mikami, T., 51–52, 58, *111*
Mikoshiba, K., 243, *270*
Miller, I. J., 121, 138, 144, 155, 162–164, 166–168, 179, *183, 192, 195, 201–202, 206, 208, 210–211, 217*
Minami, A., 239, 243, *270*
Miranda, C., 48, *110*
Mirgorodskaya, P., 230, *268*
Miroshnikova, L. I., 131, 138–139, 143–144, 156, *189, 192–195, 203*
Mischenko, V. V., 48, *110*
Mitsunobu, O., 51–52, 58, *111*
Miura, N., 234–237, 246, *269*
Miura, Y., 228, 236–245, 262, 264, *267, 269–270*
Miyamoto, S., 66, 74–76, 96, *113*
Miyazaki, H., 48, 57–58, 89–90, 100, *110, 112*
Mobley, R. C., 223, *265*
Mody, K. H., 169, *211*
Mohal, N., 85–86, *114*

Mollet, J. C., 146, *196*
Mollion, J., 126, 128, 166, 168, *185–186*, *211*
Molloy, F., 162, *207*
Monde, K., 228, 234–237, 246, *267*, *269*
Monroe, M. E., 228, *267*
Montaño, M. N. E., 140, 162–163, 166, *193*, *206–210*
Moore, R. J., 228, *267*
Morava, E., 225, *266*
Moreau, S., 126, 128, 166, *185–186*
Morett, E., 48, *110*
Morgan, E., 162, *206*
Morgan, K. C., 121, 135, *183*, *191*
Morishita, T., 158, *204*
Mori, T., 116, *180*
Mormann, M., 230, *268*
Morrice, L. M., 155–156, 162, *202–203*
Morris, E. R., 117, 144–145, 162, 167, *180*, *195*, *210*
Morris, H. R., 223–224, *265*
Morvan, H., 128, 162, 174–175, *186*, *206*, *216*
Motoki, Y., 170, *213*
Mou, H., 171, *214*
Mountfort, D. O., 156, *203*
Mouradi-Givernaud, A., 146, 162, 167, 174–175, *196*, *206*, *210*, *216*
Mourão, P. A. S., 169–170, *212*
Mueller, G. P., 145, *195*
Mukai, L. S., 120, 122, *182*
Müller, K. M., 179, *217*
Mullis, K. B., 221, *265*
Munro, S. L. A., 130, 153, 155, 166–167, 179, *188–189*, *200*, *217*
Murano, E., 120, 127, 150, 163, *182*, *185*, *198*, *207*
Murao, S., 48, *110*
Murphy, H. N., 48, *110*
Mussio, I., 117, *181*
Myslabodski, D. E., 124, 126, 150, *184*, *198*

N

Nagahori, N., 236, 239, 241, 243–245, 261, *269–271*
Nagasawa, K., 138, *192*
Nagasawa, N., 171, *214*
Nakagawa, H., 131, *189*, 228, 234–237, 246, *267*, *269*
Nakagawa, Y., 170, *213*
Nakajima, H., 47, *109*
Nakajima, M., 47, 58, 100, *109–110*, *112*
Nakamura, Y., 119, *181*
Nakanishi, K., 236–238, 262, 264, *269*
Nakano, M., 236, 239–242, *270*
Nakano, S., 58, *112*
Nakashima, H., 170, *213*
Nakasone, Y., 162, *206*
Nakata, H., 230–231, 233, *269*
Nakayama, T., 48, *110*
Nakazawa, K., 51, 53, 59, *111*
Namy, J. L., 54, *111*
Naruchi, K., 230–234, *269*
Nascimento, M. S., 168–169, *211*
Nash, R. J., 66, *113*
Nataf, J., 227, *266*
Naughton, A. B. J., 51, 101, *111*
Navarro, D. A., 138, 141–142, 165, 168, *193–194*, *209*
Nebylovskaya, T. B., 166, *210*
Neely, W. B., 145, *195*
Neill, K. F., 163, 179, *208*
Neish, A. C., 165, *209*
Neiss, T. G., 131, *190*
Nelson, W. A., 130, 163, 178–179, *189*, *208*, *217*
Nerinckx, W., 121, *184*
Neuffer, J., 77, *114*
Neushul, M., 170, *213*
Newman, R. H., 120, *182*
Nielsen, M. L., 230, *268*
Nihei, S., 119, *181*
Niikura, K., 234–237, 246, *269*
Nikapitiya, C., 169, *212*
Nilsson, J., 248, *270*
Nilsson, M., 163, *208*
Niñonuevo, M. R., 140, 166, *193*
Nishimura, S.-I., 219–265, *267*, *269–271*
Nishioka, M., 121, *184*
Nishio, O., 161, *206*

Nishiyama, S., 70–72, *113*
Nisizawa, K., 118, 173, *181*, *215*
Niu, X., 171, *214*
Nivall, P., 118, *181*
Nolan, T., 61, *112*
Noll, T., 230, *269*
Nomoto, K., 48, *110*
Nomura, K., 159, 171, *204*, *213*
Nonami, H., 121, 132–133, 148–149, 172, *184*, *190*, *197*, *215*
Nørgaard, L., 146, *196*
Norman, R. O. C., 54, *111*
Norwood, T. J., *42*
Noseda, D. G., 170, *212*
Noseda, M. D., 117, 132–133, 135, 140, 143, 148–149, 155, 163, 166, 169–170, 172–173, *181*, *190–191*, *193–194*, *207–208*, *211–212*, *215*
Nose, Y., 130, 168, *188*
Nott, K. P., *42–43*
Novotny, M. V., 228, *267*
Ntarima, P., 121, *184*
Nunes, A., 146, *196*
Nunn, J. R., 121, 129–131, 133, 135, 138, *183*, *187–189*, *192*
Nyval-Collen, P., 155, *202*
Nyvall Collen, P., 160, *205*

O

Ochiai, K., 243, *270*
O'Connor, P. B., 234, *269*
Oda, H., 159, *204*
Oestgaard, K., 160, *205*
Ogamo, A., 170, *213*
Ogawa, S., 47–48, 50–51, 53, 59, 69–70, 72–73, 92–94, 102–103, *109–114*
Ogura, K., 59, *112*
Ohara, K., 148, *198*
Ohira, S., 82, *114*
Ohkuma, Y., 47, *109*
Ohno, M., 177, *217*, 241, *270*
Ohtsuka, M., 51, 53, 59, *111–112*
Ohyabu, N., 230–234, *269*

Oka, S., 225, *266*
Okazaki, M., 173, *215*
Okazaki, T., 47, *110*
Olsen, C. E., 118, *181*
Olsen, J. V., 230, *268*
Onami, T., 66–67, *113*
Ondarza, M., 129, *188*
O'Neill, A. N., 126, 130, 133, 135, 168, *184*, *189*, *191*, *210*
O'Neill, S. A., 177, *217*
Onraët, A. C., 162–163, *206*
Opoku, G., 169, *211*
Orologas, N., 120, *182*
Orsato, A., 117, *181*
Oscarson, S., 51–52, *111*
Ossola, R., 248, *271*
Ott, F. D., 179, *217*
Overkleeft, H. S., 106, *114*
Overman, L. E., 66, *113*
Oyama, Y., 119, *181*
Ozaki, H., 118, *181*
Ozaki, M., 236–238, 262, 264, *269*

P

Pachler, K. G. R., 149, *198*
Packer, N. H., 225–226, *266*
Pagba, C. V., 163, *207*
Painter, T. J., 117–118, 122, 124, 127–128, 133, 138, *180*, *184–185*, *191–192*
Pan, C. L., 133, *190*
Pandit, U. K., 106, *114*
Pang, P.-C., 223–224, *265*
Panico, M., 223–224, *265*
Pan, Y. T., 47–48, *109–110*
Paoletti, S., 127, 150, 160, *185*, *198–199*, *205*
Pao, Y.-L., 234, *269*
Paper, D. H., 170–171, *213–214*
Papo, T., 227, *266*
Paranha, R. G., 135, *191*
Parent, C. A., 130, *189*
Parish, C. R., 171, *213*
Parolis, H., 121, 129–130, 135, 138, *183–184*, *187–188*, *192*

AUTHOR INDEX

Parolis, L. A. S., 121, *184*
Parro, V., 155, *202*
Parry, S., 223–224, *265*
Parsons, M. J., 165, *209*
Pastuszak, I., 47–48, *109–110*
Paterson-Beedle, M., *42*
Patier, P., 169, *212*
Patron N. J., 118, *181*
Paulsen, H., 234, *269*
Paulson, J. C., 225, *266*
Pavia, A. A., 54, 57, *111*
Pearce, L., 66, 70, *113*
Peat, S., 118, *181*
Pedersen, C., 149, *198*
Pedersen, H., 149, *198*
Pedersen, M., 118–119, 140, 159, *181–182, 193, 204*
Pehk, T., 162, 166, *206, 210*
Peltoniemi, H., 228–229, *267*
Penman, A., 126, 133, 143, 145, 166–167, *185, 191, 195*
Penninkhof, B., 131, *190*
Peppelman, H. A., 150, 166, *199*
Percival, E., 142, 174, 176, *194, 216*
Percival, E. G. V., 142, *194*
Percival, E. L., 116, *180*
Percy, A. E., 177, *217*
Pereira, L., 146, 150, 166, *196, 199*
Pereira, M. G., 169–170, *212*
Pereira, M. S., 169–170, *212*
Perlin, A. S., 142, 168, 174, *194, 210, 216*
Pernas, A. J., 128, *186*
Pesheva, P., 243, *270*
Peter-Katalinic, J., 261, *271*
Petersen, R. V., 146, *196*
Peterson, A., 248, *270*
Petitou, M., 220, *265*
Petrescu, A. J., 234, *269*
Peyron, M., *42*
Phelps, D. W., *42*
Pichon, R., 117, *181*
Pickering, T. D., 163, *207*
Pickmere, S. E., 165, *209*
Picton, L., 168, *211*

Piculell, L., 117, 128, *180, 186*
Pierens, G. K., *42*
Pirie, N. W., 129, *187*
Pitteri, S. J., 230, *268*
Plastino, E. M., 163, *207*
Pleasance, S., 248, *271*
Poinas, A., 161–162, *205*
Polne-Fuller, M., 156, *202*
Polyakova, A,M., 171, *214*
Pomin, V. H., 169, *211*
Potin, P., 140, 150, 156–157, 160–162, 169, 171, *193, 199, 203–205, 212, 214*
Potter, G. F., *42*
Prado-Fernández, J., 146, 150, *196, 198*
Prado, H., 163, *208*
Prakash, O., 234, *269*
Prasad, K., 162–163, *206, 208*
Pratt, M. R., 221, *265*
Price, R. G., 132, 145, *190, 195*
Prodolliet, J., 128, *186*
Prome, J. C., 130, 152, *188*
Pujol, C. A., 122, 163, 168–170, 172–173, *184, 207, 211–213, 215*
Purandare, A., 73, 101, *113*
Purchase, R., 54, *111*
Pushpamali, W. A., 169, *212*
Putaux, J.-L., 119, *181*

Q

Qian, W.-J., 228, *267*
Qiao L., 54, *111*
Qi, H., 162, *207*
Qiu, S.-J., 228, *267*
Qiu, X., 169, *211*
Quatrano, R. S., 117, *180*
Quemener, B., 121, 128, 131, 133, 170, *183, 186, 189, 213*
Quesenberry, M. S., 246, *270*
Quimener, B., 173, *215*

R

Raab, A. W., 66–67, *113*
Rama, J., *42*

Ramavat, B. K., 162–163, 169, *206, 208, 211*
Ramshaw, I. A., 171, *213*
Ranzinger, R., 225, *266*
Rarick, J., 221, *265*
Rashed, N., 46, *108*
Rathbone, E. B., 149, 152, *198*, 200
Ray, B., 122, 163, 170, 172–173, *184, 207, 212–213, 215*
Recalde, M. P., 172–173, *215*
Rechter, M. A., 135, 144, 147, *192, 195–196*
Reed, R. H., 117, *180*
Rees, D. A., 116–117, 121–122, 124, 126, 131, 133, 135, 138, 140, 143–145, 160–162, 165–167, *180, 183–185, 189, 191–193, 195, 205–206, 209*
Reis, C., 230, *268*
Reis, R. L., 171, *214*
Rej, R. N., 142, *194*
Relleve, L. S., 171, *214*
Renkonen, R., 228–229, *267*
Renn, D. W., 130, *189*
Ren, S., 148, *197*
Ren, W., 166, *210*
Revol, J. F., 129, *188*
Reynaers, H., 128, *186*
Rhinehart, B. L., 47, *109*
Ribeiro-Claro, P. J. A., 146, *196*
Richard, C., 156–157, *203*
Richard, O., 161–162, *205*
Richards, K. S., 42
Rimbert, M., 170, *213*
Rinaudo, M., 153, 159, 163, *200, 207*
Rincones, R. E., 140, *193*
Ritamo, I., 228–229, *267*
Rivas, P., 146, *196*
Rivero-Carro, H., 130, *188*
Rizzo, R., 120, 127, 150, *182, 185, 198*
Robal, M., 166, *210*
Roberts, A. W., 120, *182*
Roberts, E., 120, *182*
Roberts, M. A., 128, *186*
Robertson, B. D., 48, *110*
Robertson, B. L., 162–163, *206*
Roberts, S., 48, *110*

Robinson, K. M., 47, *109*
Rocha, H. A. O., 168–169, *211*
Rochas, C., 127–129, 146, 153, 156–157, 159–160, 162, 169, *185–186, 188, 195, 200, 203–204, 206, 212*
Rocha, S. M., 146, *196*
Roden, L., 174, *216*
Rodrigueza, M. R. C., 166, *210*
Rodriguez-Fernandez, M., 55, *111*
Rodríguez, M. C., 168, *211*
Rodriguez-Vásquez, J. A., 146, *196*
Roepstorff, P., 225, 230, *266, 268*
Rogerson, A., 156, *202*
Roleda, M. Y., 162, 163, *206, 208*
Rollema, H. S., 149–150, 166, *198–199*
Romero, J. B., 163, 166, *207, 210*
Rondeau-Mouro, C., 121, *183*
Roongsawang, N., 48, *110*
Rose, F. A., 145, *195*
Ross, A. G., 120, *182*
Rossi, G., 227, *266*
Royyuru, A. K., 234, *269*
Rudolf, B., 146, 150, 166, *196, 199*
Rudolph, B., 117, 131, 161–162, 167, *180, 190, 205, 210*
Rüetschi, U., 248, *270*
Ruiz, G., 174–175, *216*
Rupitz, K., 73, 101, *113*
Rusig, A.-M., 117, *181*
Russell, I., 121, 130, 138, *183, 188, 192*

S

Sacktor, B. S., 47, *109–110*
Sadamoto, R., 230–231, 233, *269*
Saiki, R. K., 221, *265*
Saiki, S., 171, *214*
Saito, H., 228, *267*
Saito, T., 170, *213*
Sakaguchi, H., 225, *266*
Sakai, S., 139, *193*
Sakai, T., 48, *110*
Sakamoto, N., 139, *193*
Sakamoto, T., 230–234, *269*

Sakata, T., 159, *204*
Sames, D., 234, *269*
Sampietro, A. R., 155, 159, *202*
Sandford, P. A., 177, *217*
Santos, G. A., 126, 130, 166, *184*, *189*
Santo, V. E., 171, *214*
Sarkar, K. K., 168, *211*
Sarwar, G., 159, *204*
Satake, H., 47, *109*, 230, 232, 234, *268–269*
Sato, A., 47, *109–110*
Satoh, A., 119, *181*
Sato, K., 58, *112*, 139, *193*
Sato, Y., 132, *190*
Saunders, G. W., 179, *217*, 719
Sauriol, F., 142, *194*
Schachner, M., 243, *270*
Scharf, S. J., 221, *265*
Schauer, R., 248, *270*
Schittenhelm, W., *41*
Schroender, M. J., 230, *268*
Schubert, M., 138, *192*
Schulz, B. L., 225–226, *266*
Schwarz, J. B., 234, *269*
Schwertz, A., 121, *183*
Scolaro, L. A., 170, *212*
Scott, J., 161, *206*
Seccal, M., 146, *196*
Sedgwick, R. D., 223, *265*
Seepersaud, M., 65, *112–113*
Segre, A., 150, *199*
Sekkal, M., 146, *196*
Selby, H. H., 116, *179*
Semenova, L. R., 119, *181*
Sen, A. K. Sr., 168–169, *211*
Serra, H., 228, *267*
Shabanowitz, J., 230, *268*
Shacklock, P. F., 165, *209*
Shanmugam, M., 169, *211*
Shaposhnikova, A. A., 138–139, *192*
Sharon, N., 248, *270*
Sharpless, K. B., 66–67, *113*
Shashkov, A. S., 116, 121, 129–130, 143–144, 150–155, 163–167, *180*, *183*, *187–188*, *199–201*, *210*

Sheath, R. G., 179, *217*
Sheng, W., 171, *214*
Sherman, S., 234, *269*
Shibata, T., 155, *202*
Shiga, Y., 119, *182*
Shigeiri, H., 155, *202*
Shi, H., 171, *214*
Shimaoka, H., 228, 236–242, 262, 264, *267*, *269–270*
Shimizu, H., 228, 230–231, 233, 243, *267*, *269–270*
Shimoda, C., 47, *109*
Shimonaga, T., 119, *181*
Shing, T. K. M., 62, *112*
Shinkawa, T., 228, *267*
Shin, K. S., 166, *209*
Shinohara, Y., 228, 236–245, 250–252, 258–260, 262, 264, *267*, *269–271*
Shin, T., 48, *110*
Shiomoto, K., 168, *211*
Shiozaki, M., 48, 51, 57–58, 66, 74–76, 87, 89–90, 96, 99–101, 106–107, *110–114*
Shi, Q., 176, *216*
Shirasaka, N., 48, *110*
Shi, S. S., 170, *213*
Shi, X., 171, *214*
Shobier, A. H. S., 46, *108*
Shomer, I., 160, *205*
Shulman, R. G., 47, *109*
Siddhanta, A. K., 162–163, 168–169, *206*, *208*, *211*
Sihlbom, C., 230, *269*
Sikkander, S. A., 150, 155, *199*
Silva, F. R. F., 168–169, *211*
Silva, P., 177, *217*
Simon, B., 176, *216*
Simon-Colin, C., 117, *181*
Simpson, F. J., 121, *183*
Simpson, P. R., 129–130, 164, *187*
Singh, S. K., 138, *192*
Siriwardena, A., 53, *111*
Skea, G. L., 156, *203*
Sletmoen, M., 148, 150, 160, *197*, *199*
Slootmaekers, D., 128, *186*

Smidsrød, O., 128, *186*
Smietana, M., 141, 148, *194*, *198*
Smirnova, G. P., 119, *181*
Smith, B. N., 119, *181*
Smith, C., 66, 70, *113*
Smith, D. B., 116, 128, 133, 165, 168, *180*, *210*, 248, *271*
Smith, M. S., 223–224, *265*
Smith, R. D., 228, *266–267*
Snajdrova, L., 221, *265*
Soedjak, H. S., 128, *187*
Soerbo, B., 132, *190*
Solo-Kwan, J., 174, *216*
Solov'eva, T. F., 166, *209–210*
Solovyova, T. F., 171, *214*
Sombret, B., 146, *196*
Somerville, J. C., 142, *194*
Sommerfeld, M. R., 120, 162, 176, *182*, *207*, *217*
Sommsich, I. E., 171, *214*
Song, H., 171, *214*
Song, J., 171, *214*
Sopkova-de Oliveira Santos, J., 117, *181*
Sørensen, E. S., 230, *268*
Sørensen, K. F., 230, *268*
Sousa, A. M. M., 146, 163, *196*, *208*
Sousa-Pinto, I., 140, *193*
Souza, B. W. S., 163, *208*
Spedding, H., 145, 195
Spichtig, V., 128, *186*
Stadnichuk, I. N., 119, *181*
Stahl-Zeng, J., 248, *271*
Stancioff, D. J., 126, 130, 133, 145, *185*, *189*, *191*, *195*
Standing, K. G., 228, *267*
Stanley, N. F., 124, 126, 130, 133, 145, *184–185*, *189*, *191*, *195*
Stenutz, R., 154, *201*
Stephen, A. M., 131, 133, 149, 152, *189*, *198*, *200*
Stevenson, D. E., 130, 163, 166, *188*, *207*, *210*
Stevenson, T. T., 130–131, 143, 162, 164, 178, *188*, *190*, *206*
Stewart, G. R., 48, *110*

Stewart, R. C., *42*
Stoffel, S., 221, *265*
Stoloff, L., 177, *217*
Storer, R., 66, *113*
Stortz, C. A., 117, 123–124, 128–130, 132, 138, 141–143, 153, 155, 164–168, *180*, *184*, *186–188*, *190*, *193–195*, *200–201*, *209*
Strebler, J.-L., 54, 57, *111*
Strecker, G., 174–175, *216*
Stults, J. T., 225, *266*
Sturgeon, R. J., 129, *187*
Suami, T., 70–72, *113*
Suarez, P., 163, *207*
Subba Rao, P. V., 163, *208*
Sugiyama, J., 120, *182*
Sullivan, D. M., 132, *190*
Sumi, T., 131, *189*
Sun, D., 228, *267*
Sundararajan, P. R., 119, *182*
Sundeloef, L. O., 128, *186*
Sun, L., *43*, 176, *216*
Sun, Y., 171, *213*
Suryanarayanan, G., 234, *269*
Sussman, A. S., 47, *109*
Sutton-Smith, M., 225, *266*
Suzuki, T., 159, *204*
Svensson, S., 129, 143, 146, *187*, *196*
Swan, B., 120, *182*
Sworn, G., 128, *186*
Syka, J. E. P., 230, *268*
Syn, N., 121, *183*

T

Tabak, L. A., 221, *265*
Tabet, E. A., 81–82, *114*
Tachibana, Y., 234, *269*
Tada, M., 236–238, 262, 264, *269*
Tadano, K., 70–72, *113*
Tajiri, M., 228, *267*
Takada, R., 58, *112*
Takahashi, N., 228, *267*
Takahashi, S., 47, *109–110*

Takamatsu, Y., 47, *110*
Takano, R., 129–130, 138–139, 162, 164, 168–169, *187–188*, *192–193*, *206*, *208*, *211*
Takayama, T., 66, 74–76, 96, *113*
Takegawa, M. Y., 239, 243, *270*
Takegawa, Y., 228, 230–234, 242–245, *267*, *269–270*
Takemoto, H., 230–231, 233, *269*
Takimoto, A., 241, 246–249, *270*
Takisada, M., 159, *204*
Tako, M., 162, *206*
Talarico, L. B., 170, *212*
Tal, J., 176, *216*
Tamada, M., 171, *214*
Tamura, J., 57, *111*
Tanaka, K., 223, *265*
Tandang, M.-R., 242, *270*
Taniguchi, N., 221, *265*
Tanzawa, K., 47, *109*
Taoka, M., 228, *267*
Tate, M. E., 68, 70–72, *113*
Ta, T.-V., 139, *193*
Taylor, R. J. K., 57, *111*
Teixeira, D. I. A., 163, *208*
Tekoah, Y., 175, *216*
Ten Hagen, H. K. G., 221, *265*
Terada, M., 225, *266*
Terashima, M., 239, 243, *270*
Tetler, L. W., 223, *265*
Teyssandier, I., 227, *266*
Thanawiroon, C., 150, 155, *199*
Therkelsen, G. H., 116, *179*
Thevelein, J. M., 47, *109*
Thion, L., 157, *203*
Thiruppathi, S., 163, *208*
Thomas, C. B., 54, *111*
Thomas, D. N., 117, *181*
Thom, D., 117, 145, 162, *180*
Tiller, C., 166, *210*
Tipson, R. S., 140, *193*
Tischer, S. F., 170, *212*
Tishchenko, V. P., 121, *183*
Tissot, B., 223–224, *265*

Titlyanov, E. A., 163, 166, *207*, *210*
Titlynova, T. V., 166, *209*
Tobacman, J. K., 150, 155, *199*
Todo, S., 236–238, 262, 264, *269*
Toffanin, R., 120, 127, 150, *182*, *185*, *198*
Togame, H., 230–231, 233, *269*
Toida, T., 139, 150, 155, *193*, *199*
Tojo, E., 146, 150, *196*, *198*
Tokuyasu, T., 138, *192*
Toomik, P., 162, *206*
Tripodi, G., 162, *206*
Tromp, R. H., 150, 166, *199*
Trono, G. C., 163, *207*
Trost, B. M., 82, 84, *114*
Truus, K., 162, 166, *206*, *210*
Tsekos, I., 120, *182*
Tsuda, S., 234, *269*
Tsuji, K., 82, *114*
Tsukayama, K., 173, *215*
Tsunehiro, J., 159, 171, *204*, *213*
Tsuzuki, M., 119, *181*
Tulio, S., 163, 170, *208*, *212*
Turquois, T., 146, *196*
Turvey, J. R., 118, 120–122, 129–130, 138, 140, 142–144, 149, 155–156, 159, 162–164, 173, 176, *181*, *183*, *187*, *192*, *195*, *198*, *202–204*, *208*, *215*
Tuvikene, R., 162, 166, *206*, *210*
Tyler, A. N., 223, *265*
Tyler, J. A., 42

U

Ubukata, O., 87, 106–107, *114*
Uchida, C., 47–48, 50, 69–70, 72–73, 92–94, 102–103, *109–110*, *113–114*
Uchiyama, H., 170, *213*
Ueberheide, B., 230, *268*
Uematsu, R., 228, *267*
Ugo, V., 227, *266*
Urrets-Zavalia, E., 228, *267*
Urrets-Zavalia, J., 228, *267*

Urzúa, C. C., 121, 177, *183*, *217*
Usami, S., 159, *204*
Usov, A. I., 115–179, *180–185*, *187*, *189–200*, *202–203*, *207–209*, *214–215*
Ustumi, S., 242, *270*

Vollmer, C. M., 161, *206*
Volobujeva, O., 166, *210*
von der Lieth, C.-W., 225, *266*
von Holdt, M. M., 129, *187*
Vonthron-Sénécheau, C., 117, *181*
Vreeland, V., 161, *205–206*

V

Vaher, M., 162, 166, *206*, *210*
Vakhrushev, S. Y., 261, *271*
Valente, A. P., 169–170, *212*
van Boeckel, C. A., 220, *265*
van Cauwelaert, F., 146, *196*
Van Dael, H., 146, *196*
van der Meer, J. P., 132, 159, *190*
van de Velde, F., 146, 149–150, 166, *196*, *198–199*
Vansteenkiste, J., 227, *266*
van Wiltenburg, J., 106, *114*
Varet, B., 227, *266*
Varkevisser, F. A., 128, *186*
Varki, A., 248, *270*
Vasella, A., 46, 57, *108*, *112*
Vasseur, J.-J., 148, *198*
Vasyanina, L. K., 172, *215*
Vattuone de Sampietro, M. A., 155, 159, *202*
Vedasiromoni, J. R., 169, *211*
Vederas, J. C., 54, *111*
Vella, C., 170, *213*
Verdus, M. C., 146, *196*
Vergunova, G. I., 164, *209*
Viana, A. G., 140, 163, *194*, *207*
Viebke, C., 128, *186*
Viet, M. T. P., 153, *200*
Vignon, M. R., 162, *206*
Villanueva, R. D., 140, 163, 166, *193*, *207–210*
Vincendon, M., 153, 159, *200*
Vinogradov, E. V., 154, *200–201*
Viola, R., 118, *181*
Viron, B., 227, *266*
Vliegenthart, J. F. G., 129, *187*
Voelter, W., 152, *200*
Vogl, H., 171, *214*

W

Waaland, J. R., 165, *209*
Wada, Y., 228, *267*
Wadouachi, A., 168, *211*
Waki, H., 223, *265*
Waki, I., 230, *268*
Walsh, G., 227, *266*
Walter, J. A., 165, *209*
Wang, C., 176, *216*
Wangen, B. F., 160, *205*
Wang, F., 171, *214*, 228, *267*
Wang, J., 171, *214*
Wang, K., 170, *213*
Wang, L., 228, *267*
Wang, Q., 170, *213*
Wang, S. C., 170, *213*
Wang, Y., 159, 166, *204*, *210*
Wang, Z. T., 170, *213*
Warren, R. A. J., 48, *110*
Watanabe, C., 225, *266*
Watanabe, J., 170, *213*
Watanabe, K., 129–130, 167, *188–189*
Watanabe, S., 47, 50, 103, *109*
Watkin, D. J., 66, 70, *113*
Watson, P. J., *42*
Watt, D. K., 177, *217*
Way, G., 66, *113*
Weigl, J., 159, 161, *204*
Weinberger, F., 171, *214*
Weinstein, J., 176, *216*
Weiss, R. H., 54, *111*
Welder, D. G., *41*
Welti, D. H., 130, 146, 152, *188*, *196*
Welty, D., 150, *198*
Wen, Z. C., 132, 159, *190*
Wessels, H. J. C. T., 225, *266*

West, J. A., 165, 176, *209*, *217*
Weston, A., 48, *110*
Wevers, R. A., 225, *266*
Weykam, G., 117, *181*
Whalen, E., 146, 178, *196*
Whang, I., 169, *212*
Whistler, R. L., 116, *179*
Whitehouse, C. M., 223, *266*
Whitehouse, L. A., 121, 179, *183*
Whyte, J. N. C., 120, 133, 145, *182*, *191*
Widmalm, G., 154, *201*
Wilkins, J. A., 228, *267*
Wilkins, M. R., 225, *266*
Willaume, C., 121, *183*
Williams, E. L., 120–122, 138, 164, *183*, *192*
Williams, L. J., 234, *269*
Williamson, F. B., 144, 155–156, 159, 161–162, *195*, *202–204*
Williams, S. C. R., *41*
Williams, T. P., 144, *195*
Winchester, B. G., 46, 66, *109*, *113*
Wing, D. R., 243, *270*
Withers, S. G., 73, 101, *113*
Witvrouw, M., 170, *212*
Wohlgemuth, J., 228, *267*
Wold, J. K., 142, *194*
Wong, C.-H., 161, *205*
Wong, H., 155, 161–162, 166–167, *201*, *205*, *210*
Wong, K. F., *41*, 140, 161, *193*
Wong, S. C., 225, *266*
Wong, S. F., 223, *266*
Wood, J. A., 118–119, *181*
Wormald, M. R., 234, *269*
Wormser-Shavit, E., 47, *109*
Wright, J. J., *42*
Wright, J. L. C., 121, *183*
Wrobleski, S. T., 82, 84, *114*
Wu, S.-I., 230, *268*
Wu, Y., 161, *206*

X

Xia, X., 131, *189*
Xie, B., 234, *269*
Xing, D., *42*
Xin, H., 171, *213–214*
Xu, W., 171, *214*
Xu, Z., 162, 171, *207*, *213–214*

Y

Ya Elashvili, M., 163, *208*
Yamada, K., 225, *266*
Yamada, T., 170, *213*
Yamagishi, T., 69–70, 73, 92–94, 102–103, *113–114*
Yamaguchi, K., 131, *189*
Yamaguchi, M., 239, 243, *270*
Yamamotom, N., 170, *213*
Yamashita, M., 223, *266*
Yamashita, T., 239, 243, *270*
Yamauchi, Y., 228, *267*
Yamaura, I., 155, *202*
Yang, B., 148, 166, *197*, *210*
Yang, J., 248, *271*
Yan, X., 133, 171, *190*, *214*
Yaphe, W., 116, 123, 126–130, 146, 150–153, 155, 157, 159–162, 166, 178, *180*, *184–186*, *188–189*, *195*, *198*, *200*, *202–205*
Yarotsky, S. V., 116, 120–121, 151–153, 166–167, 171–173, *180*, *182–183*, *199–200*, *214–215*
Yates, J. R. III., 225, *266*
Yeh, H. J. C., 103–104, *114*
Ye, M., 228, *267*
Yermak, I. M., 166, *209–210*
Ye, Z., 139, *193*
Yi, E. C., 228, *266*
Yokoi, T., 164, 168, *208*
Yokoyama, A., 119, *181*
Yoon, H. S., 179, *217*
Yoshida, K., 139, *193*
Yoshida, S., 228, *267*
Yoshida, T., 223, *265*
Yoshida, Y., 223, *265*
Yoshiike, R., 87, 106–107, *114*
Yoshimura, J., 58, *112*
Yoshimura, Y., 230, *269*

Yoshizawa, Y., 159, 171, *204*, *213*
Young, D. B., 48, *110*
Young, K. S., 150, 157, *198*
Youngs, H. L., 176, *217*
Yuan, H., 171, *214*
Yuasa, H., 57, *111*
Yueping, S., 171, *214*
Yu, G., 148, 150, 155, 166, 171, *197*, *199*, *210*, *214*
Yuming, Y., 48, 72–73, 92, *110*
Yu, P., 171, *214*
Yu, S., 118–119, 140, *181–182*, *193*
Yu, W., 148, 159, *197*, *204*
Yvin, J.-C., 169, *212*

Z

Zablackis, E., 161, *205*
Zablakis, E., 126, 166, *184*
Zagorski, M., 48, *110*
Zaia, J., 148, *197*
Zamze, S., 243, *270*
Zanetti, F., 150, *198*
Zanlungo, A. B., 130, *189*
Zemke-White, W. L., 177, *217*
Zhang, H., 171, *214*, 228, *266–267*
Zhang, J., 166, 171, *210*, *214*
Zhang, Q., 162, 171, *207*, *214*
Zhang, W., 171, 175, *214*, *216*
Zhang, Y., 148, *197*
Zhang, Z., 171, *213–214*
Zhao, C., 234, *269*
Zhao, T., 162, *207*
Zhao, X., 148, 166, *197*, *210*
Zhao, Z., 162, *207*
Zhong, H.-J., 128, *186*
Zhou, G., 171, *213–214*
Zhou, J., 228, *267*
Zhou, L., 228, *267*
Zhou, Y., 228, *267*
Zhou, Z., 170, *213*
Zhu, P., 133, *190*
Zibetti, R. G. M., 140, 173, *194*, *215*
Zibetti, R. M. G., 140, *194*
Zillig, P., 48–49, 54, *110*
Zinoun, M., 140, 161, *193*
Zolotarev, B. M., 147, *196*
Zou, H., 228, *267*
Zubarev, R. A., 230, *267–268*
Zúñiga, E. A., 170, *212–213*

SUBJECT INDEX

A

ABCh. *see* N-(2-aminobenzoyl)cysteine hydrazide
Agar group, polysaccharides
  agaran–agarose hybrid, 163
  algal species, 164
  calcareous algae, 164
  galactan structures, 162
  gelling properties, 163
  Gracilaria species, 162–163
  neutral regular agarose, 162
  porphyrans, 162
  sulfated polysaccharides, 165
Aldol–Wittig type of reaction, 74
Automated glycan analysis
  development, 264–265
  glycobiology and glycotechnology
    biosynthetic pathway, glycoproteins, 221–222
    synthesis and purification, 222
  glycoblotting technology and SweetBlot machine, 262
  glycomics technologies
    glycoblotting method, 234–261
    mass spectrometry, 223–234
    tryptic digests, 234
  glycoproteins and GSLs, 220
  GlycoWorkbench and SysBioWare, 261
  N-glycan profiling and ratios, 263–264
  proteomics and glycomics
    glycosylation, 220–221
    ORFs and PCR amplification, 221

C

Carrageenan, polysaccharides
  Callophyllis hombroniana, 166–167
  Gigartina lanceata, 167
  KCl-soluble fraction, 166
  partial reductive hydrolysis, 167
  samples, 166
  types, 165

D

DL-hybrids
  agarans and carrageenans, 168
  carrageenan-type structure, 169
  polymeric molecules, 168
  polysaccharides, 168

E

ECD. *see* Electron-capture dissociation
Electron-capture dissociation (ECD)
  O-glycosylation site identification, 230–231, 234
  spectra, 231, 233

F

Ferrier reaction, 58, 59
Fischer's phenylhydrazone reaction, 234–236
Floridean starch, storage carbohydrates
  amylolytic enzymes, 118, 119
  elimination reaction, 119
  enzymatic formation, 118
  low-molecular glycosides, 117
  structure and chirality, 117–118

G

Glycoblotting method
  ABCh and BlotGlycoABC™, 236–239
  analysis, O-glycans, 242–244
  BlotGlyco H beads
    preparation and SEM view, 239, 240
    protocol, 239, 241
  Fischer's phenylhydrazone reaction, 234–236
  human IgG N-glycans, 236, 237
  human-milk osteopontin
    BOA-labeled O-glycans, 243, 245
    O-glycomics, 243, 244
  mouse P19C6 cells vs. neural cells, 239–240, 243
  N-glycans expression levels, 240–241
  plant glycoproteins, 241–242
  reverse
    AFP, fibrinogen and recombinant human EPO, 247–248
    chromatography-based enrichment processes, 246
    fragment-ion counts, 259, 260
    glycan heterogeneity and glycosylation sites, 246, 247
    GSL-aldehydes, 261
    human cerebrospinal fluid, 248
    mouse-serum PA-labeled sialyl glycopeptides, 252–258
    MRM, 258–260
    MS/MS spectrum, PA-labeled sialyl glycopeptide, 252, 259, 260
    nonsialylated glycopeptides, 259, 261
    sialic acid-focused glycoproteomics, 250–252
    sialylated glycopeptides, 246–247
    terminal sialic acids, 246, 248
  trans-iminization reactions, 239, 242
Glycofragment mass fingerprinting (GMF), 225, 226
Glycosphingolipids (GSLs)
  aldehydes, 261
  biological roles, 220
GlycoWorkbench, 261

Glycuronans
  galactose and glucuronic acid, 173
  sulfated galactans, 173–174
GMF. see Glycofragment mass fingerprinting
GSLs. see Glycosphingolipids

H

Hall, L. D.
  Cambridge period, 1984–2005
    biofilm reactors, 37
    calcium alginate, 38
    Experimental NMR Conference (ENC), 38–39
    food materials, 36–37
    human diseases, 36
    in vivo and ex vivo imaging, 36
    MRI, 35–36
    zero-quantum NMR methods, 34–35
  graduate students, UBC, 34
  NMR spectroscopy, carbohydrate analysis, 12–13
  sugars, pyranose ring, 12
    angular dependence, coupling constants, 18
    Carr–Purcell–Meiboom–Gill spin-lock sequence, 25–26
    2-C-diacetoxythalliation, D-galactal triacetate, 22
    chemical microscopy, 30
    chemical shift-resolved tomography, 33
    chitosan, 31–32
    ^{13}C NMR spectra, 23
    ^{13}C T_1 values, sugars, 27–28
    cyclic phosphates, 20
    decoupling, 24
    1,2:3,4-di-O-isopropylidene-α-D-galactopyranose, 19–20
    2D NMR methods, 31
    electrophilic addition, fluorinated species, 16–18
    europium, thulium and praseodymium, 25
    ^{19}F–^{19}F couplings, 19
    fluoro sugar period, 17

^{19}F NMR, fluorinated carbohydrates, 15
glycopyranosyl fluorides, 16
glycosyl fluorides, 15
^{1}H–^{1}H dipole–dipole relaxation mechanism, 26
^{1}H spin–lattice relaxation times, 26
internuclear double resonance (INDOR), 24
interproton distances, 29
^4J couplings, 18–19
J-multiplets, 30
Karplus dependence, 14
methoxymercuration, acetylated glycals, 21–22
nitroxide adducts, ESR spectroscopy, 33
noise decoupling technique, 23
nuclear magnetic double-resonance methods, 19
paramagnetic ions, 32
pentopyranosyl fluorides, 16
radiofrequency (RF) carrier, 23
scientific literature, 14–15
shift reagents, 25
stereospecific effects, anomeric protons, 27
syn-diaxial and vicinal gauche effect, 27
tin coupling constants, 21
T_1-null method, 28–29
zeugmatography technique, 30
HILIC. *see* Hydrophilic-interaction chromatography
huEPO. *see* Human erythropoietin
Human erythropoietin (huEPO)
   glycan heterogeneity, 222, 227
   glycoproteomics technique, 247–249
Hydrophilic-interaction chromatography (HILIC)
   glycopeptide separation, 228, 234
   reverse glycoblotting method, 246

M

MALDI. *see* Matrix-assisted laser-desorption ionization
Mass spectrometry (MS)
   description, 223
   disaccharides characterization, 147
   glycomics
      cross-ring fragments and permethylation, 225, 227
      elucidation, glycan compositions, 225
      GlycosidIQ platform, 225
      GMF procedure, 225–226
      MALDI-based ionization, 223, 225
      N-and O-glycans, 223
      protocol, 223, 224
   glycoproteomics
      ECD–MS/MS spectrum and analysis, 230–232
      electron-induced dissociations, 230
      HILIC and lectin-affinity chromatography, 228
      huEPO, 222, 227
      O-and N-glycosylation sites, 231, 233–234
      periodinate-based nonspecific oxidation, 227–228
      purified human plasma serotransferrin, 228, 229
   methyl groups, 146
   post-source decay (PSD) mode, 148
   prompt fragmentation, 149
   red algal galactans, 147
   soft-ionization methods, 148
   sulfated oligosaccharides, 147–148
Matrix-assisted laser-desorption ionization (MALDI)
   mass mapping, 223
   plate, 262
   TOF and TOFMS analysis
      asialo-glycopeptides, 247
      crude tryptic digests, human IgG, 236, 237
      N-glycan profiling, 263
      O-glycans, 243, 244
Methylation analysis
   alkoxide, 142–143
   D-galactose residues, 142
   polysaccharide, 143
   red algal galactans, 143

Montreuil, J.
  glycomics, 5
  human milk glycoproteins, 5
  lactotransferrin discovery, 5
  pentose-containing nucleic acids, 4
MRM. *see* Multiple-reaction monitoring
MS. *see* Mass spectrometry
Multiple-reaction monitoring (MRM)
  channels, 259
  experiments, 252, 259
  glycoblotting-assisted assays, 250, 252
  reverse glycoblotting-assisted assays, 259–261
  technology, 248, 250

N

N-(2-aminobenzoyl)cysteine hydrazide (ABCh), 237, 238
Noise decoupling technique, 23
Nuclear Overhauser enhancement spectroscopy (NOESY), 30, 31

O

Open reading frames (ORFs), 221
ORFs. *see* Open reading frames

P

Partial acid hydrolysis
  agarose and 3,6-anhydrogalactose, 133–134
  borane-4-methylmorpholine, 133
  carrageenans autohydrolysis, 132
  galactans, 132
  glycosidic bonds, 138
  oligosaccharides, 135–137
  partial depolymerization, 135
  pyruvic acid residues, 135
  structural analysis, agar, 133
PCR. *see* Polymerase chain reaction
Polymerase chain reaction (PCR)
  amplification, 221
  glycanspecific enrichment, 234

R

Red algae polysaccharides
  agars and carrageenans, 116
  algal systematics and polysaccharide composition
    acid hydrolysis, 179
    advantages, 178
    chemical structures, 177
    molecular genetics, 179
    NMR spectroscopy, 178–179
    reductive hydrolysis procedure, 178
  cellulose, mannans and xylans
    acid hydrolysis, 120
    biosynthesis, 120
    cellulose preparations, 120
    chemical structure, 121
    conchocelis and generic phases, 120
    matrix-assisted ultraviolet laser, 121
    relative insolubility, 119–120
    water-insoluble, 122
  chemical investigation, 116
  galactans structural analysis, 116–117
  gelling galactans, 117
  glycuronans, 173–174
  storage carbohydrates, floridean starch, 117–119
  sulfated galactans
    agar group, 162–165
    biological activity, 169–171
    carrageenan group, 165–167
    chemical methods, 129–145
    DL-hybrids, 168–169
    enzymatic and immunological methods, 155–161
    information and nomenclature, 122–127
    isolation, 127–128
    physicochemical methods, 145–155
  sulfated mannans
    anticoagulant action, 173
    aqueous sodium chloride, 172
    mannose and xylose, 171–172
  unicellular and freshwater heteropolysaccharides, 174–177

SUBJECT INDEX

Freezing–thawing procedure, 127
Retrosynthetic analysis, trehazolin
  functionalized cyclopentitol, 49, 50
  thiourea derivative, 49

S

Spin-echo correlated spectroscopy (SECSY), 30, 35
Sulfated galactans
  biological activity
    anticoagulant activity, 169
    bacterial lipopolysaccharides, 171
    carragenans, 169
    exhibit antioxidant activity, 171
    inhibitory effect, 170
    matrix-degrading enzymes, 170–171
    potent antiviral activity, 170
  chemical methods
    composition analysis, 129–132
    desulfation, 138–142
    methylation analysis, 142–144
    partial acid hydrolysis, 132–138
    periodate oxidation, 144
  DL-hybrids, 168–169
  enzymatic and immunological methods
    agarolytic enzymes, 155–159
    carrageenans, 159–160
    enzymes releasing sulfate, 160–161
    immunological methods, 161
  information and nomenclature
    biological source, 124
    carrageenans, 124–125
    desulfation procedures, 127
    "deviating" structures, 123
    disaccharide repeating units, 122
    physicochemical properties, 122
    red algal galactans, 123–124
    shorthand terminology system, 126
    structural analysis, 126–127
  isolation
    algal biomass treatment, 127
    analytical procedures, 128
    carrageenans fractionation procedure, 128

gel-forming ability, 127
Freezing–thawing procedure, 127
size-exclusion chromatography, 128
physicochemical methods
  mass spectrometry, 146–149
  nuclear magnetic resonance spectroscopy, 149–155
  vibrational spectroscopy, 145–146
polysaccharides
  agar group, 162–165
  carrageenan group, 165–167
Sulfated mannans
  anticoagulant action, 173
  aqueous sodium chloride, 172
  mannose and xylose, 171–172
Swern oxidation, 54–55
SysBioWare, 261

T

Trehalase inhibitors
  enzymes, 46
  glycoconjugates, 46
  sugar isothiocyanate fragment, synthesis
    azido derivative, 51
    β-elimination, unstable dithiocarbamide triethylamine salt, 51, 52
    benzylation and mesylation, 51–52
    D-galactose pentaacetate, 51, 53
    metal isothiocyanate, 50
    S_N2-type azidation, 51, 53
    Wittig–Horner–Emmons type of reaction, 51, 52
  trehazolamine fragment synthesis
    acylated oxazolidinones, 81–82
    analogues, 53, 54
    cis-2-butene-1,4-diol, 82–85
    cyclopentadiene, 77–81
    D-arabinose, 65–66
    D-glucose, 54–60
    D-mannitol, 66
    D-mannose, 60–63
    D-ribonolactone, 73–77
    D-ribose, 63–65
    3-hydroxymethylpyridine, 85–87

Trehalase inhibitors (cont.)
  myo-inositiol, 68–73
  norbornyl derivatives, 85
  sugar 1,5-lactones, 66–68
  trehazolin
    biosynthesis, 48–49
    description, 47
    natural occurrence and characterization, 47–48
    retrosynthetic analysis, 49–50
    synthesis (see Trehazolin synthesis)
  Trehazolamine fragment synthesis
    acylated oxazolidinones, 81–82
    analogues, 53, 54
    cis-2-butene-1,4-diol, 82–85
    cyclopentadiene, 77–81
    D-arabinose
      p-methoxybenzyl (PMB) group, 65–66
      2,3,5-tri-O-benzyl D-arabinose, 65
    D-glucose
      5-epi-trehazolin, 57, 59
      epoxidation, 55
      Ferrier reaction, 59
      hydrogenation, azido group, 57
      open-chain O-methyloxime derivative, 54, 55
      reductive carbocyclization, 55, 57
      regiospecificity, 60
      silylation and benzylation, 57, 58
      Swern oxidation, 54–55
      $ZnCl_2$ treatment, 57
    D-mannitol, 66
    D-mannose
      deacetylation and monoacetonation, 60
      free-radical cycloisomerization, 61, 62
      O-benzylhydroxylamine, 61
      regio-selective benzylation, 60–61
      trehazolamine diastereoisomeric analogue, 61, 62
      triethylborane plus triphenyltin hydride-mediated carbocyclization, 63
    D-ribonolactone, 73–77
    D-ribose

      anhydrous ethanol treatment, 63–64
      free-radical cyclization, 63
      N,N-dimethylformamide (DMF), 65
    3-hydroxymethylpyridine, 85–87
    myo-inositiol
      acid hydrolysis, 72–73
      cis-hydroxylation, 72
      deacylation, 69
      exo-alkene, spiroepoxide, 69, 70
      free hydroxyl group oxidation, 71
      osmium tetraoxide treatment, 70
      penta-N,O-acetyl derivatives, 70, 71
      periodate oxidation, 68–69
      peroxyacid oxidation, exo-methylene group, 73
    norbornyl derivatives, 85
    sugar 1,5-lactones
      glycosidases inhibition, 68
      hydrogenation, 67–68
      2-iodo-5-formyl-1,5-lactone and precursor synthesis, 66, 68
  Trehazolin synthesis
    anomeric center, trehazolamines
    amino alcohols, 93, 95
    aminocyclitol condensation, 90
    benzyl groups removal, 96, 98
    bromodeoxy and epoxy analogues, 99
    coupling, 92, 93
    cyclization, 99, 100
    cyclohexyl analogue, 99–101
    diastereoisomer, 92, 94
    5-epi-trehazolin, 90, 92
    HCl, MeOH, 96, 97
    hydrogenolysis, 89
    intramolecular cyclization, 87
    N-alkyl and N-aryl ureas, 89, 91
    oxazoline formation, 96, 98
    replacement, aminocyclitol, 93, 95
    2,3,4,6-tetra-O-benzyl-1-deoxy-α-D-glucopyranosyl isothiocyanate, 89, 90
    trehalostatin diepimer, 96
    trehazolin tribenzyl ether, deprotection, 89, 91

biosynthesis, 48–49
deoxynojirimycin analogues
   1-deoxynojirimycin-trehalamine fused
      compounds, 106, 107
   desilylation, tetrasilylated compound, 107
   disubstituted thiourea, 108
   N-(tert-butoxycarbonyl)glycine, 106, 107
1-thiatrehazolin
   acetal hydrolysis, 104
   intramolecular nucleophilic
      displacement reaction, 104, 105
trehazolamines
   α and β anomers, 103, 105
   6-isothiocyanoglucose derivative,
      coupling, 100, 102
   maltose-type trehazolin derivatives,
      101, 103
   oxazolidinone hydrolysis, 101, 104
   pseudodisaccharides, inhibitory
      activities, 100
   2,3,4-tri-O-benzyl-6-deoxy-
      6-fluoro-α-D-glucopyranosyl
      isothiocyanate, 102, 105

U

Unicellular and freshwater
heteropolysaccharides
Apophloea lyallii, 177
medullary cells, 176
molecular structure, 174–175
NMR spectroscopy, 175
oligosaccharides, 175
osmotic and ionic conditions, 176
partial acid hydrolysis, 174
polyacrylamide gel electrophoresis, 177
viscosity, 176

W

Wittig–Horner–Emmons type of reaction,
   51, 52

X

Xylans
   acid hydrolysis, 120
   biosynthesis, 120
   cellulose preparations, 120
   chemical structure, 121
   conchocelis and generic phases, 120
   matrix-assisted ultraviolet laser, 121
   relative insolubility, 119–120
   water-insoluble, 122

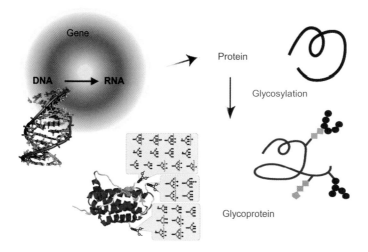

PLATE 1 Biosynthetic pathway of glycoproteins. No one can predict individual glycoforms and the microheterogeneity of glycoproteins, because glycan biosynthesis is not a template-driven posttranslational modification. Recombinant human erythropoietin, the glycoprotein drug produced by mammalian cells, has highly complex N-glycan heterogeneity. (Original figure made by the author).

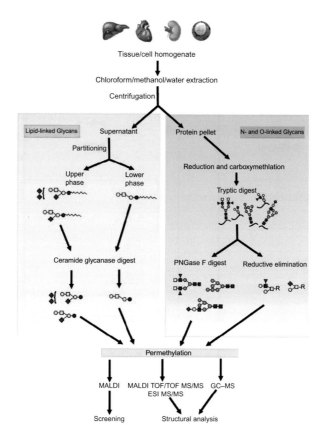

PLATE 2 Schematic representation of MS strategies for glycomics as reported by Dell *et al*. This figure is adapted from ref. 12.

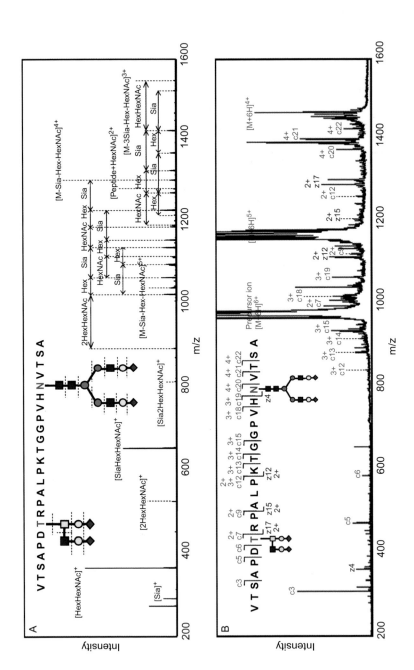

PLATE 3 MS/MS spectra of the 6+ charge state (m/z 982.43) of a synthetic MUC1 glycopeptide carrying N- and O-glycans. (A) CID, (B) ECD spectra. This figure is adapted from ref. 75.

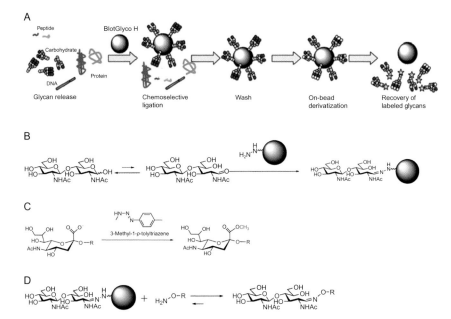

PLATE 4 The glycoblotting protocol, using BlotGlyco H beads. (A) Schematic representation of a practical glycoblotting protocol employing BlotGlyco H beads. The streamlined chemical reactions performed ''on-bead'' are composed of (B) chemical ligation of glycans by hydrazide groups, (C) methyl esterification of carboxyl group of sialic acid residues,[87] and (D) release of desired glycans as labeled derivatives by trans-iminization reaction[89,90] with designated aminooxy compounds. (This figure was made by the author.)

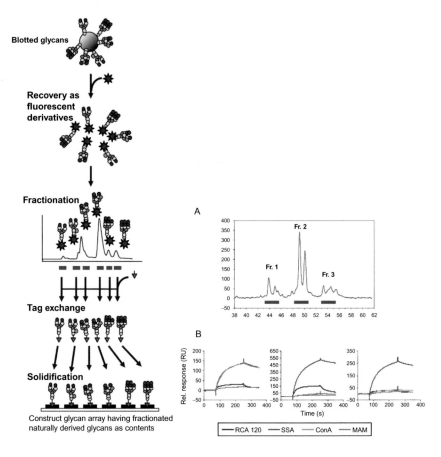

PLATE 5 Schematic diagram showing the sequential multiple-tag conversion, based on trans-iminization of N-glycans enriched by glycoblotting and the construction of a glycan array. (A) Normal-phase chromatography of N-glycans enriched by glycoblotting from human α_1-acid glycoprotein, labeled by transiminization with an anthraniloyl hydrazine derivative. Frs. 1, 2, and 3 correspond to di-, tri-, and tetrasialylated N-glycans. They were then subjected to a tag-exchange reaction with biotin hydrazide for immobilization on the streptavidin-coated surface. The interactions with various lectins were monitored by surface plasmon resonance. (B) Sensorgrams showing interactions of the biotinylated N-glycans of Fr. 1 (left), Fr. 2 (middle), and Fr. 3 (right) with 4 lectins (*Sambucus sieboldiana* lectin (SSA), *Maackia amurensis* lectin (MAM), *Ricinus communis* agglutinin 120 (RCA$_{120}$), and concanavalin A (Con A)). This figure is adapted and arranged from ref. 89.

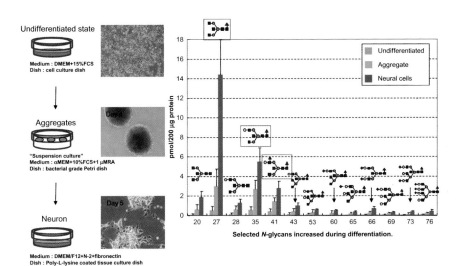

PLATE 6 Differentiation of mouse P19C6 cells to neural cells, and N-glycan changes as monitored by the glycoblotting method. Undifferentiated cells were stimulated by retinoic acid, and the cells were subjected to the glycoblotting protocol optimized for cellular glycomics. It was demonstrated that the expression level of various bisecting-type N-glycans increased drastically during cell differentiation to neural cells.[91] Three glycoforms circled by red lines are known to exist in the mouse brain system.[92,93] (This figure was made by the author.)

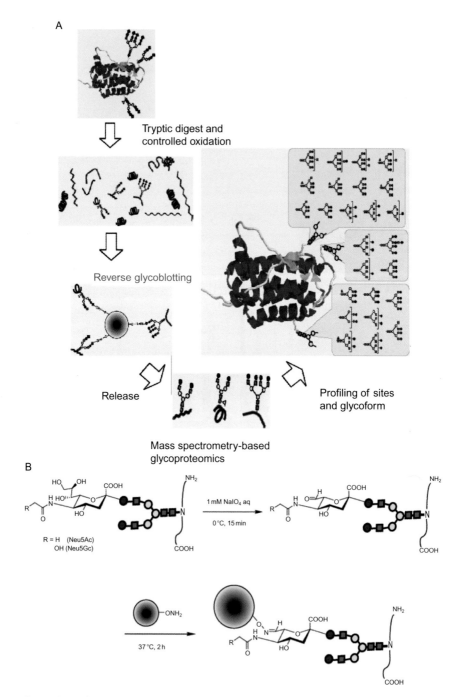

PLATE 7 The first "reverse glycoblotting" protocol reported in ref. 97. (A) Schematic representation, showing the procedure for glycoproteomics, profiling both the glycan heterogeneity and sites of glycosylation. (B) General conditions for selective oxidation and reverse glycoblotting of the terminal sialic acids. This figure is adapted from ref. 97.

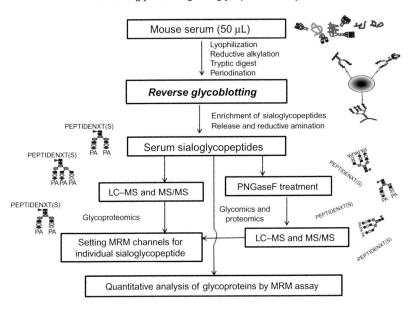

PLATE 8 (Continued)

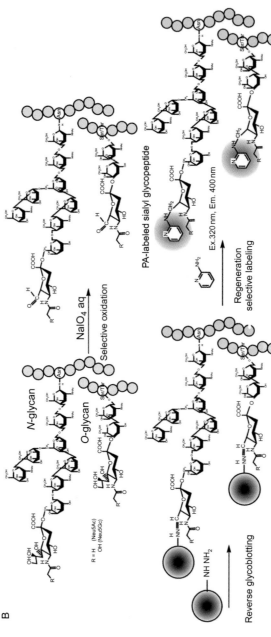

PLATE 8 A novel strategy of "sialic acid-focused" glycoproteomics. (A) A workflow of the comprehensive approach of the reverse glycoblotting-assisted MRM assays. (B) An improved protocol for reverse glycoblotting, release, and fluorescence tagging of enriched glycopeptides. These figures are adapted and arranged from ref. 113.

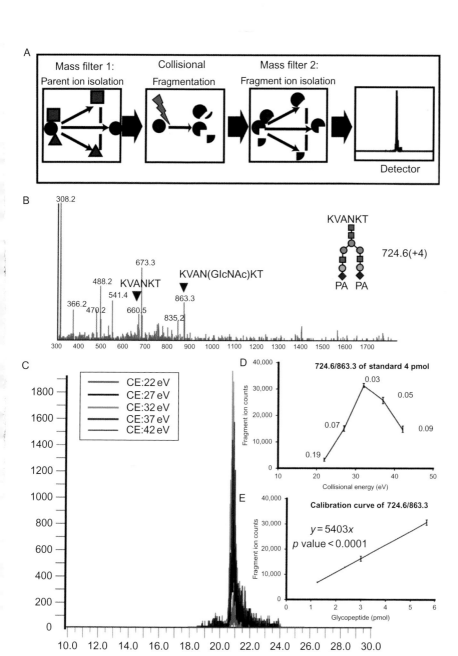

PLATE 9 Schematic diagram of MRM assays and a typical procedure for MRM setting. (A) The general concept of MRM. (B) The MS/MS spectrum of PA-labeled sialyl glycopeptide from egg yolk. (C) The MRM chromatogram under different collision energies (CE) at 22, 27, 32, 37, and 42 eV for 724.6/863.3 as Q1/Q3. (D) Relationship between fragment-ion counts versus CE for MRM assay [the numerical value means each coefficient of variation (CV)]. (E) Relationship between fragment-ion counts versus the amount of PA-labeled glycopeptide. This figure is adapted from ref. 113.

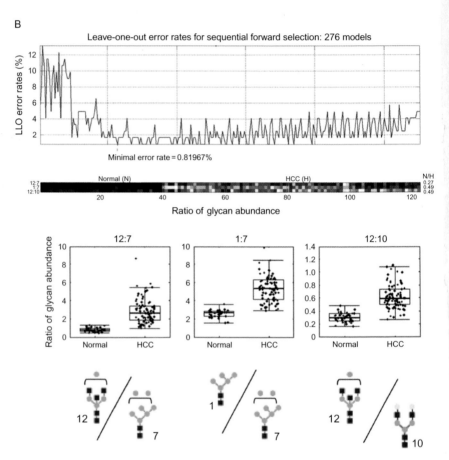

PLATE 10 Large-scale glycomics performed by means of the automated glycan-processing machine "SweetBlot" in a test with HCC patients (80 samples) and healthy donors (28 samples). (A) The raw mass spectra were subjected to compositional analysis and structural elucidation. (B) upper panel; process of the feature-subset selection in which sequential addition of the most significant N-glycans ratios at each step, testing a total of 276 models, resulted in a minimal error rate of 0%; lower panel; heat-map view of the selected three N-glycan ratios as features for the classification (brighter color indicates higher ratio of glycan abundance), and box-plot expression of the selected features (ratios of N-glycan abundance and oligosaccharide structures are shown in the figure). Parts of these figures are adapted from ref. 88.